CP
OB/GYN

Clinical Perspectives in Obstetrics and Gynecology

Series Editor:

Isaac Schiff, M.D.

CP
OB/GYN

Clinical Perspectives in Obstetrics and Gynecology

Series Editor: Isaac Schiff, M.D.

Published Volumes:

Shoupe and Haseltine (eds.): *Contraception (1993)*
Lorrain (ed.): *Comprehensive Management of Menopause (1993)*
Gonik (ed.): *Viral Diseases in Pregnancy (1994)*
Flamm and Quilligan (eds.): *Cesarean Section: Guidelines for Appropriate Utilization (1995)*

Forthcoming Volumes:

Reindollar and Gray (eds.): *Molecular Biology for the Obstetrician-Gynecologist*

Published Volumes (*Series Editor:* The Late Herbert J. Buchsbaum, M.D.):

Buchsbaum (ed.): *The Menopause (1983)*
Aiman (ed.): *Infertility (1984)*
Futterweit: *Polycystic Ovarian Disease (1984)*
Lavery and Sanfilippo (eds.): *Pediatric and Adolescent Obstetrics and Gynecology (1985)*
Galask and Larson (eds.): *Infectious Diseases in the Female Patient (1986)*
Buchsbaum and Walton (eds.): *Strategies in Gynecologic Surgery (1986)*
Szulman and Buchsbaum (eds.): *Gestational Trophoblastic Disease (1987)*
Cibils (ed.): *Surgical Diseases in Pregnancy (1990)*
Collins (ed.): *Ovulation Induction (1990)*
Sanfilippo and Levine (eds.): *Operative Gynecologic Endoscopy (1990)*
Altchek and Deligdisch (eds.): *The Uterus (1991)*

Cesarean Section

Guidelines for Appropriate Utilization

Bruce L. Flamm
Edward J. Quilligan
Editors

With 24 Illustrations

Springer-Verlag
New York Berlin Heidelberg London Paris
Tokyo Hong Kong Barcelona Budapest

Editors:
Bruce L. Flamm, M.D., Research Chairman, Kaiser Permanente, Southern California Permanente Medical Group, Riverside, CA 92505, USA

Edward J. Quilligan, M.D., Professor Emeritus, Department of Obstetrics and Gynecology, University of California-Irvine Medical Center, Orange, CA 92668, USA

Series Editor:
Isaac Schiff, M.D., Chief of Vincent Memorial Gynecology Service, Women's Care Division of the Massachusetts General Hospital and the Joe Vincent Meigs Professor of Gynecology, Harvard Medical School, Boston, MA 02114, USA

Library of Congress Cataloging-in-Publication Data
Cesarean section: guidelines for appropriate utilization / [edited by] Bruce L. Flamm, Edward J. Quilligan.
p. cm.—(Clinical perspectives in obstetrics and gynecology)
Includes bibliographical references and index.
ISBN-13: 978-1-4612-7556-5 e-ISBN-13: 978-1-4612-2482-2
DOI: 10.1007/978-1-4612-2482-2
1. Cesarean section. 2. Surgical indications. 3. Cesarean section — Prevention. I. Flamm, Bruce L. II. Quilligan, Edward J., 1925– . III. Series.
[DNLM: 1. Cesarean Section. WQ 430 C4216 1995]
RG761.C486 1995
618.8'6—dc20 94-29533

Printed on acid-free paper.

Softcover reprint of the hardcover 1st edition 1995

Production managed by Bill Imbornoni; manufacturing supervised by Jeff Taub.
Typeset by Best-set Typesetter Ltd., Hong Kong.

9 8 7 6 5 4 3 2 1

Preface: Defining the "Appropriate" Cesarean Section Rate

I wrote an article for *Contemporary Obstetrics and Gynecology* in 1983 entitled "Making inroads against the C-section rate."[1] In this article, I examined the various factors responsible for the relative rapid increase in the cesarean section rate in the United States between 1970 and 1983, when it rose from 5.5% to 18%. I suggested that it was possible to achieve a cesarean section rate between 7.8% and 17.5%, and here I would like to quote directly from the article: "If a hospital were to pursue a conservative policy on section, for the usual obstetric population, what section rate should it expect?" This frequently asked question is hard to answer because hospital obstetric populations differ so markedly. *However, hospital staffs should examine their cesarean rates frequently, indication by indication.* Purely as a guide (emphasis mine), the table on page vi lists some high and low values. These values were 2%–4% for failure to progress, 2%–6% for repeat cesarean section, 1.3%–3.5% for breech and abnormal lie, 1.5%–3% for fetal distress, and 1% for third-trimester bleeding.

In 1992, the cesarean section rate in the United States was 22.6%, a somewhat slower increase in the past decade than the very rapid rise of the 1970s. The rate seems to vary widely between states, with a low of 16.3% and a high of 28.4%, between hospitals with lows below 15% and highs above 40%,[2] and even between doctors in the same hospital with similar practices, from 10% to 17%.[3] This wide variability depends primarily on the section rate for failure to progress in labor, fetal distress, and repeat cesarean sections. Frequently those physicians who have a very high cesarean section rate do not encourage vaginal births after a previous cesarean section, are not active in their management of desultory labors, and are too ready to label fetal distress when they see a pattern that is only mildly abnormal. To some, these physicians are guilty of greed, ignorance, and fear. In my personal opinion, greed plays a very small role in the cesarean rate. While it is true that most physicians charge more for a cesarean section than a

Cesarean section rates

Indication	Percentage	
	Low	High
Failure to progress	2.0	4.0
Repeat cesarean section	2.0	6.0
Breech and abnormal lie	1.3	3.5
Fetal distress	1.5	3.0
Third-trimester bleeding	1.0	1.0
Totals	7.8	17.5

From Quilligan,[1] by permission of *Contemporary Obstetrics and Gynecology*.

vaginal delivery, I have yet to meet a physician who would do something they believed would harm their patient even if they were paid ten times as much for a section. On the other hand, there are fears and misconceptions. I have heard many doctors say "I have never been sued for a section I did, but I have been sued for the section I did not do." The fear of not having performed a section in my opinion is real, although difficult to prove, and until the public can be educated that cesarean section delivery cannot eradicate fetal death and damage, this fear will remain and will be responsible for some unnecessary cesarean sections. Bruce Flamm and I hope this book will correct misconceptions that have been responsible for many unnecessary cesarean sections.

I am still frequently asked the same old question: What is an ideal cesarean section rate? I still give an answer similar to the 1983 answer, perhaps somewhat modified. Every hospital that has an obstetric service should have some committee that examines every cesarean section that is performed in that hospital and determines whether it was indicated or not. If it was not indicated, then the physician who performed the section should be educated as to why it was not indicated. Only through repeated educational efforts will individual physicians lower their section rates. If the hospital rate remains high, then it seems reasonable that the licensing body in the state has an obligation to examine the records of that hospital. Perhaps the rate is justified; if not, education of the entire staff is in order. How do I define "high" rate? If the hospital rate is persistently above the level statewide, it would certainly seem reasonable to look at the records.

Another way to look at hospital performance is to look at a segment of the cesarean section rate that depends primarily on the philosophy of the physicians and their education as well as their patients' education, the elective cesarean rate or its reverse, the rate of vaginal birth after cesarean section (VBAC). The VBAC rate should be on an upward trend toward 50% and certainly at the national level of 25%.

What about the other major factors in fetal distress and failure to progress in labor? Electronic fetal heartrate monitoring has been blamed for the rapid increase in cesarean section rate for fetal distress; however, it is not the monitoring but the interpretation of the data that is at fault. Unfortunately, many of the tracings that have resulted in cesarean section show a temporary decrease in fetal oxygenation and are not indicative of fetal distress. It must be kept in mind that the heartrate pattern shown by the monitor has a high false-positive rate for fetal distress; therefore, other measures such as fetal scalp stimulation, fetal acoustic stimulation, and fetal scalp sampling must be employed to rule out false-positive cases. Committees should determine whether these measures are being used if the section rate for fetal distress exceeds 2%.

Further, let us not forget our educational efforts not only to get physicians and nurses to correctly interpret patterns and use ancillary diagnostic tests but also to educate the public that not all abnormal patterns are indicative of ongoing brain damage but may be the result of damage that has occurred days or months before the onset of labor. Failure to progress in labor would seem to be the most difficult figure to determine; however, some important concepts can help guide a committee examining individual cases. It is paramount that accurate records be kept of uterine activity, cervical change, and fetal presenting part descent.

The active management of labor, described later in this volume, has been very successful in keeping a very low cesarean section rate at the National Maternity Hospital in Dublin, Ireland. In my opinion, the two most important parts of that program are ensuring that the patient is in labor on hospital admission and the prompt use of oxytocin when the patient falls off the labor curve. Both these factors require close observation of the patient, and the concept of one-on-one nursing or midwife care is obviously important.

The key I would like to stress is education, education at every level—the patient, the nurse, and the physician. It is the only effective method to lower the cesarean section rate, but to be effective education must be repetitious. I wish I could say that simply reading this book will lower the cesarean section rate; it will not. However, this volume will give you a series of suggestions and the scientific rationale behind those suggestions. If followed, these ideas will help you to decrease your cesarean section rate.

E.J. Quilligan, M.D.

References

1. Quilligan EJ. Making inroads against the C-section rate. Contemp Obstet Gynecol 1983;Jan:221–225.

2. Gabbay M, Wolfe SM. Unnecessary cesarean sections: curing a national epidemic. Washington, DC: Public Citizens Health Research Group, May 1994.
3. Dermott RK, Sandmire HF. The South Bay cesarean section study. II: The physician factor as a determinant of cesarean birth rates for failed labor. Am J Obstet Gynecol 1992;166:1799–1806.

Contents

Contributors

Frank H. Boehm, M.D.
Maternal-Fetal Division, Department of Obstetrics and Gynecology, Vanderbilt University Medical Center, Nashville, TN 37232, USA

Peter Boylan, M.D.
National Maternity Hospital, Holles Street, Dublin 2, Ireland

T.J. Broadhead, M.B., Ch.B.
Department of Obstetrics, Queen's Medical Centre, Nottingham NG7 2UH, UK

Douglas Brown, Ph.D.
Department of Obstetrics and Gynecology, Louisiana State University School of Medicine in New Orleans, New Orleans, LA 70112, USA

Zane A. Brown, M.D.
Professor of Obstetrics and Gynecology, University of Washington School of Medicine, Seattle, WA 98195, USA

Luis A. Cibils, M.D.
Mary Campau Ryerson Professor of Obstetrics and Gynecology, University of Chicago, Chicago Lying-In Hospital, Chicago, IL 60637, USA

Richard H. Demir, M.D.
The Center for Human Reproduction and The Foundation for Reproductive Medicine, Inc., The Department of Obstetrics and Gynecology, Mount Sinai Hospital Medical Center, Chicago, IL 60610, USA

Thomas E. Elkins, M.D.
Department of Obstetrics and Gynecology, Louisiana State University School of Medicine in New Orleans, New Orleans, LA 70112, USA

BRUCE L. FLAMM, M.D., F.A.C.O.G.
Research Chairman, Kaiser Permanente, Southern California Permanente Medical Group, Riverside, CA 92505, USA

EMANUEL A. FRIEDMAN, M.D., SC.D.
Professor Emeritus of Obstetrics and Gynecology, Harvard Medical School, One Lincoln Plaza, 20K, New York, NY 10023, USA

LAURA B. GARDNER, M.D., M.P.H., PH.D.
President, Axiomedics Research, Inc., 300 Third Street, Suite 4, Los Altos, CA 94022, USA

NORBERT GLEICHER, M.D.
The Center for Human Reproduction and The Foundation for Reproductive Medicine, Inc., The Department of Obstetrics and Gynecology, Mount Sinai Hospital Medical Center, Chicago, IL 60610, USA

JANICE R. GOINGS, R.N., C.N.M.
Coordinator of Nurse-Midwifery Services, Kaiser Permanente Medical Center, Riverside, CA 92505, USA

CORNELIA R. GRAVES, M.D.
Maternal-Fetal Division, Department of Obstetrics and Gynecology, Vanderbilt University Medical Center, Nashville, TN 37232, USA

JOHN C. HOBBINS, M.D.
Professor of Obstetrics and Gynecology, Chief of Obstetrics, University of Colorado Health Sciences Center, Denver, CO 80262, USA

D.K. JAMES, M.A., M.D., F.R.C.O.G., D.C.H.
Department of Obstetrics, Queen's Medical Centre, Nottingham NG7 2UH, UK

STEPHEN A. MYERS, D.O.
The Center for Human Reproduction and The Foundation for Reproductive Medicine, Inc., The Department of Obstetrics and Gynecology, Mount Sinai Hospital Medical Center, Chicago, IL 60610, USA

DAVID A. MILLER, M.D.
Department of Obstetrics and Gynecology, LAC/USC Women's Hospital, Los Angeles, CA 90033, USA

JEANNE B. NOVAS, M.D.
The Center for Human Reproduction and The Foundation for Reproductive Medicine, Inc., The Department of Obstetrics and Gynecology, Mount Sinai Hospital Medical Center, Chicago, IL 60610, USA

Richard H. Paul, M.D.
Department of Obstetrics and Gynecology, LAC/USC Women's Hospital, Los Angeles, CA 90033, USA

Jeffrey P. Phelan, M.D., J.D.
Co-Director, Maternal-Fetal Medicine, 1030 South Arroyo Parkway, Suite 110, Pasadena, CA 91105, USA

Edward J. Quilligan, M.D.
Professor Emeritus, Obstetrics and Gynecology, University of California Irvine Medical Center, College of Medicine, Orange, CA 92668, USA

Hung N. Winn, M.D.
Associate Professor of Obstetrics and Gynecology, Director, Division of Maternal-Fetal Medicine, St. Louis University School of Medicine, Chief of Obstetrics, St. Mary's Health Center, St. Louis, MO 63117, USA

Introduction

In 1993, President Bill Clinton rode into Washington on the crest of a massive "health care reform" wave. The reform did not occur in the way the Clinton administration had envisioned, but major changes have indeed been taking place. In what has been termed an "orgy of measurement," everything we do in medicine is now being observed and evaluated. In such an environment it would be naive to expect America's most common operation to escape careful scrutiny. In a recent *Journal of the American Medical Association* article, Donald Berwick, M.D., points out,[1] "Physicians can themselves achieve the purposes of health system reform. In fact, we should; all of the visible alternatives are worse." He lists eleven worthy aims of health care reform, including the reduction of cesarean rates to pre-1980 levels.

The one million cesareans performed each year in the United States represent a fivefold increase in 20 years. Recent reports claiming that fully half these operations are unnecessary[2,3] have fanned the flames of the cesarean controversy. Surely this estimate is exaggerated, but just as surely we could be doing better. Quality improvement experts know that variation is often a warning sign. When one hospital has a 15% cesarean rate while a similar hospital in the same town has a 45% rate, it is only natural for someone to ask, why? When 2,000 babies are delivered annually at each of these hospitals, but only 300 cesareans are performed at one while 900 are "required" at the other, sooner or later someone is going to ask why?

There is no easy answer to this question. Patient risk-factors and demographics may explain part of the variation but not all of it. Some would point an accusing finger at us obstetricians and claim that it's our fault. Naturally I'm a bit biased, but I don't think we deserve all the blame. Women continue to request or even demand cesarean operations. If we don't comply, in the current legal climate we clearly run the risk of being sued for medical malpractice if there is any problem with a delivery. Experts remind us that physicians generally prevail in such cases; this, however is of little solace to the obstetrician caught in the clutches of a grueling five year legal battle.

Some would claim that the escalating cesarean rate is a financial issue and that paying obstetricians less for cesarean deliveries would quickly turn things around. This is faulty logic. Cesareans cost more than vaginal births, but only a small fraction of the additional cost goes to doctors. Hospitals do in fact charge much more for cesarean births but this does not explain rising cesarean rates. If cesareans could be done more cheaply, it is doubtful that their popularity would wane. In fact, the opposite might prove to be true.

Legal and financial issues aside, parents and obstetricians really share a common goal. Both want every pregnancy to culminate with the birth of a perfect baby to a healthy mother. Many patients and some physicians have begun to believe that ever-increasing cesarean rates will lead us closer to this goal, but statistics do not support this contention. Many nations with the lowest perinatal mortality rates also have the lowest cesarean rates. Likewise, many U.S. hospitals have very low cesarean rates as well as low perinatal mortality rates.

It is often said that "where there's a will there's a way." This book shows the *way* in which cesarean rates can be safely lowered. Several of our authors have found methods to lower their cesarean rates to 15% (or less) while maintaining excellent maternal and perinatal outcomes. However, the tools provided in this book can only work effectively if there is first a *will* to use them. Willingness to change is a prerequisite for improvement.

Finally, it should be stressed that cesarean delivery is not an evil that needs to be eliminated. As we strive to attain reasonable cesarean rates we must never forget that cesarean operations, when appropriately utilized, have saved the lives of many mothers and countless babies.

Bruce L. Flamm, M.D., F.A.C.O.G.

References

1. Berwick, DM. Eleven worthy aims for clinical leadership of health system reform. JAMA 1994;272:797–802.
2. Silver L, Wolfe SM. Unnecessary cesarean sections: how to cure a national epidemic. Washington, DC: Public Citizens Health Research Group, 1989, 1994.
3. Steinbrook, R. Half the cesarean operations in U.S. called unnecessary. Los Angeles Times. January 27, 1989, Part 1. p. 1.

1 Cesarean Delivery in the United States: A Summary of the Past 20 Years

Bruce L. Flamm

This chapter summarizes what has happened to cesarean rates in the United States during the past 20 years and outlines some of the factors responsible for the changes that have taken place.

United States Cesarean Statistics: 1970 to Present

A dramatic change in the U.S. national cesarean rate has taken place during the past 20 years (Figure 1.1). Obstetricians trained recently, accustomed to 20%–25% cesarean section rates, may find it difficult to comprehend that during most of this century the U.S. cesarean delivery rate remained consistently below 4%. In 1970 there were 3.7 million births and the national cesarean rate reached 5%. By 1975, the rate reached 10%.[1] In a span of 5 years, the overall rate had doubled. By 1978 the rate had reached 15%, three times what it had been just 8 years earlier. The number of cesareans was growing by 50,000 additional operations each year, and 500,000 cesareans were being performed annually.[2] In 1979 the National Institutes of Health (NIH) assembled a 19-member Task Force on Cesarean Childbirth. In 1981, the results of the task force investigation and their recommendations were published in a 537-page document entitled "Cesarean Childbirth."[3] The report received widespread publicity but had little apparent effect on the rising cesarean rate.[4,5] By 1983 the national rate reached 20%.[6,7] The revised national birth certificate data for 1990, first reported in 1993 because of the lag time in processing millions of records, indicates that the rate seems to have stabilized at about 23%.[8,9] A 23% national rate means that approximately 1 million cesarean operations are now performed in the United States each year (Figure 1.2). Preliminary data from the National Hospital Discharge Survey (NHDS) also indicate that the national cesarean rate is stabilizing at about 23.5%.[10] The NHDS report estimated that 977,000 cesareans were performed in 1990, compared to 966,000 in 1991. As is explained on page 3, these numbers are so similar that they may not represent a significant difference. The NHDS report also points out that 8.5% of women who had a previous cesarean delivered vaginally in 1986 compared with 24.2% in 1991. This dramatic increase in the vaginal birth after cesarean (VBAC) rate has clearly been a major factor in the recent leveling-off of the overall cesarean rate.

How Is the National Cesarean Rate Determined?

This question appears to be trivial, but it is not. It would seem to be a simple matter of dividing the total number of cesareans by the total number of births. However, at least until the past few years, the total number of

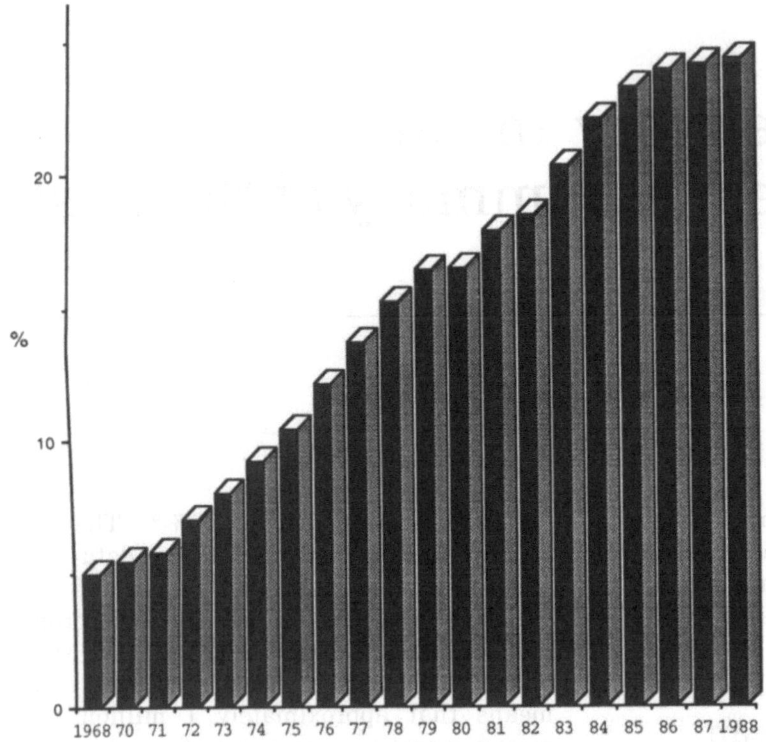

FIGURE 1.1. Cesarean rate in the United States, 1968–1988.

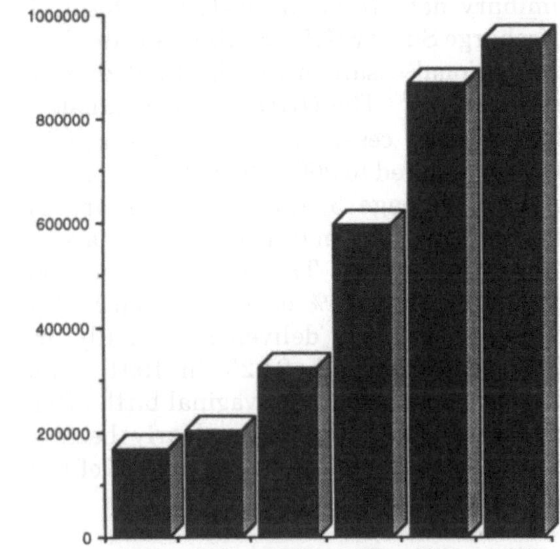

FIGURE 1.2. Number of cesarean operations in the United States, 1965–1990.

cesarean deliveries was unknown and hence the national rate could not be calculated directly. Until 1989 the national cesarean rate was estimated by extrapolation from a carefully selected sample population. The NHDS is a survey that has been conducted each year since 1965 by the National Center for Health Statistics (NCHS).[11] Each year the charts of several hundred thousand patients discharged from hundreds of short-stay hospitals are abstracted. Some of these patients have had admissions for obstetric care. International Classification of Disease (ICD) codes indicating cesarean delivery are tabulated and extrapolated to apply to the entire national population. Before 1989, this was the only method of determining national cesarean statistics. In 1989, the U.S. Standard Certificate of Live Birth was revised to include specific details on the method of delivery.[8] The revised birth certificate now indicates whether the birth was by cesarean and, if so, specifically whether it was a primary or repeat operation. If the birth is vaginal, the revised birth certificate also indicates if it was a vaginal birth after prior cesarean delivery. The new birth certificate is a major step forward in obtaining accurate national statistics. For 1989, data on method of delivery were available for 45 states and the District of Columbia but were not available for Louisiana, Maryland, Nebraska, Nevada, and Oklahoma. Although 95% of the birth certificates from reporting areas indicated the method of birth, 5% did not. Hence, at least for 1989, calculation of national rates from the revised birth certificate data still involved some degree of extrapolation.

The two methods (NHDS and revised birth certificate) for determining the U.S. national cesarean delivery rate have been compared. Analysis of data from the revised birth certificate for 1989 yielded a national cesarean rate of 22.8%. The NHDS data for 1989 indicated a 23.8% rate. Because about 4,000,000 births occurred in 1989 in the United States, the NHDS method indicated that about 40,000 more cesareans were performed than were detected by the revised birth certificate. Alternatively, the NHDS data may have overestimated the number of cesareans by 40,000 procedures. In any case, considering the huge number of births involved, the correlation between these two methods is remarkably good. It should be emphasized that, because we do not know the *exact* cesarean rate, small changes in the rate from one year to the next may not be statistically significant. In other words, we are not absolutely certain if the 1989 national rate is closer to 22.8% (birth certificate data) or 23.8% (NHDS data).

The Cesarean Controversy

With the possible exception of abortion, no other issue in women's health care has created more controversy in the past decade than cesarean operations. A sampling of the dozens of cesarean-related articles that have appeared in newspapers and magazines across the United States in the past few years are listed in Table 1.1.

A glance at the titles of these articles demonstrates that rising cesarean rates have generated a great deal of concern. While the fact that the U.S. cesarean rate has increased dramatically is indisputable, the question of whether this change is favorable or deleterious is not so straightforward. During the past several years people all over America have expressed greatly different opinions. Some women have openly stated that they believe current cesarean rates are ridiculously high and that this rise has occurred because obstetricians will seek out any excuse to perform the operation. But many other women have stated that they believe

TABLE 1.1. Selected media coverage of cesarean rates

Year	Source	Title of article
1987	LA Times	C-Sections: Are There Too Many?
1988	LA Times	"Runaway" C-Section Rates Reflect "Crisis"
1988	LA Times	Hospital Plan Reduces Use of Cesareans
1989	LA Times	Half the Cesarean Operations in the U.S. Called Unnecessary
1990	Physicians Weekly	Cesareans: A Million a Year
1990	USA Today	Doctor's Age Key in Decision on Cesareans
1990	OBC Management	C-Sections Under Fire
1991	Parenting	The Cesarean Rage
1991	Press Enterprise	Corona Hospitals Rank High in Number of C-Section Births
1991	Wall Street J	C-Sections Tied to Economic Factors in Study
1991	Consumer Reports	Too Many Cesareans
1991	Associated Press	Private Hospitals Record High Repeat Cesarean Rate
1992	Expecting	What's Behind the Rise in Cesareans?
1992	American Baby	Can You Avoid a Cesarean?
1992	Parade Magazine	When Is a Cesarean Really Necessary?
1993	Associated Press	Feds Say Doctors Perform Too Many Cesareans
1993	Good Housekeeping	C-Sections: Of the Million Done Each Year, Fully Half Aren't Needed

cesarean delivery is wonderful and that vaginal birth is far more painful and probably outmoded.

Even among obstetricians there is no consensus. Some doctors believe that cesarean rates greater than 10% reflect unnecessary surgery and others honestly believe that even a 25% rate is too low. There are actually some physicians who believe that *all* babies should be delivered by cesarean. In the words of one such doctor,[12]

> Why is the trend toward more and more cesarean sections? I submit that, in this day of advanced anesthesiological and surgical techniques, cesarean section is both easier and safer than vaginal delivery. Why should the *modern* woman undergo the sweaty, gut-wrenching ordeal of labor that may last twelve to twenty-four hours or more? Why must she face the frantic rush to the operating room for a stat section when the fetal heart rate drops? In summary, I think the goal should be for all women to give birth by cesarean section. I will bet that in 50 years, 90% will.

At the opposite end of the spectrum from the doctor just quoted are the members of two national organizations who strongly oppose rising cesarean rates. C/SEC (Cesarean Support Education and Concern) was founded in 1973 after the U.S. cesarean rate had increased from 5.2% to 7.8% in the previous 3 years. The Cesarean Prevention Movement (CPM), which later became the International Cesarean Awareness Network (ICAN), was formed in 1982 as the national cesarean rate increased to 18%. The use of the word "movement" in the group's original name conjured up visions of hostile confrontations. This was unfortunate, because all the members I have met are caring individuals who volunteer hundreds of hours each year to help women develop realistic expectations about childbirth. The CPM eventually grew to 75 chapters in 35 states and claimed to be counseling more than 35,000 women each year. Although most obstetricians were probably unaware of their existence, these two organizations probably contributed significantly to the current leveling-off of the national cesarean rate at about 23% to 24%.

Prominent obstetricians have also taken a strong stand against rising cesarean rates. The editor of an obstetric journal stated,[13]

> In our specialty, critical scrutiny has lately focused on one area: the runaway cesarean rate. The problem has now caught the attention of the news media, and public criticism has become increasingly severe. This is not to argue that the low cesarean rates of the 1960s would be appropriate in 1986. But for almost one in four babies to be delivered operatively is a national disgrace.

The cesarean controversy has been fueled by a report listing individual cesarean rates for more than 2400 hospitals across the United States and claiming that more than half the operations were unnecessary.[14] Clearly, the vast majority of obstetricians in the United States would adamantly disagree with this conclusion. However, even those obstetricians who believe that current rates are reasonable may be surprised by the report's finding that, at some American hospitals, more than 50% of all births are now by cesarean. A book coauthored by Dr. Sidney Wolfe, director of the Washington-based Health Research Group, contends that 3 million American women have already had un̲necessary cesareans and that the number is growing by almost half a million cases each year.[15] Again, many experts would strongly disagree with this contention. But it would be difficult to argue with the conclusion that *some* unnecessary operations are being performed.

Does Cesarean Section Prevent Cerebral Palsy?

Those who defend current cesarean rates point out that perinatal mortality rates have fallen as cesarean rates have increased. Their opponents observe that decreased perinatal mortality rates can be readily explained by improved antepartum care and great advances in neonatal intensive care. More than a decade ago, this issue was already being debated on an international level. A report from Parkland Hospital in Dallas,

Texas (U.S.A.) claimed that their liberal use of cesarean sections, 18% for 1983, was associated with decreased intrapartum deaths and neonatal seizures when compared to National Maternity Hospital in Dublin, Ireland.[16] However, the Dublin group pointed out that for a valid comparison of uncommon events, larger numbers of births needed to be studied. When data for 1982 and 1984 were included, almost four times as many cesareans were performed in Dallas but no significant difference in perinatal mortality was demonstrated between the two hospitals.[17] Moreover, when the study interval was expanded to include more than 20,000 births at each hospital, there was actually a higher incidence of neonatal seizures in Dallas. The cesarean rates in Dublin during the study interval ranged from 4.2% to 6.0% while the rates at Dallas ranged from 17.3% to 18.0%. The Dublin group concluded that, "The dilemma facing contemporary obstetrics is how to continue to justify the massive increase in cesarean births when there is not convincing evidence of benefit to the child." The authors went on to say that "the high incidence of cesarean section in the United States is not justified by results and that by analogy with Dublin the same perinatal mortality rate could be achieved with one third the number of operations performed."

The controversy over whether the increased use of cesarean section has improved fetal outcome has not been resolved. As one author pointed out, "The contention over the years has been that cesarean sections decreased cerebral palsy and other neurologic problems; however, this has not been proved. Nevertheless, the reason for the sharp increase in cesarean births appears to be in many cases for protection of the fetus."[18] In a prospective study designed specifically to address this issue, unselected patients were divided into two groups.[19] One group had obstetric management directed at minimizing the use of cesarean section, and the other was a routine management group. After 2 years the cesarean rate on the first service was 5.7% while the rate on the routine management service was 17.6%. Despite the threefold difference in cesarean rates, perinatal mortality and morbidity rates were not significantly different on the two services.

A recent publication from the National Institutes of Health reviewed the literature of the past 25 years in an attempt to determine if cesarean delivery had been found to prevent cerebral palsy.[20] The conclusion of the review was that children born by cesarean have no documented reduced risk of childhood neurologic problems or cerebral palsy. The authors also compared the rates of cerebral palsy in nations with very different cesarean rates. In spite of cesarean rates that varied from 7% to 22%, all areas reported cerebral palsy rates of 1.1 to 1.3 per 1000 neonatal survivors born weighing more than 2500 g.

What Caused the Cesarean Rate to Quintuple?

While there is widespread agreement that the U.S. cesarean rate has increased dramatically during the past 20 years, there is no consensus as to *why* it happened. It is clear that no single factor can explain why 200,000 cesareans were deemed to be adequate in 1970 while almost 1,000,000 were performed in 1990. An increase in the overall birthrate does not provide an explanation because approximately 3.7 million births occurred in each of these 2 years. Some of the reasons put forth to explain the rising cesarean rate are outlined in Table 1.2.

The exact contribution of each of these factors to the overall rise in the cesarean rate would probably be impossible to determine. However, it is clear that some factors have had more impact than others. This point will be of more than academic interest to physicians and hospitals interested in lowering their overall cesarean rates; those factors that have had the most impact on raising cesarean rates may also have the greatest potential to reduce them.

What individual factors seem to have contributed most to rising cesarean rates? Data from the NHDS showed that the overall U.S.

TABLE 1.2. Some factors that have contributed to increased cesarean rates

Medical-legal factors
Higher malpractice premiums
Higher malpractice awards
Perceived increased risk of litigation
Advertising by malpractice attorneys
Social factors
Childbearing by older women
The "premium" baby concept
The rise and fall of "natural" childbirth
Medical factors
Management of breech presentation
Management of multiple gestations
Management of genital herpes
Management of postdates pregnancy
Management of the extremely small fetus
Concerns about macrosomia and shoulder dystocia
Decreased use of forceps and vacuum
Epildural analgesia
Fetal monitoring
Repetitive factors
Cycle of one cesarean leading to repeat operations
Convenience factors
Physician related
Patient related
Monetary factors
Hospital compensation for cesarean versus vaginal birth
Physician compensation for cesarean versus vaginal birth
Miscellaneous factors
Perceived increased safety of surgery
Physician experience during training

cesarean rate increased from 16.5% in 1980 to 22.7% in 1985.[6] More than 90% of this 5-year increase appeared to be related to three factors. Repeat cesareans accounted for 48%, dystocia for 29%, and fetal distress for 16% of the increase. Analysis of these three indications will provide a good starting point for anyone interested in the evaluation of cesarean section rates. However, many of the issues presented in Table 1.2 are highly intertwined. For example, a discussion of repeat cesarean versus trial of labor certainly involves social and convenience factors as well as medical factors. Likewise, a discussion of fetal distress would not be complete without considering potential medical-legal issues. Unfortunately, in contemporary obstetrics a discussion of almost any possible indication for cesarean would not be complete without consideration of medical-legal ramifications. Cesarean-related discussions almost always end up focusing on lawyers and lawsuits. Murphy's law states that if anything can go wrong, it will. The obstetrical corollary is that if anything goes wrong, you will be sued. To this corollary it is often added that if you did not do a cesarean you will lose the case. This is commonly rephrased as "the only cesarean I've ever been sued for is the one I didn't do." Many American physicians believe that the single most effective way to reduce unnecessary cesareans would be tort reform that eliminates frivolous lawsuits.

References

1. Hibbard CT. Changing trends in cesarean section. Am J Obstet Gynecol 1976;125:798.
2. Petitti D, Olson R, Williams RL. Cesarean section in California—1960 through 1975. Am J Obstet Gynecol 1979;133:391.
3. National Institutes of Health. Report of a consensus development conference. Bethesda, Maryland: National Institutes of Health, Department of Health and Human Services, 1981. (DHHS publication No. 82-2067.)
4. Winkler JD, Kanouse DE, Brodsley L, Brook RH. Popular press coverage of eight National Institutes of Health consensus development topics. JAMA 1986;255:1323–1327.
5. Gleicher H. Cesarean section rates in the United States: the short-term failure of the National Consensus Development Conference in 1980. JAMA 1984;252:3273–3276.
6. Taffel S, Placek P, Liss T. Trends in the United States cesarean section rate and reasons for the 1980–85 rise. Am J Public Health 1987;77:955–959.
7. Taffel SM, Placek PJ. One-fifth of 1983 US births by cesarean section. Am J Public Health 1985;75(2):190–192.
8. Ventura S, Heuser R, Arbetha J, Dunn T. Advance report of new data from the 1989 birth certificate. US Dep Health Hum Serv Mthly Vital Stat Rep 1992;40(12):6–7.
9. Ventura S, Taffel S, Mathews T. Advance report of maternal and infant health data from the birth certificate, 1990. US Dep Health Hum Serv Mthly Vit Stat Report 1993;42(2):1–31.

10. Health objectives for the nation. Rates of cesarean delivery—United States, 1991. Morb Mortal Wkly 42(15):285–289.
11. Graves JG. National hospital discharge survey annual summary. Rockville, Maryland: National Center for Health Statistics, 1989. (NCHS Publication 89-1760.)
12. Overhulse PR. The cesarean section rate. JAMA 1990; 264:971.
13. Queenan JT. The cesarean section rate: out of sight, but not out of mind. Contemp Obstet Gynecol 1986;10:7–8.
14. Silver L, Wolfe SM. Unnecessary cesarean sections: how to cure a national epidemic. Washington, DC: Health Research Group, 1989.
15. Wolfe SM, Jones RD. Women's health alert. Reading, Massachusetls: Addison-Wesley, 1990.
16. Leveno KJ, Cunningham FG, Prithchard JA. Cesarean section: an answer to the House of Horne. Am J Obstet Gynecol 1984;63:485–490.
17. O'Driscoll K, Foley M, MacDonald D, Stronge J. Cesarean section and perinatal outcome: response from the House of Horne. Am J Obstet Gynecol 1988;158:449–452.
18. Harvey GA. Current status of cesarean births. Obstet Gynecol Annu 1984;13:71–82.
19. Porreco RP. High cesarean section rate: a new perspective. Obstet Gynecol 1985;65:307–311.
20. Scheller J, Nelson K. Does cesarean delivery prevent cerebral palsy or other neurologic problems of childhood? Obstet Gynecol 1994; 83:624–630.

2 Worldwide Utilization of Cesarean Section

T.J. Broadhead and D.K. James

Cesarean section (CS) is one of the most common major operations performed around the world. In this chapter we attempt to answer the following questions:

What is the variation in cesarean section rates worldwide?
Have there been any changes in these rates?
What are the possible explanations for any differences?
Are differences in cesarean section rates reflected in differences in outcome?

Worldwide Cesarean Section Rates

There are considerable variations in CS rates both within and between different countries. However, at a national level, relatively few countries routinely provide population-based data on the frequencies of CS.[1,2] In some countries, such as the United States, where most deliveries occur in hospitals, regional statistics probably provide a reasonable estimate of national performance, but it would be unwise to make this assumption for every country. From the information that is available on national rates, we have compiled an overview that is summarized in Table 2.1. The difficulties in obtaining that information are as just stated, and although it has been possible to obtain statistics from a large number of developed countries, few data are available on the use of CS in third-world countries.[3] Figure 2.1 shows the CS rates per 100 hospital deliveries in 13 selected countries for 1983, the most recent year in which rates were available for comparison.

The highest reported CS rate was in Brazil, at 32% in 1986.[4] More recent figures available for the United States show a rate of 25% in 1988.[5,6] These high levels are in marked contrast to those of Jamaica (5.5% in 1989)[3] and Czechoslovakia (7% in 1986).[4]

Within Europe, rates also differ widely. In England and Wales, the rate was 12.1% in 1989[7] whereas in Italy it was 17.5% in 1987.[7] Some of the lowest rates in Europe have been reported in the Netherlands (see Table 2.1), which favors midwife and general practitioner care of low-risk women. In addition there is a high proportion of home deliveries, with a third of births occurring outside hospitals, and 42% of women being cared for solely by midwives.[7] In general, lower rates are found in Eastern Europe countries. For example, Czechoslovakia had a rate of 7% in 1986[4] and Hungary, a rate of 10% in 1985.[4]

It is not possible to give a national figure for Australia because not all the states provide information. However, the most recent figure for Western Australia was 16.9% in 1987.[8] In comparison, New Zealand had a CS rate of 10% in 1985.[4]

In developing countries, surprisingly high rates have been reported, including 32% for Brazil in 1986,[4] 27% for the Mexico City

TABLE 2.1. Overview of worldwide cesarean section rates (%)

Country	Year 70	71	72	73	74	75	76	77	78	79	80	81	82	83	84	85	86	87	88	89	90	91	92	Reference source
U.S.A.	5.5	5.8	7.0	8.0	9.2	10.4	12.1	13.7	15.2	16.4	16.5	17.9	18.5	20.3		23.0		24.0	25.0					1, 4, 5, 6, 10, 11, 14, 19, 41, 42, 42
Canada	5.7	6.4	7.2	8.0	9.0	9.6	10.8	12.1	13.9	15.1	15.9					19.0			20.0					1, 4, 5, 10
England	5.0	5.2	5.3	5.3	5.6	6.0	6.5	7.4	7.5	8.5	9.0	9.3	10.1	10.1		10.0			15.0	12.1			13.0	1, 4, 5, 10, 37
Scotland	5.9	5.8	7.0	8.0	9.2	10.4	12.1	13.7	15.2	16.4	16.5	17.9	18.5	20.3	13.5									1, 4, 38
Puerto Rico																29.0								4
Bavaria										11.3	11.2	11.7	12.2	13.2		15.0								1, 4
Denmark	5.7		5.9		7.2	7.5	8.0	8.8		9.0		11.7	11.7	12.8		13.0		12.1						1, 4, 39
Italy											11.2	12.1	13.0	14.5	15.7	15.8		17.5						4, 7, 18, 19
Portugal	6.1	7.2	8.3	7.9	8.7	8.9	8.4	9.1	9.1	10.3	9.5	9.4	9.8	12.9				13.0						1, 4
Greece														13.0					16.7					4, 39
Sweden				5.5	6.8	7.8	9.6	10.9	11.6	11.7	12.1	12.4				12.0								1, 4
Spain																12.0								4
Norway	2.2	2.4	2.5	3.0	3.6	4.1	5.0	6.4	7.3	8.0	8.3	8.7	9.0	9.4		12.5								1, 4, 14, 38
Switzerland																	11.0							4
New Zealand	3.9		3.9	5.2	5.0		5.2					9.7	9.8			10.0								1, 4
Jamaica															4.3	5.8	8.1	4.6	4.5	5.5				3
W. Australia	4.2	4.1	4.6	5.1	7.3	8.2	8.7	9.9	11.7	13.0	11.2	11.8	12.5	13.3	13.9	15.1	15.6	16.9						1, 8
S. Australia												16.9					19.0							40
Victoria														14.5				16.4						40
Tasmania							6.0												12.9					40
Brazil					15.0						31.0		27.9				32.0							4, 17
East Africa												0.02												9
Netherlands	2.1	2.1	2.2	2.5	2.6	2.7	2.9	3.5	3.8	4.3	4.7	4.9	5.3			10.0						7.9		1, 4, 37
Hungary	6.2	6.7	6.9	7.0	6.7	6.5	6.6	6.9	7.4	7.6	8.0	8.6	9.2	9.5		10.0		10.2						1, 4, 39
Japan						6.4	6.5	6.8				8.0				7.0								1, 4
Czechoslovakia	2.3	2.3	2.4	2.5	2.6	2.8	3.1	3.4	3.7	4.0	4.4	4.7	5.2	6.0			7.0							1, 4
Austria												6.5	7.0	7.5										1
Belgium										7.2		7.4	8.0	8.1										1
Finland	6.0	6.9	7.2	7.9	8.1	8.2	8.8	10.0	10.9															1, 39
France			6.1				8.5			11.9		10.9						10.2	14.4					1
Saudi Arabia											5.4	6.0	5.4	7.5	9.9									15
Israel										3.9														39

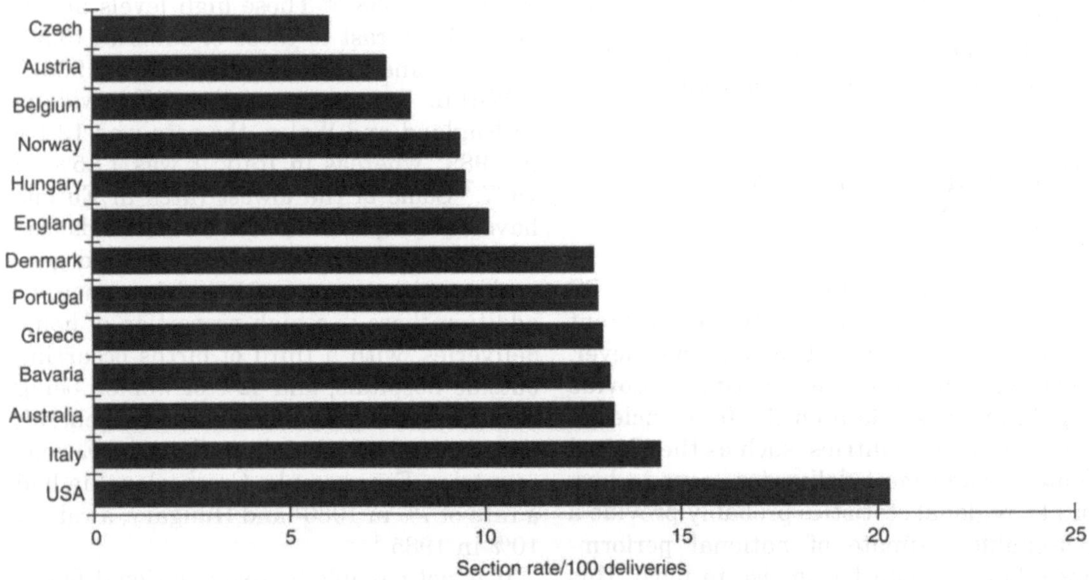

FIGURE 2.1. Cesarean section rate per 100 deliveries in 13 selected countries, 1983 (extracted from Table 2.1).

metropolitan area,[3] and 27% for Puerto Rico.[3] Few data are available for African countries but some reports have suggested extremely low rates. Indeed some authors have speculated that because the rates are so low, a number of necessary operations are not being performed.[9]

Trends in Worldwide Cesarean Section Rates

Although Table 2.1 and Figure 2.1 demonstrate that worldwide rates in CS differ, Table 2.1 also illustrates that its use throughout the western world has risen consistently in almost all countries during the past 20 years.[7] For example, it is the most commonly performed operation in the United States, where the rate was 16.5% in 1980[1] and rose to 25% by 1988.[5,6] This is perceived as a major public health problem and a cause for increasing concern because of the associated higher mortality, higher morbidity, and hence higher cost in comparison to vaginal deliveries. Increasing CS rates could place a serious burden on the health care programs of developing countries in which resources tend to be scarce.[3]

Figure 2.2 shows the trends in CS rates per 100 deliveries in seven selected countries from 1970 to 1992. The rates within these countries for any one year differ, but it can be seen that there is an upward trend in delivery by CS. Apart from Brazil, from which only patchy information is available, the United States has maintained the highest CS rate for the period shown. The Canadian trend is similar, as one might expect given the similarity in obstetric practice between these countries. These two countries have been consistently the "league leaders" during the past two decades.

Reasons for National Differences

Why are there such wide differences in the utilization of CS among different countries? The reasons can be classified as medical and nonmedical (Table 2.2).

Medical Reasons

In the United States between 1971 and 1982, four indications were responsible for most deliveries by CS: previous CS, breech presentation, fetal distress, and dystocia.[10]

Previous Cesarean Section

In the 1970s, dystocia and previous CS were the main factors responsible for the overall

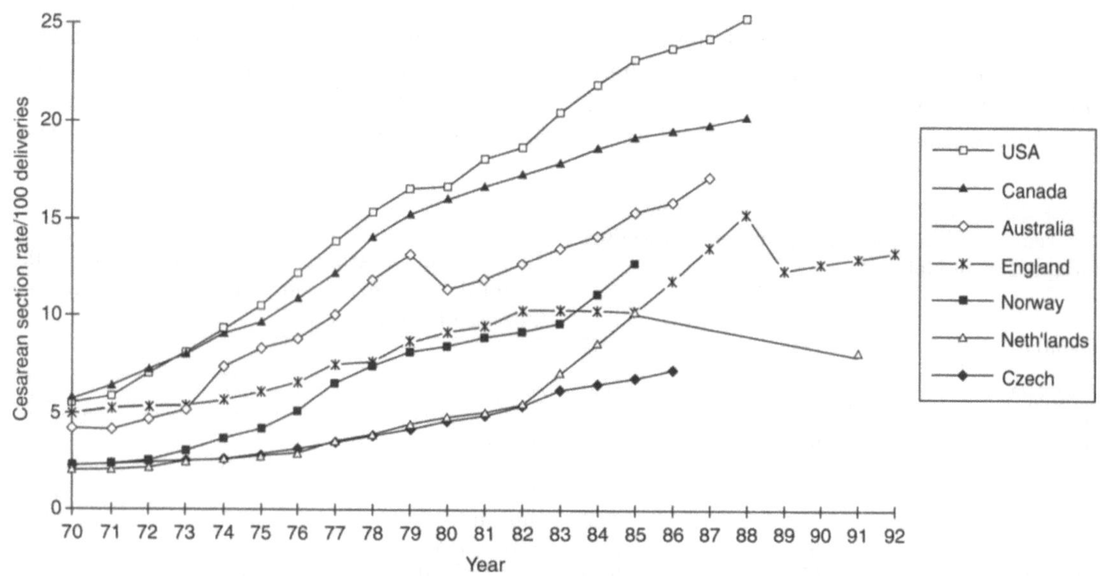

FIGURE 2.2. Trends in cesarean section rate per 100 deliveries in seven selected countries, 1970–1992 (see Table 2.1).

TABLE 2.2. Reasons for differences in rates of cesarean sections

Medical	Nonmedical
Previous cesarean	Maternal age
Breech presentation	Socioeconomic factors
Fetal distress	Cultural reasons
Cephalopelvic disproportion	Hospital resources
Dystocia	Geographical location
Low birthweight and preterm delivery	Private practice
	Litigation
	Physician factor

TABLE 2.4. The number of repeat cesareans as a percentage of the total number of operations (adapted from Notzon et al.[1])

Country	Rate
Australia	37
Bavaria	20
France	28
Hungary	8
New Zealand	11
Norway	23
Scotland	43
U.S.A.	35

rate increases.[10] However, by the early 1980s previous CS had become the principal factor behind the continued rise. Some reports have suggested that the single most important indication in determining the overall rate is previous CS.[10] Nevertheless, there are differences between countries in the number of women who have repeat CS (Table 2.3). Large differences also exist in the number of repeat CS performed as a percentage of the total CS rate (Table 2.4).

These differences among countries in the number of women undergoing repeat CS reflect different obstetric practices. The lowest proportion of vaginal births after previous CS is found in North America, where the figures are 6.6% for the United States and 5.7% for Canada.[4] Cragin's dictum of 1916 tends to be adopted there: "once a CS, always a CS."[5] However, various studies show that rates of 40%–80% for vaginal delivery after CS are possible.[11] Given the concern that is being expressed about the clinical and financial impact of rising CS rates,[3] a critical review of policies regarding management of women with a previous CS seems advisable.

Breech Presentation

The management of breech presentation, both preterm and term, continues to be controversial.[12] It is not surprising that countries differ widely in their use of CS for breech presentation (Table 2.5). Indeed there is probably as much variation again within countries. All countries illustrated show a rise in CS rate for breech presentation. The highest was 93.3% in Sweden, and the lowest was in the Netherlands at 34.8%. In the United States, the rate has risen from 67.2% in 1980 to 80.4% in 1985, despite the

TABLE 2.3. Percentage of women with a previous cesarean section undergoing repeat procedure (adapted from Notzon[4])

	Year	
Country	1980	1985
Australia		86.0
Brazil	80.0	
Bavaria	66.7	61.0
Canada	96.2	94.3
Hungary	67.4	68.3
Netherlands	41.1	44.6
New Zealand	87.7	81.8
Norway	57.4	53.5
Scotland	61.3	61.7
U.S.A.	96.6	93.4

TABLE 2.5. Cesarean section rates per 100 hospital deliveries for breech presentation (adapted from Notzon[4])

	Year	
Country	1980	1985
Bavaria	67.1	79.8
Canada	55.0	69.2
Denmark	55.8	75.4
Hungary	32.7	38.7
Netherlands	24.5	34.8
New Zealand	41.5	45.9
Sweden	91.9	93.3
U.S.A.	67.2	80.4

recommendations of the U.S. Cesarean Birth Task Force in 1980.[13] This group issued guidelines for selecting patients in whom a vaginal delivery of a full-term breech presentation could be anticipated:

1. Estimated fetal weight less than 3.58 kg
2. Normal pelvic dimensions
3. Frank breech without an extended head
4. Delivery by an experienced obstetrician

In contrast to these recommendations, others have suggested that all breeches be delivered by CS. For example, in one study the argument of the protagonists is that the risks of vaginal delivery to the baby are greater than the risks to the mother with elective CS. The incidence of intrapartum and neonatal death associated with vaginal birth was 0.83% compared with 0.03% in those born by CS. In addition, the numbers of low Apgar scores and neonatal intubation were doubled in babies born vaginally or by emergency CS compared to elective CS.[12] Others have calculated that the excess risk of neonatal death attributable to vaginal delivery of the term breech was about 4 per 1000.[12]

The reasons for the rise in rates of CS for breech presentation are complex. Part of the explanation may be a reflection of reduced training opportunities. New obstetricians are less often exposed to vaginal delivery and are unlikely to feel competent to attempt such a delivery, thus a spiral of falling vaginal delivery rates ensues. This in itself leads to more CS, with a previous section being the indication, as outlined earlier. In addition, medical-legal considerations may contribute to the rising CS rate for breech presentation.

Fetal Distress

Given the wide variation in criteria used for the diagnosis of "fetal distress," one might expect a wide variation in the rates of CS for this indication. Surprisingly therefore, with the exception of the United States, the variation among countries for this complication was not so great as with other indications, with the incidence being 20%–30% (Table 2.6). Further, there has not been a consistent rise with time.

Fetal heart rate (FHR) monitoring began in the early 1970s. By 1980, almost 48% of mothers of live births were electronically monitored. However, this produced well-documented increased rates of CS for fetal distress.[14,15] Admittedly, as experience increased the CS rate became less. Further, the addition of fetal blood sampling as an adjunct to FHR monitoring increases the specifity of diagnosis of hypoxia and limits the CS rate.[13]

Obviously, the availability of these methods for monitoring the fetus during labor differs among countries, which may explain some of the small differences. Where there is greater use of fetal monitoring, more frequent diagnosis of fetal distress is made. However, it may be that this complication is being overestimated. For example, 25% of all CS in France in 1982 were for fetal distress but 8%–10% of Apgar scores were normal at delivery.[13] The difficulties in the identification of intrapartum hypoxia have been well discussed,[13] and until methods of fetal monitoring with greater accuracy are available it is unreasonable to expect a fall in the number of CS performed for fetal distress.

Cephalopelvic Disproportion

The diagnosis of cephalopelvic disproportion (CPD) appears to be uniform for most countries. With the exception of the Nether-

TABLE 2.6. Cesarean section rates per 100 hospital deliveries for fetal distress (adapted from Notzon[4])

	Year	
Country	1980	1985
Bavaria	27.3	37.9
Canada	24.3	24.2
Denmark		21.6
Hungary	25.5	27.0
Netherlands	18.8	21.5
New Zealand	29.5	29.8
U.S.A.	66.9	55.7

lands (79%) and Denmark (71%), CS rates are in excess of 85% for all countries (Table 2.7). The only concern about drawing too many conclusions about these figures is that the accuracy of diagnosis of CPD is likely to be inconsistent among the various reports. Indeed, some of the CS performed for CPD could possibly result from dystocia.

Dystocia

Dystocia as a complication leading to CS delivery is a significant contributor. However, the term is vague and does not necessarily represent a single pathology. For example, inefficient uterine activity and CPD can both lead to the diagnosis. While there are inconsistencies in the way the term is used and hence the diagnosis made, there are two additional reasons why CS rates for dystocia may vary.

First, the diagnosis is more common following induced labor.[16] Indeed, some authors have argued that CS following induction with such an outcome is more correctly termed failed induction.[17] Induction rates vary widely worldwide, and this may contribute to some of the variation in CS rates from dystocia.

Second, the use of augmentation of labor is not uniformly practiced by all obstetric units. The protagonists argue that their lower CS rates result in large part from this approach.[13] While the size of the effect is debatable, there is sound evidence that use of this form of management will significantly reduce CS rates.[13]

TABLE 2.7. Cesarean section rates per 100 hospital deliveries resulting from cephalopelvic disproportion (CPD) (adapted from Notzon[4])

	Year	
Country	1980	1985
Bavaria	75.5	86.3
Canada	89.2	94.0
Denmark	77.8	70.5
Hungary		96.5
Netherlands	75.6	79.2
New Zealand	87.6	89.6
Sweden	93.0	92.4
U.S.A.	95.7	98.6

Low Birthweight and Preterm Delivery

Improvement in neonatal outcome, largely resulting from advances in neonatal intensive care, especially in the areas of ventilatory support and surfactant replacement, has altered obstetric practice.[13] However, it is tempting to speculate that some of the improvement is caused by the greater use of steroids antenatally where preterm delivery is contemplated.[13] Few would have performed CS for fetal reasons before 32 weeks gestation 10–15 years ago, but now CS for severe fetal compromise earlier than that gestation time is commonplace. Once again, however, there appears to be wide variation between countries in their use of CS for this indication (Table 2.8).

The differing rates presumably reflect the standards of neonatal facilities that are available in various countries. However, within individual countries, rates are highest among women with very low birthweight infants (<1500 g) and become lower as birthweight increases. In this context, it is of note that at the other end of the scale, CS rates are again increased for birthweights over 4 kg.[18]

Nonmedical Reasons

Maternal Age

Childbearing patterns also differ among countries. In western society, there is a trend for women to have fewer children and begin childbearing at an older age.[11]

TABLE 2.8. Cesarean section rates per 100 hospital deliveries for low birthweight babies (<2500 g) (adapted from Notzon[4])

	Year	
Country	1980	1985
Bavaria	21.2	35.1
Denmark	29.7	41.2
Hungary	10.8	15.8
Sweden	42.5	44.2
U.S.A.	27.3	

In a 20-year U.S. study, CS rates were consistently highest in women of 35 years and older (Table 2.9).[11] However, while the overall rate over two decades increased fivefold, the lowest increase (by a factor of 3.8) was seen in the oldest age group.

The finding of an increased risk of CS delivery with increasing maternal age appears to be a relatively consistent conclusion of several studies.[11,18–20] The reasons proposed for this are not clear, but include an increased risk of malpresentation and a decreasing proportion of occipitoanterior positions in older mothers.[20] Others have suggested the increased risk results from increased dysfunctional labor patterns in older women.[20] However, these obstetric/medical factors are unlikely to be the sole explanation. Undoubtedly in some cases, obstetricians may have a lower threshold for performing CS on these older women because their pregnancies are considered to be "precious,"[18] that is, the likelihood of having further pregnancies is low.

Socioeconomic Status

Several reports from Brazil,[21] Italy, and the United States have suggested there is a positive correlation between socioeconomic status and the use of CS, with women of higher social class having higher rates. For example, a survey conducted between 1982 and 1983 in Los Angeles County, California, showed that women with a median family income of more than $30,000 had a primary CS rate of 23%, as compared with 13% among those with a median family income under $11,000 (Figure 2.3).[22] It was concluded that rates of primary CS vary directly with socioeconomic status and that this association cannot be accounted for by differences in maternal age, parity, birthweight, race, ethnic group, or complications of pregnancy or childbirth. However, this observation may not solely be the responsibility of the obstetrician. The phenomenon of women requesting or even demanding CS is well recognized.[23]

Cultural Factors

Cultural factors can influence CS rates. For example, Brazil has one of the highest rates of CS in the world. Economic factors are important, with rates increasing with family income. In addition, there is strict adherence to "once a CS always a CS." However, arguably the most important reason for Brazil's high rate is that many CS are performed to carry out an intraoperative sterilization. In Brazil, for religious reasons, sterilizations are only allowed under very special circumstances, for example, when two independent physicians certify that future pregnancies would be contraindicated. In addition, there is no reimbursement by the national health insurance program. To get around these factors, doctors and families usually agree that the sterilization should be carried out during a CS. The doctor may then charge an extra fee and give other reasons for the operation in the hospital case notes.[21]

One study of 5960 mothers in southern Brazil from 1982 to 1986 reported that 9.4%

TABLE 2.9. Cesarean section rates by age of mother in the United States, selected years 1965–1985 (data from Taffel et al.[11])

		Age of mother (years)				
Year	All ages	<20	20–24	25–29	30–34	≥35
1985	22.7	16.1	21.2	22.9	26.6	30.7
1980	16.5	14.5	15.8	16.7	18.0	20.6
1975	10.4	8.4	9.0	11.1	13.6	15.0
1970	5.5	3.9	4.9	5.9	7.5	8.3
1965	4.5	3.1	3.5	4.3	6.4	7.9
Factor of increase	5.0	5.2	6.1	5.3	4.2	3.8

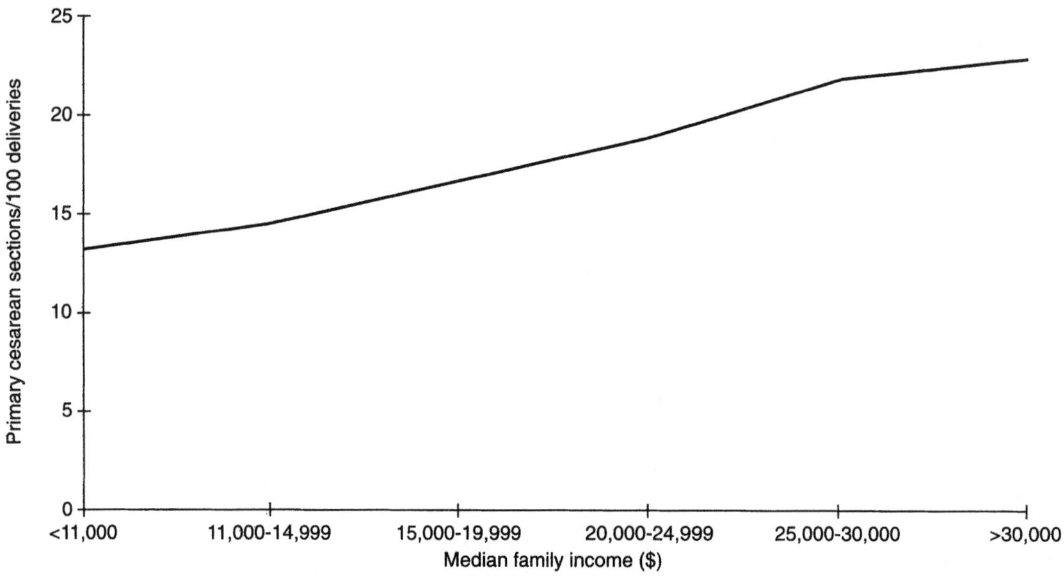

FIGURE 2.3. Rates of primary cesarean section in Los Angeles County (California, U.S.A.) according to median family income. (Adapted from information appearing in *The New England Journal of* Medicine. See Gould et al. [22])

of all women undergoing CS were sterilized. However, among women who had had their first child by a vaginal delivery and had undergone a CS for the second delivery, almost one-third (31%) were sterilized. For those with two previous normal births and a third by CS, 68% were sterilized. This increased to 80% when a fourth child was delivered by CS after three normal deliveries.[21]

In contrast, the rate of CS in Saudi Arabia is much lower. Figures from the Maternity and Childrens Hospital, Riyadh, show a rate of 3.9% in 1979, which had increased to 9.9% by 1984. Cultural reasons have been suggested to be responsible in part for this low rate. Vaginal delivery is particularly important in Saudi Arabia because of the social preference for a large family: the average size is six or seven children. Because repeated CS have the potential to limit family size, use of CS may have serious social repercussions such as marital breakdown or remarriage by the husband.[15]

Hospital Resources and Status

The resources that are available at a particular hospital and whether it is a community or a teaching hospital may have a bearing on CS rates. Some reports have shown that fetal distress is three times more likely to be diagnosed at a teaching hospital than in a community hospital.[24] This in part reflects the fact that teaching hospitals handle more women with high-risk pregnancies and may have greater availability of fetal monitoring. Further, community hospitals may lack the resources for an emergency CS that are thought to be required for management of a trial of labor in those women who have had a previous CS.[24] Although these valid reasons might explain why teaching hospitals have higher CS rates, there is still a suspicion that those practicing in such institutions may have an inherent tendency to prefer CS when faced with a choice.

Other factors include the availability of neonatal facilities and an epidural service. The effect of epidural analgesia on labor remains controversial. Some have reported that an on-demand epidural service does not affect the primary CS rate[25] while others have claimed that epidurals increase it.[26]

Clearly, these types of resources are likely to be extremely limited in developing countries, and as such it is easy to see how these factors would result in a lower CS rate.

Geographical Location

The location of hospitals and their proximity to the community is unlikely to be significant in developed countries, but in third-world areas it can have a significant influence on the rate of CS. For example, extremely low rates have been reported at rural hospitals in East Africa. In fact, rates are so low (25 per 100,000) as to imply that numerous obstetric disasters are occurring in remote areas because of difficulty of access to medical facilities and lack of availability of this surgical procedure.[9] Vesicovaginal fistulas and childhood disability from cerebral trauma or anoxia at birth are just two likely examples of such problems.[9]

Private Practice

Source of payment for obstetrical care is related to CS use. This is particularly the case in the United States, where women who are privately insured have the highest CS rates.[26–28] Studies have shown similar patterns in Brazil and Italy.[18,29] However, this strong relationship has not been demonstrated for all countries with their differing health care systems.

A Californian study in 1986 reported on 461,066 deliveries.[29] Women with private insurance had the highest rate (29.1%), and successively lower rates were observed for women covered under different schemes.[29] Similar reports have come from Brazil,[30] one of which showed that 75% of deliveries of private patients were by CS compared to less than 25% for indigent patients. Medical indications were unable to explain these variations. In an attempt to decrease this rate, the government insurance program subsequently equalized the reimbursement fees for vaginal delivery and CS.

Financial considerations certainly play an important role. It may be that the obstetrician is subjected to pressure from their private patient to perform a CS. Alternatively, it may be that the performance of CS provides obstetricians with incentives in terms of greater financial reward and less inconvenience.

Litigation

The practice of "defensive" medicine is increasing, particularly in the United States where litigation is high. Rising malpractice claims have been documented by the American College of Obstetricians and Gynecologists (ACOG). In 1985, 73% of ACOG Fellows, in response to a survey, reported that one or more professional liability claims had been filed against them, compared with 67% in 1983.[11] Undoubtedly litigation is an important factor in CS rates. In a survey of 306 consultants asking for their reasons for the rise in CS rates,[7] litigation was the leading reason given by 42% of British obstetricians for the rise in U.K. CS rates and by 84% of American obstetricians for the rise in the United States.[7] Despite its importance in these countries, litigation has little relevance in other countries, such as Italy, where malpractice claims are a relatively uncommon phenomenon.[18] If litigation continues to rise, its impact on CS rates will increase, because obstetricians are more likely to deliver by this method rather than attempt even a slightly difficult vaginal delivery at the risk of a malpractice claim.

Physician Factor

Apart from the factors already described, arguably the most important influence is that of the obstetricians themselves, who make the decisions to deliver a baby by CS. Many medical and nonmedical factors influence the decisions to perform a CS birth. Obstetricians will differ in their interpretation of these factors according to their training and experience. In any given case, some may decide that CS is necessary, while others may not.

The Green Bay study of 1986–1988 analyzed 1030 CS deliveries performed on singleton pregnancies by 11 different obstetricians practicing during this time. Rates varied from 5.6% to 19.7% for the different physicians and were not attributable to patient obstetric risk factors, socioeconomic status, service status, or duration of the physician practice.[31] Other such studies have shown similar results.[6]

Cesarean Section Rates and Outcome

It has been clearly demonstrated that CS rates have continued to rise in recent years. What effect has this had on the outcome of pregnancy?

The traditional measure used to assess outcome is perinatal mortality (PNM). Perinatal mortality rates have decreased in recent years, and it seems logical to apply a cause-and-effect relationship between these and increasing CS rates. The relationship between CS rates and PNM rates for 14 countries is illustrated in Figure 2.4. The data suggest that frequencies of delivery do not contribute much to variation in perinatal mortality rates. For example, in 1985 the United States had a CS rate of 23%, with a PNM ratio of 10.8. This is in contrast to Sweden, which had a lower CS rate of 12% but also a lower PNM ratio of 7.3 (Figure 2.4). Clearly, perinatal mortality must be more closely linked with other characteristics that vary among the nations.[2] This was well demonstrated in 1983 by O'Driscoll and Foley, who compared the PNM and CS rates between Dublin and the United States from 1965 to 1980. Perinatal mortality improved along parallel lines for this period, but while the CS rate was increasing from 5% to 15% in the United States, that in Dublin remained remarkedly constant at less than 5%.[32] Many studies have confirmed this and failed to show a strong correlation between CS and PNM rates.[18,31–34]

Factors other than increased CS rates clearly are responsible for the improved PNM rates. It is possible that the regionalization of perinatal care, better techniques for fetal monitoring, and improving neonatal care have a greater impact. In addition, it has also been suggested that a reduction in neonatal deaths from respiratory distress syndrome has had a larger affect on PNM in recent years than has the mode of delivery.[35]

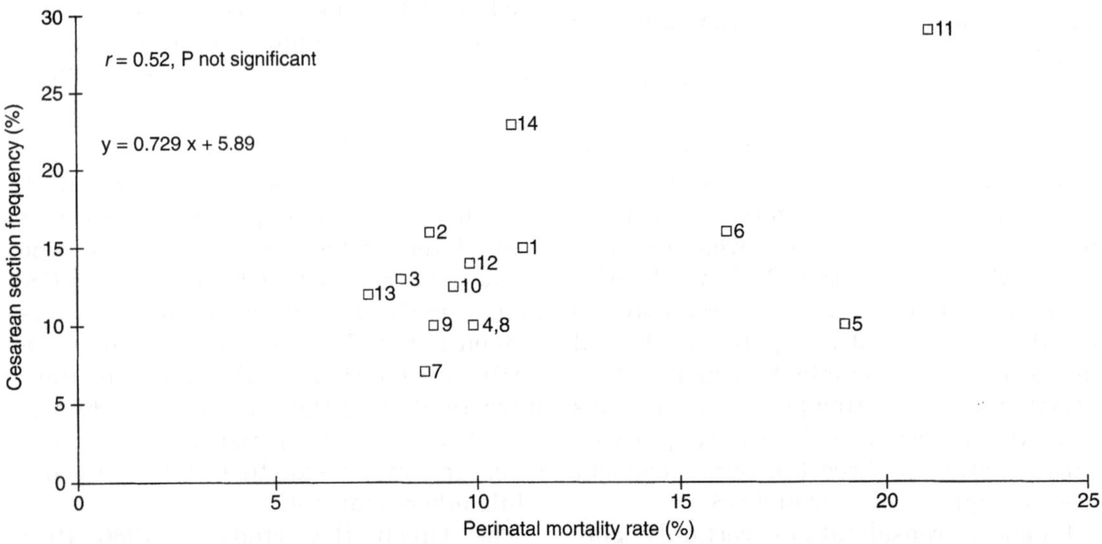

FIGURE 2.4. Perinatal mortality rates compared to cesarean rate frequency in 14 countries, 1985. *Key*: 1, Australia; 2, Canada; 3, Denmark; 4, England; 5, Hungary; 6, Italy; 7, Japan; 8, Netherlands; 9, New Zealand; 10, Norway; 11, Puerto Rico; 12, Scotland; 13, Sweden; 14, United States. (Adapted from Notzon.[4])

To assess the effectiveness of CS with respect to outcome, perhaps it would be useful to look more specifically at outcome (mortality and morbidity) for specific indications rather than overall PNM rates. Thus, for example we should study the benefit of CS for fetal distress and outcome measures relating to fetal asphyxial insult, or we should study the benefit of CS for repeat CS and outcome measures relating to fetal risk and maternal complications.

Conclusions

Worldwide CS rates differ greatly. The only consistent finding is that they are increasing in every country. The reason for the differences and the rise in CS rates are many and complex, reflecting medical and nonmedical factors.

What should be the ideal rate? "How long is a piece of string?"[36] There is no answer to that question; however, that is no reason for complacency. When looking at CS rates for our own institution or discussing the optimum mode of delivery for a specific mother, there are certain guidelines that will avoid excessively high figures.

We should be vigilant to avoid socioeconomic, emotional, and age-related factors that might influence our decisions, although it is difficult to argue with a mother who demands a CS. We should guard against using different standards with private patients. We should resist attempts to make our decisions litigation driven. In developing countries, mechanisms should be in place for transfer of women in labor who might benefit from CS where no such facility exists locally.

In women with previous single CS, we should consider vaginal delivery as an option in subsequent pregnancies. In a woman with a breech presentation we should not automatically think of a CS as the only delivery option. We should seek improved and more accurate methods or diagnosing intrapartum fetal hypoxia. Certainly our CS rates for fetal distress diagnosed by FHR abnormality will be lower if we use fetal scalp pH sampling as an adjunct.

Finally, our CS rates for "dystocia" will be lower if we avoid unnecessary inductions and adopt a policy of augmentation of labour.

Acknowledgments. We are grateful to Mrs. L. Straw and Dr. D. Sahota for their help in the preparation of this chapter.

References

1. Notzon FC, Placek PJ, Taffel SM. Comparisons of national CS rates. N Engl J Med 1987;316:386–389.
2. Bergsjo P, Schmidt E, Pusch D. Differences in the reported frequencies of some obstetrical interventions in Europe. Br J Obstet Gynaecol 1983;90:628–632.
3. Webster LA, Daling JR, McFarlane C, Ashley D, Warren CW. Prevalence and determinants of CS in Jamaica. J Biosoc Sci 1992;24:515–525.
4. Notzon FC. International differences in the use of obstetric interventions. JAMA 1990;264:3286–3291.
5. Lomas J. Holding back the tide of caesareans. Br Med J 1988;297:569–570.
6. Goyert GL, Bottoms SF, Treadwell MC, Nehra PC. The physician factor in caesarean birth rates. N Engl J Med 1989;320:706–709.
7. Savage W, Francome C. British CS rates: have we reached a plateau? Br J Obstet Gynaecol 1993;100:493–496.
8. Read AW, Waddell VP, Prediville WJ, Stanley FJ. Trends in CS in Western Australia 1980–1987. Med J Aust 1990;153:318–323.
9. Nordberg EM. Incidence and estimated need of CS, inguinal hernia repair, and operations for strangulated hernia in rural Africa. Br Med J 1984;289:92–93.
10. Anderson GM, Lomas J. Recent trends in CS rates in Ontario. Can Med Assoc J 1989;141: 1049–1053.
11. Taffel SM, Placek PJ, Liss T. Trends in the United States CS rate and reasons for the 1980–85 rise. Am J Public Health 1987;77: 955–959.
12. Thorpe-Beeston JG, Banfield PJ, St. G Saunders NJ. Outcome of breech delivery at term. Br Med J 1992;305:746–747.

13. Derom R, Patel NB, Thiery M. Implications of increasing rates of CS. In: Studd J, ed. Progress in obstetrics and gynaecology, Vol. 6. New York: Churchill Livingstone, 1987: 175–194.
14. Borthen I, Lossius P, Skjaerven R, Bergsjo P. Changes in frequency and indications for CS in Norway 1967–1984. Acta Obstet Gynecol Scand 1989;68:589–593.
15. Chattopadhyay SK, Sengupta PB, Edress YB, Lambourne A. CS; changing patterns in Saudi Arabia. Int J Gynaecol Obstet 1987;25:387–394.
16. Gibb DMF, Cardozo LD, Studd JWW, Cooper DJ. Prolonged pregnancy: is induction of labour indicated? A prospective study. Br J Obstet Gynaecol 1982;89:292–295.
17. Bergsjo P, Bakketeig LS, Eikhom SN. Case-control analysis of post-term induction of labour. Acta Obstet Gynecol Scand 1982;61: 317–324.
18. Signorelli C, Elliott P, Cattaruzza MS, Osborn J. Trend of CS in Italy: an examination of national data 1980–1985. Int J Epidemiol 1991;20:712–716.
19. Parazzini F, Pirotta N, La Vecchia C, Fedele L. Determinants of CS rates in Italy. Br J Obstet Gynaecol 1992;99:203–206.
20. Peipert JF, Bracken MB. Maternal age: an independent risk factor for caesarean delivery. Obstet Gynecol 1993;81:200–205.
21. Barros FC, Vaughan JP, Victoria CG, Huttly SRA. Epidemic of CS in Brazil. Lancet 1991; 338:167–169.
22. Gould JB, Davey B, Stafford RS. Socioeconomic differences in rates of CS. N Engl J Med 1989;321:233–239.
23. Hall MM. When a woman asks for a CS. Br Med J 1987;294:201–202.
24. Anderson GM, Lomas J. Explaining variations in CS rates: patients, facilities or policies? Can Med Assoc J 1985;132:253–256.
25. Gribble RK, Meier PR. Effect of epidural analgesia on the primary caesarean rate. Obstet Gynecol 1991;78:231–234.
26. Thorp JA, Parisi VM, Brylan PC, Johnstone DA. The effect of continuous epidural analgesia on CS for dystocia in nulliparous women. Am J Obstet Gynecol 1989;161:670–675.
27. Stafford RS. Alternative strategies for controlling rising CS rates. JAMA 1990;263: 683–687.
28. Bertollini R, DiLallo D, Spaden T, Perucci C. CS rates in Italy by hospital payment mode: an analysis based on birth certificates. Am J Public Health 1992;82:257–261.
29. Stafford RS. CS use and source of payment: an analysis of California hospital discharge abstracts. Am J Public Health 1990;80:313–315.
30. Janowitz B, Nakamura M, Lins FE, Brown ML, Clopton D. CS in Brazil. Soc Sci & Med 1982;16:19–25.
31. DeMott RK, Sandmire HF. The Green Bay CS study: 1. The physician factor as a determinant of caesarean birth rates. Am J Obstet Gynecol 1990;162:1593–1602.
32. O'Driscoll K, Foley M. Correlation of decrease in perinatal mortality and increase in CS rates. Obstet Gynecol 1983;61:1–5.
33. Sanchez-Ramos L, Kaunitz AM, Peterson HB, Martinez-Schnell B, Thompson RJ. Reducing CS at a teaching hospital. Am J Obstet Gynecol 1990;163:1081–1088.
34. Taylor U, Zentay Z, Ganesh V, Apuzzio J, Murphy G, Iffy L. Rates of CS and neonatal mortality. Aust NZ J Obstet Gynaecol 1992; 32:203–205.
35. Rydhstrom H, Ingemarsson K, Ohrlander S. Lack of correlation between a high CS rate and improved prognosis for low-birthweight twins (<2500 g). Br J Obstet Gynaecol 1990;97:229–233.
36. Chamberlain G. What is the correct CS rate? Br J Obstet Gynaecol 1993;100:403–404.
37. Treffers P, Pel M. The rising trend for caesarean birth. Br Med J 1993;307:1017–1018.
38. Notzon FC, Bergsjo P, Cole S, Irgens LM, Daltveit AK. International collaborative effort (ICE) on birth weight, plurality, perinatal and infant mortality: IV. Differences in obstetrical delivery practice: Norway, Scotland and the United States. Acta Obstet Gynecol Scand 1991;70:451–460.
39. Stephenson PA, Bakoula C, Hemminki E, et al. Patterns of use of obstetrical interventions in 12 countries. Paediatr Perinatol Epidemiol 1993;7:45–54.
40. Foote AJ, Giles WB. Review of obstetric operative intervention rates. Asia-Oceania J Obstet Gynaecol 1992;3:195–198.
41. Anderson GM, Lomas J. Determinants of the increasing caesarean birth rate. Ontario

1979 to 1982. N Engl J Med 1984;311:887–892.
42. Shiono PH, McNellis D, Rhoads GG. Reasons for the rising caesarean delivery rates. Obstet Gynecol 1987;69:696–700.
43. Halpin GJ, Rose E, Shapiro E. Trends in CS rates. N J Med 1989;86:867–873.

3
Dystocia and "Failure to Progress" in Labor

EMANUEL A. FRIEDMAN

Too often invoked as the rationale for cesarean section, dystocia is a major (if not the principal) contributor to the phenomenal growth in the cesarean section rate during the past several decades in the United States and, to a greater or lesser extent, in every other industrialized nation in the world. There is no denying that our current cesarean section rate is excessively and unacceptably high, and there is no longer any logical argument to support it. Based on the most recent available national data (1991), the Centers for Disease Control (CDC) reports that the national cesarean rate has apparently leveled off at 23.5%.[1] The United States and Canada previously led the world in the frequency of cesarean deliveries,[2] and we can take little solace in the news that our current rates are exceeded only by Brazil and Puerto Rico.

In more specific and telling terms, the number of unnecessary cesarean sections, according to the recent CDC statement,[1] was 349,000, 36.1%, of the 966,000 done in 1991. There was some good news in this announcement to the effect that women who had previously had a cesarean section were being delivered vaginally (vaginal birth after cesarean, VBAC) more often; the frequency of successful VBACs rose to 24.2% from 20.2% in 1990, a small but nonetheless gratifying advance. However, the falling rate of repeat cesareans coupled with the unchanging overall rate has to reflect the bad news that, notwithstanding all efforts to reverse the trend, still more primary cesarean sections are being done annually, not fewer.

The national cost of these unnecessary cesarean sections, as calculated by the CDC solely from the differential hospital costs of cesarean versus vaginal delivery, is more than 1 billion dollars annually. It deserves to be emphasized, moreover, that the 1–2 billion-dollar estimate for unnecessary cesareans does not begin to take into account the economic toll in lost wages, prolonged convalescence, augmented home care needs, and the well-documented increase in the number and severity of major and minor complications so commonly associated with cesarean delivery, to say nothing about the medical, physical, psychologic, and societal repercussions that are likely to affect the patient and her family and which cannot be assigned a dollar value.

This information comes at a time when the economic impact of medical care is foremost in the minds of the public and the government. The pervading concern has come about as a consequence of the perception that our current health care system is too expensive and inefficient, and that it should be feasible to provide the same or better care at equal or less cost to more people (conceivably, to everyone) by some form of governmentally controlled health care system. The perception is bolstered by the fact that more than 12% of our gross national product goes for health care, far in excess of other nations in which the health care is at least comparable in quality and distribution. Given the general dissatisfaction with our current economic environment, it is clear that increasingly intense interest will be focused on modifying and ameliorating health care practices in those areas of medicine that have already been defined as potentially problematic. Obstetrician-gynecologists, having been identified as practicing a discipline in which "unnecessary" surgical procedures such as cesarean section and hysterectomy are being done at what is widely regarded as an excessive rate (regrettably, with some justification), are therefore especially vulnerable.

Given this situation, there is an obvious urgent need for us to address the problem objectively and unemotionally to help understand the issues, determine their root causes, consider potential solutions, and develop means for correcting them. This chapter addresses the single condition recognized to be among those leading the list of factors contributing the current high cesarean section rate, namely, dystocia. The only other competing factor in terms of incidence is repeat cesareans, which is variously encountered in about the same frequency or slightly more often.

The quantitative impact of dystocia on the increase in the frequency of cesarean sections has been put at between 28% and 43%.[3-9] In perspective, these relative incidence figures take on greater relevance when it is realized that many surgical interventions for dystocia yield babies of no more than normal weight and size.[10] Still more impressively, more than half (56.5% in one report[11] and 66% in another[12]) of cesarean sections in nulliparas are done for dystocia.

Thus, one-third to one-half or more of the growth in the cesarean rate can be attributable to this factor alone, and the combination of repeat cesareans and dystocia accounts for two-thirds to three-quarters of the increase. It should be clear from these data that if it were possible to reduce the frequency of cesarean sections undertaken for just these indications alone, so that these two factors would be used by obstetricians to justify abdominal delivery only when truly indicated, the resulting reduction in the cesarean section rate would be substantial.

What Is Dystocia?

Dystocia is etymologically derived from the Greek root *tokos*, childbirth, and the prefix modifier *dys*, meaning hard, bad, or difficult. Thus, the term is merely a nonspecific descriptor of a difficult parturition. Its use without further definitional characterization, therefore, is at best suspect. The suspicion that the term may be inappropriately used to describe a given labor, and thereby justify an unnecessary operative intervention, is magnified by the not-infrequent application of the term on the basis of a strictly subjective (if not overtly erroneous) interpretation of the events of labor.

As the shadow of doubt looming over the merit of a diagnosis of dystocia has become increasingly appreciated and publicized, admonitions have been forthcoming to abandon it or to define it in clinically objective terms so as to yield diagnostic relevance.[13-16] As a consequence, substitute language has been introduced. The term now most widely used in lieu of dystocia is its clinical euphemism, "failure to progress." Substitution of this new term for dystocia has apparently occurred widely, but it has occurred without concomitant introduction of measures to ensure that it would be clearly defined in specific terms, documented when used, and applied meaningfully to any given case and only in those in which it truly reflects the existence of a labor process that is not only nonprogressive, but one that really warrants aggressive intervention by cesarean section.

In a few institutions where compliance with these stringent constraints occurred, the intended effect of reducing the number of unnecessary abdominal deliveries from the inappropriate use of the diagnostic term dystocia did materialize.[17] Programs to monitor the quality of labor care, introducing use of labor curves (see following) or equivalent labor definitions to describe the specific type of labor aberration, plus internal surveillance of management practices supplemented with external reviews, proved capable of reducing the cesarean section rate by a considerable degree without adversely affecting outcome results. Where educational programs for doctors, nurses, and the lay public were tried alone, no impact was discernible, and the cesarean section rate continued to rise.[18] Overall, however, very little has occurred nationwide to reduce the cesarean section rate.[4] Needless to say, merely changing the terminology, which appears to be all that took place in most communities, could not have been realistically expected to have had any material impact.

Nonspecific Nature of Failure to Progress

A gravida who fails to progress in labor may be exhibiting one of a number of different conditions ranging from patently innocuous to potentially very serious.[19–21] It is necessary to recognize that failure to progress comprises several dissimilar and quite unrelated disorders, each of which must be differentiated from the others and managed in rather specific ways according to management principles which are appropriate for each abnormality and may prove inappropriate, if not outright counterproductive or even harmful, if applied to the treatment of the others.[22]

To diagnose a given laboring patient as having dystocia or failure to progress, and to proceed forthwith to undertake the major surgical procedure of cesarean section without first differentiating which one of these several conditions she may actually be experiencing, cannot be considered acceptable. Conducting one's practice in this manner would be no more correct than, for example, undertaking an exploratory laparotomy for appendectomy in every adult with abdominal pain without suitable clinical and laboratory evaluation to both confirm the diagnosis and rule out nonsurgical conditions.

To evaluate labor progression in such a woman, to determine whether she actually has a labor abnormality, and to identify its specific type if present, one must be familiar with the normal course of labor, have a working knowledge of the limits of the normal labor pattern, and be able to recognize the kinds of possible disorders. Moreover, after a specific disorder is diagnosed, it becomes necessary for the health care provider to undertake a timely assessment to determine what is the likely underlying cause of the disorder.

It is only with this type of information in hand, namely, a defined disorder with an identifiable probable etiology, that one can proceed with an appropriate management program tailored to meet the needs of that particular patient and her fetus. To achieve this worthwhile goal, the attendant must know about the kinds of conditions that can cause each of the different types of labor aberrations and thus investigate for them. He or she must know what therapeutic options are available and appropriately applicable to the specific labor disorder and the concurrent conditions prevailing in that patient. Moreover, the practitioner must know what can be expected prognostically from each condition in regard to delivery outcome and result to mother and fetus, thereby to be able to choose among them intelligently and with the best interests of both mother and fetus in mind.

Graphing the Labor Course

To distinguish among the several conditions responsible for the situation in which a laboring patient appears not to be making progress, the simple expedient of graphing the labor course should prove very useful. It has long been common knowledge that normal labor almost always follows a definable pattern over time.[22] Aside from consideration of the requirements for careful assessment of maternal and fetal status (matters of great importance not considered in this chapter, which is limited to the diagnosis and management of dystocia and therefore addresses the labor course only), the complexity of the clinical observations used to try to interpret the dynamic changes that occur during labor encompass those that focus on the contractility pattern, the changes that can be expected to occur in cervical dilatation and effacement, and those anticipated for fetal descent through the birth canal.

It is obvious that uterine contractions are the sine qua non of labor; little will occur over the course of labor without them. Nonetheless, the obstetrician's ability to utilize the pattern of contractions, that is, their intensity, frequency, and duration or the wave pattern generated by continuous intrauterine pressure measurements, as clinically relevant data for purposes of assessing labor progress has thus far proved to be of essentially no clinical value in assessing whether a given labor is normal.[23,24] Similarly, while information about effacement of the cervix is sometimes helpful in other ways, such as in determining whether a patient is in true labor or perhaps is likely to be a suitable candidate for labor induction, it does not help solve the clinically important concerns over whether a given labor is normal or not.

The clinician's role can be greatly simplified by concentrating on the two remaining factors, namely cervical dilatation and fetal descent, both of which can be considered to represent the effective end result of those labor forces acting to cause the labor to progress along normal lines so that the mouth of the womb is opened sufficiently to allow the fetus to exit from the uterus and to be eventually delivered to the outside world. Plotting the course of cervical dilatation against the time that elapses after labor begins (or is assumed to have begun) and fetal station on a similar time basis permits one to evolve a graphic representation of the labor course. Tracing the pattern can be done by the use of a simple square-ruled grid drawn on paper or, as is the case in many obstetric units around the world, a preprinted blank graphic form incorporated into the patient's hospital chart for this purpose.

The form consists of a graph with the horizontal axis divided into hours. The vertical axis is usually constituted into 10 equal divisions, scaled in two ways simultaneously: on the left border by cervical dilatation, in centimeters, from 0 to 10; and on the right border by fetal station from −5 to +5 according to the convention that designates station by the level in centimeters to which the forward leading edge of the fetal presenting part has descended relative to the referent plane ("zero" station) of the ischial spines at the midplane of the maternal pelvis (negative numbers signify stations cephalad to the ischial spines and positive numbers, caudad).

If the older but less preferable system of dividing the pelvis into thirds is used instead, so that the range of station designations is −3 to +3, it is recommended that suitable adjustments be made in the scale to ensure that the recordings are comparable to those made using the centimeter scale and the evolving descent curves can be evaluated accordingly by the same criteria. The adjustment can be made graphically or by simple arithmetical calculation (one station on the thirds scale being equivalent to 1.67 cm on the centimeter scale).

For greatest utility, one should ordinarily begin the timing scale on the horizontal time axis (at the lower left corner of the grid) for a given patient with the onset of her labor, by convention that time when she reports that the contractions were perceived by her and became regular, acknowledging that in due course it may become clear (by hindsight only) that she may not actually have been in labor at all at that time. This will mean that there will often be a period of time, extending from the onset of labor until the patient is first seen and examined, during which the graph will necessarily remain blank; while this might seem an ill-advised waste of space in the chart, it may prove useful to help determine the duration of the initial portion of the labor process and to remind all attendant personnel that that portion of the labor occurred outside their surveillance. While events during the prodromal or latent phase labor may seldom prove to be clinically important, recent data suggest that the disorder of prolonged latent phase, heretofore deemed innocuous, may actually contribute to adverse fetal outcome.[24]

The results of each successive vaginal examination for observing the degree of cervical dilatation and fetal station from the time of hospital admission until delivery are entered into the grid at the appropriate points on the graph according to when they were obtained. For example, a determination of 3 cm of cervical dilatation found at the examination done just 7 h after labor began would be entered on the seventh line from the left margin of the graph, three divisions up from the bottom. As each subsequent dilatation entry is recorded in sequence and joined to the preceding one by a straight line, it can be expected to yield an evolving pattern of cervical dilatation. The same is done to trace the descent pattern.

Graphing a labor as it progresses has the advantage of allowing attendant personnel (as well as the patient) to see the pattern of her labor as it develops. This will show whether or not she is following a pathway consistent with normal in regard to its general shape and falling within the range of normal component durations and slopes. Under ordinary circumstances, for example, all normal gravidas can be expected to trace very typically shaped patterns.

The normal dilatation curve is S shaped. It begins as a flat, essentially unchanging line at the onset of labor. This is the *latent phase* of the first stage of labor. After a period of time, dilatation accelerates, although the contractility pattern or the patient's subjective perception of her uterine contractions is unlikely to change to any degree. The upswing of the dilatation curve marks the end of the latent phase of the first stage of labor and the beginning of the *active phase*, which continues until the end of the first stage at full cervical dilatation. The dilatation continues to speed up (the *acceleration phase*, the first of the three identifiable components of the active phase) until it reaches a fairly rapid speed. Once its maximum rate has been achieved, dilatation usually continues in this fashion at that constant rate (that is, the curve assumes linearity, now in the *phase of maximum slope*) until the cervix is nearly fully dilated. The rate of dilatation then appears to slow down (the *deceleration phase*) until the cervix is fully retracted and fully dilated at the onset of the second stage.

The deceleration phase is produced artifactually in that the cervix is being pulled centrifugally at the same rate as earlier in the phase of maximum slope, but instead of dilating, it is being drawn upward by the action of the upper uterine segment. Thus, continuing dilatation appears to slow down or even cease in due course as the cervix is progressively retracted around the fetal head. The interval encompassing the acceleration phase, phase of maximum slope, and deceleration phase is the active phase of the first stage. The combination of latent and active phases comprises the first stage.

At the same time, the descent curve remains flat until the dilatation pattern is well advanced into the active phase. Descent usually begins during the phase of maximum slope of the dilatation curve, and reaches its maximum rate by the beginning of the deceleration phase or, at latest, at the onset of second stage. Once a maximum rate of descent is achieved, descent continues at this rate, proceeding linearly, until the presenting part reaches the level of the perineum.

Defining Abnormal Labor Patterns Objectively

The characteristic pattern of the normal dilatation curve thus is sigmoid in shape and the normal descent curve is hyperbolic.

When a given patient's labor pattern deviates qualitatively from these shapes in certain recognizable ways, that labor pattern can be diagnosed as being abnormal. The kinds of abnormalities that can be identified on the basis of such qualitative deviation from normal are those in which the expected linear dilatation in the active phase stops (for at least 2 h) before the the second stage is reached; this is called *secondary arrest of dilatation*. Another is the parallel cessation of descent (for at least 1 h) after the maximum slope of descent has begun; this is *arrest of descent* in the second stage. A third disorder recognized by its shape is that of *failure of descent*, readily diagnosable if the fetal descent fails to materialize at or after the time descent should already be at its maximum, that is, when the deceleration phase of dilatation begins or, at the latest, when the second stage has begun.

In addition to abnormalities based on the characteristic shapes of the dilatation and descent curves, other aberrations can be identified on the basis of the clinically valuable, quantitative attributes they exhibit. The latter are based on the statistical distribution of the normal range of durations and slopes of the several component parts of the curves. Just as the aforementioned qualitative variations in shape were able to identify several abnormalities of labor progression, so the quantitative changes serve in like manner as important indicators of the presence of others. The qualitative abnormalities thus defined include prolonged latent phase, protracted dilatation, prolonged deceleration phase, and protracted descent, as detailed in Table 3.1.

Potential Errors in Making the Diagnosis of Failure to Progress

As should now be evident from our review of the kinds of labor disorders that are so clearly diagnosable on the basis of the deviations from the normal cervical dilatation and fetal descent curves, failure to progress, or the apparent lack of progress in either cervical dilatation or fetal descent, for any finite period of time in labor may represent any of several quite different entities. The range of possibilities is listed in Table 3.2,

TABLE 3.1. Labor disorders objectively identifiable by graphic analysis with diagnostic criteria and managment options

Labor pattern	Diagnostic criteria: Nulliparas	Diagnostic criteria: Multiparas	Recommended therapy
Prolonged latent phase	>20 h	>14 h	Therapeutic rest Exceptionally, oxytocin if delay of 6–10 h is unacceptable
Protracted dilatation	<1.2 cm/h	<1.5 cm/h	Expectancy and support CS if CPD disclosed
Protracted descent	<1.0 cm/h	<2.0 cm/h	Expectancy and support CS if CPD disclosed
Secondary arrest of dilatation	>2 h arrest in active phase		Oxytocin only if CPD ruled out; CS if CPD probably present
Prolonged deceleration	>3 h	>1 h	Oxytocin only if CPD ruled out; CS if CPD probably present
Arrest of descent	>1 h arrest in second stage		Oxytocin only if CPD ruled out; CS if CPD probably present
Failure of descent	No descent in second stage		Oxytocin only if CPD ruled out; CS if CPD probably present

CS, cesarean section; CPD, cephalopelvic disproportion.

TABLE 3.2. Types of labor conditions that failure to progress may represent

Labor condition	Probability of diagnostic error	Source of diagnostic error
Normal latent phase	100%	Not an abnormality because it is found in all labors without exception
Prolonged latent phase	100%	Not an indicator of obstructed labor
Arrest disorder Secondary arrest of dilatation Prolonged deceleration phase Arrest of descent Failure of descent	50%	Unassociated with CPD in half the cases
Protraction disorder Protracted dilatation Protracted descent	70%	Unassociated with CPD in more than two-thirds of cases
Prolonged second stage associated with Epidural anesthesia Fetal malposition Maternal exhaustion Ineffective uterine contractions Poor expulsive efforts	? (Unknown)	Not all long second stages are the result of CPD; a prolonged second stage is not a sufficient indication per se for a cesarean

CS, cesarean section; CPD, cephalopelvic disproportion.

along with an estimate of the probability of making a clinical error if a diagnosis of failure to progress is invoked without further clarification and a cesarean section is done forthwith on that basis alone. Depending on the type of labor condition that actually exists in a given case, this practice will frequently prove to have been the wrong choice.

Normal Latent Phase

The condition of least significance insofar as it reflects a normal state of affairs is one that is found in the course of every labor, whether normal or not, namely the normal latent phase. During this initial part of the first stage, nothing by way of active dilatation or descent is supposed to be happening. Careful observation will sometimes demonstrate some softening in the consistency of the cervix and even some minor degrees of effacement, especially in multiparas. Effacement is usually complete in nulliparas long before labor begins. In some gravidas, again primarily multiparas, even some dilatation can be detected before the active phase begins; if this occurs at all, it takes place to a very limited extent and very slowly. A rate as high as 0.6 cm/h has been observed, but this is rare.[22]

However, it may be all too easy to be deluded into believing that the patient whose labor has not yet advanced beyond the normal latent phase is exhibiting the disorder of failure to progress for which aggressive intervention by cesarean section is indicated. This applies especially if the observer is either unaware that that is what is expected during early normal labor or, perhaps more often, does not fully appreciate that this patient is still in her latent phase. This latter is the kind of misinterpretation that can arise when the cervix of a patient, particularly a multipara, has already been dilated somewhat in the days or weeks before the onset of her labor. This is the gravida who is typically admitted to the labor unit some hours after labor begins and at that time is found to have a well-dilated cervix, say 4 cm. The impression is that her labor must be in the active phase because her cervix is so dilated. This not uncommon false impression is based on the belief held by many that the degree of dilatation defines the onset of active phase, rather than the upswing of the dilatation curve. The active phase can begin at almost any cervical dilatation, although it is admittedly most likely to be in progress by the time the cervix reaches 4 cm; however, just because the cervix is so far dilated does not perforce means that the patient is in the active phase.

Given this mindset, if serial observations over the next several hours should show that the cervix has not continued to dilate, the patient may become a source of consternation to the attendant personnel who mistakenly conclude that a labor abnormality exists in the form of failure to progress. Their misdiagnosis stems from not having recognized that, contrary to expectations, this woman was still in the latent phase of labor; indeed, she may not yet even have been in labor at all. The difficult differential diagnosis in such an instance involves distinguishing between normal latent phase and secondary arrest of dilatation; it must be addressed in this scenario. More often than not, the situation is easily resolved; that is, the correct diagnosis is made, by taking into consideration the relatively short interval that has elapsed from the onset of labor, a review of the known prelabor dilatation documented by a recent outpatient or office examination, and perhaps the relatively inconsequential contractile pattern that reflects the antecedent prodromal or early labor. In most cases, the diagnosis of the normal latent phase should be obvious merely from the history alone; the patient reports that the onset of regular uterine contractions occurred only a relatively short time ago. Knowledge about what the limits of normal for a latent phase should be (see Table 3.1) and about the fact that advanced dilatation does not necessarily mean that the patient must be in the active phase should provide the necessary insight.

If, in any given case, there is still some remaining concern about whether the patient's condition represents true secondary arrest of dilatation in the active phase, as opposed to a normal latent phase, there would then be ample justification for undertaking an evaluation for cephalopelvic disproportion. Having ruled it out by means of a careful, astute vaginal examination, one might even consider uterotonic stimulation if the contractile pattern warranted it. However, to act upon the misdiagnosis by intervening aggressively with a cesarean section under these circumstances would be clearly unacceptable. To do so, in most instances, would mean that the cesarean delivery was being done for nothing more than a normal latent phase.

Prolonged Latent Phase

A less common error is diagnosing failure to progress in the presence of a prolonged latent phase. Patients who begin labor and do not make any progress in the latent phase for many hours, passing the time limit beyond which nearly all normal gravidas will have entered the active phase, are easily recognized as meeting all the diagnostic criteria for this labor aberration. This condition, when determined to exist, responds most favorably to a program of therapeutic rest, usually in the form of some narcotic-analgesic agent administered for purposes of providing a period of 6–8 h of sleep. During this interval, uterine quiescence is usually but not always achieved. When gravidas managed in this way awaken, most will be found to have advanced well into the active phase; they can be expected to proceed to delivery uneventfully, although they are not necessarily immune from the kinds of complications that may affect any woman in labor. In fact, as stated earlier, there may some cause for concern based upon a recent report suggesting that the prognosis, both maternal and fetal, is not entirely optimistic, especially as regards the increased frequency of subsequent labor disorders, cesarean delivery, neonatal depression, and need for resuscitation.[25]

Most of the remaining patients who are rested and fail to advance into the active phase of labor after they awaken are found to be out of labor altogether; that is, their uterine contractions have subsided completely, signifying that the foregoing event represented nothing more than a long interval of false labor. Thus, the narcotic agent has served the very useful purpose of offering a means for making the differential diagnosis between true labor and false labor. It must follow from this that if cesarean section had been undertaken for a patient in this subgroup, it would have been done in a gravida who had been in false labor. It seems superfluous to add that this would be unconscionable.

In the past, morphine was advocated as the sedative of choice for patients with prolonged latent phase, and it has proved clinically effective.[22] However, a recent as-yet unpublished investigation indicates tht morphine may not be entirely safe when used for this objective (Oriol N). Oximetry studies conducted under the aegis of an obstetric anesthesia service showed that maternal oxygen saturation frequently fell after morphine was given to patients in labor, even without perceptible changes in respiratory rate or the appearance of clinically discernible cyanosis; it even fell below an acceptable 80% level in 2 of 13 women. While no adverse effects have as yet been detected among fetuses in these cases, the observation of resulting maternal hypoxemia, however mild, serves to warn against use of depressing narcotic agents in these patients, particularly if there is some concurrently ongoing or potential problem with regard to fetal oxygenation. Some consideration should therefore be given to substituting other, potentially less hazardous agents for morphine, such as barbiturates or tocolytics. Limited experience with the latter, not as yet fully evaluated or formally reported, suggest they may prove worthy for this purpose.

As to the false designation of patients with prolonged latent phase as having dystocia or failure to progress and therefore warranting cesarean section, the latter diagnoses are not supportable and the intervention assuredly cannot be condoned. Cesarean section done merely for prolonged latent phase is both inappropriate and unnecessary. Lack of progression is characteristic of this labor disorder, of course, but it does not represent a serious labor aberration in the same sense as a major labor aberration, such as true arrest of dilatation or of descent, does in terms of the kind of associated conditions that may place mother and fetus in jeopardy.

There should be little if any clinical difficulty in differentiating prolonged latent phase from any other condition in the course of labor. The diagnosis takes little diagnostic acumen; if anything, it tends to be overdiagnosed by obstetricians who appear to be impatient to medicate patients so as to give them a therapeutic trial of rest to ascertain if they are in true or false labor. They thus invoke the diagnosis before the latent phase has actually exceeded normal limits, and proceed with the therapeutic rest regimen. The end result is the same, but the gravida's labor has been unduly prolonged as a consequence of the unnecessary use of the inhibitory drug.

Arrest Disorders

The four identifiable arrest disorders (see Table 3.1) constitute the most serious of the labor aberrations that can be diagnosed objectively by means of the graphic analysis. They are serious because they occur so often in gravidas whose labor is obstructed, that is, who have insurmountable cephalopelvic disproportion (CPD). The incidence of documentable CPD among women who develop one or more of the arrest disorders is about 50%; it is much higher (clinical experience suggests it may be as high as 90% but more precise data are as yet unavailable) if the arrest occurs while oxytocin is being administered or continues uncorrected for any length of time despite uterotonic augmentation by oxytocin.

As a basic management principle, it is mandatory that gravidas who demonstrate an arrest disorder be evaluated for CPD before any decisions are made about management. If CPD is present, or cannot be ruled out, cesarean section would appear to be entirely justified. The risks of permitting such patients to continue in their labors or to administer oxytocin to offer a further trial of labor are great.[22] Nonetheless, questions are now being raised as to whether the ostensible risks are sufficiently real to warrant such a recommendation for management.[26] The American College of Obstetricians and Gynecologists (ACOG) has even issued a statement advocating the use of oxytocin for use in cases of arrest of dilatation (but not arrest of descent) if the contractions are deemed inadequate.[27]

The criteria suggested for determining adequacy of uterine contractions include consideration of an objective determination of the contractions by use of Montevideo units, which are derived by the summation of the amplitude, minus the basal tonus, of all contractions that occur in an interval of 10 min. Specifically, an adequate contractility pattern is defined as one that generates at least 200 Montevideo units (per 10-min window) over a period of 2 h in labor. However, for this measurement of contractility to be at all meaningful, it requires the use of an indwelling intrauterine pressure catheter that will yield a properly calibrated continuous strain gauge recording. The use of external tokodynamometry is valueless in this regard and must not be relied upon for purposes of evaluating the contractility. Assessment by palpation of the contractions with the examining hand is far better, even though subjective and not capable of providing a quantifiable measurement.

The lack of an adequate scientific foundation for this recommendation as to what constitutes a minimally acceptable contractility pattern needs to be addressed because it is based on data pertaining to the level of measured contractility (in Montevideo units) achieved by 90% of parturients who respond favorably to oxytocin by successful cervical dilatation and subsequent fetal descent and vaginal delivery, not on what is minimally necessary to effect dilatation and descent in patients with an arrest or protraction disorder.[29] Nevertheless, it does have the advantage of at least being objective and reproducible. Moreover, it helps to focus on a common fault in managing patients whose labors are poorly progressive or not progressive at all strictly on the basis of ineffective contraction. This fault consists of the failure to pay sufficient heed to the effort to correct the contractions for a long enough period of time to give the gravida an opportunity to effect progress if she is capable of doing so.

It should be appreciated, however, that some of these patients have developed poor contractility patterns because the uterus has evolved this pattern over time in response (assigning a teleologic purpose to it) to a physical impediment to progress, that is, to CPD. In other words, just as a hypercontractility pattern (frequent contractions of high amplitude, often with coupling of contractions) may serve as an index of uterine overreaction to obstructed labor, so may a hypocontractile pattern. In fact, both may occur seriatim in a given labor. One generally sees hypocontractility, reflected in what is clinically perceived as inadequate contractions, later in labor than hypercontractility; this perhaps reflects the exhaustion of some myokinetic mechanism over the course of time. To assume all instances of poor contractility represent some intrinsic myometrial dysfunction is just as wrong as assuming all are the result of disproportion. Evaluation of the individual patient is clearly in order to determine which she represents.

One can applaud the admonition implied in the ACOG pronouncement to the effect that cesarean section is not indicated in all such cases; at the same time, however, cesarean section may actually prove to be appropriate in some of them. Individualization is imperative. To advocate routine use of oxytocin for enhancing the contractions under those circumstances would expose those gravidas in whom the poor uterine contraction pattern signals CPD to potentially serious hazard. The practice, if complied with widely, would in effect turn the clock back to a time when oxytocin was used almost indiscriminately, and without clear understanding about what constituted its contraindications, with rather unfortunate results.

With respect to the catch-basket inclusion of all arrest disorders within the diagnostic designation of failure to progress, aside from the nomenclatural confusion it fosters there is no strong clinical argument against its use, with the proviso that it be strictly limited to true arrest disorders. There is an academic argument that can be brought to bear, however. Before the several graphic labor disorders were formally described and named, the obstetrical literature was marked by a profusion of terms (regrettably, a few are still in use in some areas around the world) to designate disorders of labor progress. A total of more than 60 were collected at one time in just the English language literature alone.[22] This Babel came about largely because of the very subjective nature of clinical evaluation of the labor phenomenon then in vogue. With the introduction of greater objectivity in labor assessment, it was no longer necessary to cloak our collective ignorance in terminologic obfuscation. Introducing newer terms at this time, without apparent need and without clear-cut and well-defined criteria does not seem justified, and should be resisted.

The question of how one determines the presence or absence of CPD deserves further consideration because it is crucial to the management of patients with arrest disorders, and to an equal degree of those with protraction disorders (see following). The only currently acceptable technique is clinical pelvimetry by means of vaginal examination for assessing cephalopelvic relations. Prenatal clinical pelvimetry is useful for identifying a woman with a small or anatomically misshapen pelvis, an admittedly rare finding today. More pertinently, pelvic examination should be done on a routine basis for every prenatal patient at the first antepartum visit and again late in the course of pregnancy for determining the architectural characteristics and capacity of the pelvis. This information is important for anticipating the mechanism of labor and identifying those women whose pelvic capacity is limited.

Data on fetal growth over pregnancy, uterine size by abdominal palpation and fundal measurement, and fetal size by ultrasonographic imaging, if indicated (by suspicion of macrosomia, small pelvis, small maternal stature, maternal diabetes, large uterus, and other relevant indicators of potential problems), will be supplemented in the course of labor by additional examination to evaluate the relationship between the maternal pelvis and the fetal head. The fetal head is obviously the "pelvimeter" of most critical relevance. How the head fits into the mother's pelvis and whether there is sufficient room to accommodate safe passage are questions that can be answered by timely, skillful, intelligent, knowledgeable, and detailed vaginal examination.

Of special importance in this examination is a dynamic assessment of the cephalopelvic relationships; that is, one that determines the ability of the fetal head to be thrust into the pelvis in response to the anticipated forces of labor. This examination, called the Mueller–Hillis maneuver (MHM),[29] is achieved by undertaking a vaginal examination during a uterine contraction with an assistant applying firm, but gentle, downward pressure against the uterine fundus (Kristellar or fundal pressure). If cervical dilatation is well advanced so that the lower uterine segment offers little resistance to caudad thrust of the fetal head, the head will ordinarily be felt to descend in the pelvis by 1–2 cm; flexion and internal rotation may also be perceived simultaneously. These are indicators of ample room for future descent and clearly demonstrate that there can be no CPD, at least down to the level to which the head descended.

If the head is fixed at its current station and does not descend at all when this examination is done while simultaneous endogenous and exogenous forces are being applied, CPD is likely to exist. A recent question has been raised as to the validity of this approach by Thorp et al.,[30] who concluded that this valuable clinical tool should be summarily abandoned because they were unable to show that it could predict future problems in the course of labor. They tested its predictive value well in advance of the time in late active-phase labor when the pelvic capacity is actually being tested by the fetal head, in effect using it as a screening device for CPD. Although the original descriptions of the method did suggest it for this purpose, that is not the manner in which it is used in clinical practice, as emphasized as long ago as 1938 by Hillis.[29] More recently, Greenhill and I[31] recommended using this maneuver in late active phase and especially in second stage, when pelvic architecture plays a major role in shaping the course of fetal descent.

The primary value of the MHM is to diagnose an obstruction to labor progress when the fetal presenting part has reached and impacted against the pelvic impediment, specifically when labor is well-advanced and particularly after the dilatation or descent pattern has signaled the problem by evolving an arrest or protraction disorder. At minimum, 5 cm cervical dilatation is required to ensure that the cervix itself offers negligible resistance to the thrust of the fetal head when fundal pressure is applied. Further, the downward pressure should be applied during a uterine contraction (not between contractions, as reported by Thorp et al.[30]) to provide a fair test of what the combination of contractile force and firm fundal pressure is capable of achieving. If the head is fixed under these circumstances, disproportion is very likely; if it descends and rotates, bony dystocia is effectively ruled out, at least at that fetal station. As undertaken by Thorpe et al.,[30] the testing of the predictive value of the MHM in early labor under conditions not designed to assess its true diagnostic capability almost guaranteed that the study would show no effect. Therefore, their recommendation to abandon the use of the MHM in clinical practice must be considered inappropriate. It is probably too soon to forsake a tried-and-true method that has been shown to be so clinically valuable for so long, particularly because there is preciously little else available in our diagnostic armamentarium to replace it. Moreover, its use has been recommended for these purposes by the ACOG.[27]

Protraction Disorders

It may seem inappropriate to include the two protraction disorders here in our discussion of the kinds of conditions that may be mistakenly called failure to progress because continuous progress is being made (by definition) as these labor aberrations evolve, albeit abnormally slow. However, by virtue of the very slow progress that occurs in cervical dilatation or fetal descent when these conditions exist, attendant personnel may unwittingly believe that no real progress is actually occurring. The error in diagnosis means that these patients will be managed much more aggressively, and therefore inappropriately, than is warranted by the actual labor pattern they exhibit.

Given the recognized degree of interobserver error in estimating cervical dilatation and fetal station, there may not be any recorded changes over time; alternatively, small recorded advances may be interpreted to represent no change at all. Therefore, a protraction disorder may exist in a gravida and remain undiagnosed. More to the point, failure to progress may be diagnosed instead. If this occurs and it leads to a management option in which intervention by cesarean section occurs, it would be most unfortunate because about 70% of such women can be delivered safely by the vaginal route.

Although little is known about what causes the typical dilatory progress seen in gravidas with protraction disorders, it is recognized that about one-third have associated CPD. This relationship alone makes the protraction disorders clinically important, but because so many affected women do not have CPD, it is worth emphasizing that uniform management by cesarean section is clearly not warranted. Knowing this, the obstetrician confronted with such a patient wisely undertakes to determine whether or not it is safe to allow vaginal delivery. If CPD is encountered, cesarean section is indeed appropriate. This does not intend to suggest that every woman with a protraction disorder should be subjected to operative delivery, however, because if careful and thorough assessment of the cephalopelvic relations shows that CPD is unlikely, a program of expectancy and support can result in a safe vaginal delivery.

The aforementioned ACOG statement recommends the use of oxytocin for these patients if the contractility pattern is inadequate, using the Montevideo unitage criteria detailed previously.[27] Although administration of oxytocin is intuitively a reasonable option, there is no documented basis for this recommendation, particularly in view of the fact that the course of dilatation and descent has been shown to be unaffected by oxytocin in cases of protraction disorder.[22] There is an apparent and as yet unexplained dissociation between the contractile pattern and the effect that the uterine contractions has with regard to the work they are able to accomplish in dilating the cervix and causing the fetus to descend. Moreover, while oxytocin does no apparent good when used for stimulating the uterus to contract in these cases, it has the potential to cause harm, especially if it results in fetal hypoxia from hypercontractility or physical trauma from excessive head deformation in the presence of CPD.

The conservative program of expectancy and support consists of detailed discussion with the patient and her significant other to educate them about what is in store for them. The labor that follows will be uncommonly long; there will be need to mobilize and provide psychologic supports; careful attention must be paid to fluid needs and electrolyte balance; analgesia and anesthesia must of necessity be stringently limited because the slow labor progress may be worsened by use of inhibitory agents to which the uterus is especially sensitive; careful maternal and fetal monitoring has to be maintained.

Labors conducted in this way can be expected to continue to advance until vaginal delivery is accomplished, provided no problems arise with regard to fetal or maternal wellbeing. Deserving of special mention is the requirement to avoid the temptation to undertake forceps delivery in these case. It is understandable that one should wish to effect delivery as the second stage drags on, but that temptation is best resisted in the interest of ensuring against the fetal injury to which these infants may be exposed even by what may seem to be an easy outlet-type forceps delivery.[32,33] It is as yet unclear why the morbidity and mortality for these offspring should be increased in association with any form of instrumental delivery, but given the observation that such is the case, one should attempt instead to seek a safer spontaneous delivery, if at all possible.

Although those parturients with obvious CPD are not permitted to undergo this management option, having been winnowed out and delivered by cesarean section earlier, those few gravidas with undetected CPD will have taken this path. Accordingly, one can expect some of the women who were treated expectantly to fail along the way. The most important indicator signaling such failure is the development of an arrest disorder, superimposed on the preexisting protraction pattern. When the labor of a patient with a protraction disorder becomes arrested, it serves as a strong portent of obstructed labor. The prognosis for a safe vaginal delivery among them is very guarded; cesarean section is an option that deserves consideration at this point without further trial of labor.

Prolonged Second Stage

For many years, it was believed that a second stage exceeding 2 h exposed the fetus to considerable hazard. This firmly held concept, probably originating early this century by intuitive logic rather than objective substantiation,[34] appears to have been given major impetus and authoritative support from a report by Hellman and Prystowsky[35] showing that the longer the second stage lasted, the greater the risk of infant mortality and maternal hemorrhage. Additional buttressing came from the observation that the probability of the fetus developing acidosis increased with progressive prolongation of the second-stage duration.[36] In a more recent and larger series of nulliparas, some with greatly prolonged second stage, Cohen[37] was able to demonstrate a contrary view to the effect that there was no discernible increase in perinatal mortality with progressive lengthening of the second stage. A number of other investigations have confirmed these observations as well.[33,38–41]

Similarly, while there is good documentation that increasing levels of lactate appear in fetal umbilical arterial blood with increasing duration of second stage, inversely related to the falling pH levels, it has been demonstrated that this partly reflects maternal lactate levels plus the accumulation and excess production of lactate in the fetus, but not necessarily from fetal hypoxemia.[42] The safe limits of this process are unknown, but continuous monitoring of the acid–base balance in the human fetus carried through the delivery process has determined that recovery from the resulting mild respiratory acidosis occurs very quickly, indicating the likelihood that its impact of the fetus is negligible or nonexistent.[43]

The recommendations for management of the prolonged second stage derived from the observations of no ill effect from long second stages include admonitions to avoid reflexively considering cesarean section when a patient's second stage extends beyond any arbitrary duration.[37] Thus, allowing patients to proceed into a long second stage is quite acceptable, even if descent is occurring at an abnormally slow rate, with the important proviso that appropriate electronic fetal heart rate monitoring is done on a continuous basis and CPD has been definitively ruled out beforehand. This is not intended to give license to ignore such gravidas. On the contrary, prolongation of the second stage may reflect the existence of one or another of the arrest or protraction disorders detailed earlier in this chapter. Correct diagnosis of the specific aberration, if it does exist, is important under these circumstances. Even more important is the evaluation that should flow from this recognition, with emphasis on disclosing any CPD that may account for the abnormal labor pattern.

Again, if one is unable to ensure that there is ample room for safe vaginal delivery, cesarean section will ordinarily be quite justified. For the rest, it is clear that vaginal delivery should be achievable. To accomplish this, careful maternal and fetal surveillance is needed to ensure that continued progress is being made, albeit slowly, and that both mother and fetus remain in good condition. If necessary and called for on the basis of an inadequate contractility pattern, oxytocin augmentation of the uterine contractions may prove beneficial as well. As long as things continue to go well, and one is reasonably certain that there is no anatomic obstruction that would make fetal descent hazardous, allowing continued labor is not only acceptable but entirely appropriate. Therefore, expectancy and careful surveillance provide a useful and apparently safe alternative to both cesarean section and midforceps delivery, the latter having been done heretofore for no better reason than that the second-stage duration exceeded an arbitrary 2-h limit.

Quantifying the Potential Impact on the Cesarean Section Rate

A plea can be made that all those cases in which failure to progress or dystocia is invoked as a diagnosis leading to cesarean delivery should instead be designated according to a more specific diagnosis based on a more objective graphic assessment of the labor pattern. If this were done, suitable evaluation would then be undertaken to determine the associated problems and the ensuing management would address the individual needs of each patient. Were this to be done in all cases, a substantial proportion of cases now subjected to cesarean section will be permitted to deliver vaginally.

Given the figures for the errors inherent in a diagnosis of failure to progress or dystocia tallied in Table 3.2 for cases in which specific labor aberrations are present, coupled with general population data for the incidence of these several conditions,[22] it should be possible to approximate the reduction in cesarean section rate attainable if obstetrical practices were modified accordingly. To quantitate the rate reduction, such a calculation would be based on the assumption that vaginal delivery will take place for all those patients who are now being delivered by cesarean section in the presence of a labor pattern characterizable as (1) the latent phase, whether normal or prolonged in duration, (2) an arrest disorder unassociated with CPD, (3) a protraction disorder in a gravida with ample cephalopelvic relations, and (4) a prolonged second stage associated with a poor contractile pattern unrelated to anatomic obstruction. This would result in definitively reducing the cesarean rate for women delivered in the latent phase by 100%, for those with arrest disorders by 50%, for those with protraction disorders by 70%, and for those with long second stages by 60% (because of the relative impact of the arrest and protraction disorders that make up this group). In total, no fewer than 65% of the cesarean sections now undertaken for failure to progress would not be done.

What might this mean in terms of its impact on the actual cesarean section rate? If we accept the current national cesarean rate as 23.5%, a figure just announced by the CDC for 1991 data,[1] and embrace the generally agreed-upon concept that as many as 40% are being done for failure to progress or dystocia, then not doing 65% of the cesarean sections now being done for these diagnoses would yield an absolute reduction in the overall rate of 6.1%. This would drop the overall cesarean section rate to 17.4%, and this does not even begin to take into account what might be attainable simultaneously by concerted efforts to deliver potential VBAC patients vaginally in larger numbers. Furthermore, if this program were to be applied at those institutions at which the cesarean section rate is now far in excess of the national average, in some approaching a startling 50%, the benefit would be even greater in relative terms.

This in fact comes very close to the national objective of 15 percent that has been stated by federal health officials as the goal for the year 2000. This worthy goal has been deemed unlikely to be achievable by the ACOG and has already been effectively abandoned by the CDC.[1] It can be seen from this brief analysis that not only is this goal within the realm of feasibility, but it should be readily reachable and, in concert with efforts to enhance the frequency of successful VBACs, easily exceeded.

The impact such a program would have on the cesarean section rate for nulliparas is even more impressive. We can apply the same simple calculations to the current cesarean frequency of 30% or more,[11] about half or more of which are being done for failure to progress or dystocia. We can derive a conservative estimate of 9.8% for the reduction attainable by the focused management principles detailed here. The result of this calculation is that the overall cesarean rate for nulliparas would fall to 20.2%. If the true relative frequency of cesarean sections done for dystocia or failure to progress is as high as 66%, as has been reported in at least one study,[12] then the absolute overall reduction would total 12.9%, leaving a section rate among nulliparas of 17.1%. If reached, this would clearly constitute a major step in the right direction.

In an attempt to achieve this some states have introduced public health care legislation to test and, more recently, to mandate such obstetric practices based on the recommendations of the NIH Consensus Development statement on cesarean childbirth, first formulated in 1980.[13] These recommendations included standardizing nomenclature by replacing the term "failure to progress" with diagnoses based on the graphic labor curve pattern or equivalent labor definitions, and utilizing a hospital quality control system composed of internal institutional reviews of management practices supplemented by external reviews to ensure compliance. In addition to requiring good chart documentation, use of the labor curves or some form of equivalent documentation, and more complete chart notes and operative notes, such an implemented program would insist on updating, modifying, or developing protocols for obstetric patient care; education of the staff and the affected patients, especially as regards VBAC and fetal monitoring; quality assurance activities with increased participation by physicians and nurses; and appropriate use of oxytocin for labor augmentation.

A program designed precisely along these lines and aimed at monitoring quality of obstetric care in a sampling of one state's obstetric units, including 176 hospitals, has actually had a clinical demonstration trial, as reported by Dillon et al.[17] The preliminary results are indeed very gratifying. The state cesarean section rate fell from 25% in 1987, just before the program was introduced, to 23.6% in 1990 and 23.4% in the first seven months of 1991. Even greater rate reductions were seen in the hospitals in which the program had been implemented, and still stronger declines among those 24 units that had been additionally subjected to external review.

Lest this be considered just the result of imposing strong rules in cosmopolitan or academic institutions, it should be noted that another similar clinical trial was conducted in a rural community hospital in Canada.[12] Criteria were introduced for the diagnosis of dystocia based upon the 1986 National Consensus Conference on Aspects of Cesarean Birth.[15] These specified, for example, that the diagnosis could not be made unless the labor was documented to have entered the active phase (designated by cervical dilatation of at least 3 cm); there was acceptance of delayed labor to allow vaginal delivery (that is, no mindset to effect delivery after an arbitrary duration of second stage), plus encouragement to use oxytocin to augment nonprogressive labor where indicated. The results of this program in reducing the cesarean section rate over a 4-year period, 1985–1989, were even better than expected. Overall, it fell from 23% to 12%, with an increase in the frequency of successful VBACs and a remarkable reduction in the rate of cesarean sections for dystocia among nulliparas by half, from 16% to 8%. Indeed, the latter accounted for most of the overall decline in the cesarean section rate in this series.

In New York state, the Department of Health has since taken the unusually enlightened step of promulgating rules making the use of failure to progress and dystocia unacceptable as diagnoses for which cesarean section could be done, and requiring all formal diagnoses of labor disorders to be based on the graphic patterns discussed earlier. It is too soon to know what the impact of a statewide program such as this will be on the cesarean section rate, but if it proves successful—and it is reasonable to predict that it will from the aforementioned pilot study results—it is likely to serve as a template for other states to follow. It is regrettable only in that it has taken such governmental action to force physicians to do what is clearly indicated and in the best interests of their patients.

Alternative and Supplementary Programs

> The most striking development in operative obstetrics during the last 25 years has been the extension of the operation of caesarean section, as a means of combating the complications of parturition . . . It would be a sad day for prospective mothers if caesarean section were to run riot and were to be indiscriminately employed for complications which can be quite as well treated by ordinary vaginal procedures.[44]

This insightful statement by Munro Kerr, the obstetric doyen of a previous generation, was written 30 years ago and is interesting because it shows how accurately today's climate in which cesarean section has "run riot" could be predicted. It was cited by Seiler[45] in a wistful proposal advocating a retrogressive reintroduction of such abandoned practices as vaginal breech delivery and midforceps for "correctable dystocia" for purposes of reducing the cesarean section rate. Although one cannot support or even take seriously Seiler's proposal to turn the clock back and encourage the resurgence of procedures that fell from favor because of their adverse consequences,[46] it does at least recognize that many of today's cesarean sections are actually done for dystocia that is clinically correctable by means other than cesarean section so that vaginal delivery can be accomplished. The solution that is recommended, however, is anachronistic.

Are there more acceptable approaches? Aside from the one just outlined and championed, a program with growing popularity is that of active management of labor as advocated by O'Driscoll et al.[47,48] Its details and merits are beyond the scope of this review (see Chapter 4). It is valid, nevertheless, to point out that much of the benefit accrued from the active management program appears to come primarily from the total commitment of the both physician and nursing staffs, the use of strict criteria for diagnosing that a patient is truly in labor (no small task), standardization of care, organization and efficient functioning of the labor unit, a strong patient enducation program, and one-on-one attendance and supervision of each gravida by a committed and knowledgeable nurse labor monitor (not the uninformed and emotionally involved significant other). While routine administration of oxytocin to women who are documented to be in labor but who do not show the requisite minimally acceptable rate of dilatation of 1 cm/h is undoubtedly important in the successes reported with this approach, it is perhaps of less relevance because, if the uterine stimulation is really limited only to those fulfilling the stated criteria, it should apply to relatively few patients.[49]

There is ample documentation to the effect that there are often subtle (and sometimes not so subtle) factors at play in the decision of doctors to undertake cesarean section for dystocia. The higher cesarean rates among private patients, whose section rates ought to be lower than those of nonprivate gravidas by virtue of the differential incidence of maternal and fetal complications, raises the specter of the physician's role in influencing the choice of delivery mode based on hidden agenda items, including such unacceptable reasons as convenience.[50–56] Taylor[57] attributed the lower rates on nonprivate services of community hospitals and in patients delivered at county, university, and military hospitals to the more objective approaches to management at those units and the shared patient care responsibility that exists there, implying oversight. He attributed the higher cesarean section rates for dystocia among private patients to the fact that prolonged labor is not tolerated by patients, that oxytocin augmentation is used less, and when used, may alarm the obstetrician because of associated fetal heart rate changes, and that labor progress may be interfered with by the epidural anesthesia that is used more often among these gravidas.[58] Patient tolerance of labor is directly related to their prelabor education and intrapartum support, an intrinsic feature of active management. Use of oxytocin has not been shown to be a real factor in determining the type of delivery in cases of "functional dystocia" as defined by Seitchik et al.,[59,60] although its benefits are widely extolled nevertheless.

The use of oxytocin for active management of the patient who is not progressing adequately does apparently serve to ensure against being deluded into considering a patient who is actually still in the latent phase as if she were exhibiting some abnormal state of dystocia for which cesarean section should be done. It does not guard completely against other abuses of latent phase management, however, specifically because by advocating routine administration of oxytocin, the occasional patient who may instead actually be in false labor will receive it and her failure to respond (to the inadvertant induction of labor) may be perceived as a labor aberration worthy of aggressive intervention by cesarean section, the very problem the use of oxytocin was trying to prevent. Moreover, if the patient actually has a protraction or an arrest disorder to account for the dilatory progress, the labor augmentation will be providing a form of therapy many will later need. Thus, at least one-half of the affected patients with arrest disorders and more than two-thirds of those with protraction disorders will perforce get the therapy that is actually indicated for them. This is not necessarily an unqualified blessing.

The cases referred to here are those who, under the program described, would in due course have been diagnosed as having the specific labor aberration; the diagnosis made, they would then have been evaluated for its cause with dynamic clinical pelvimetry and shown not to have demonstrable CPD. Because active management does not take the precaution of addressing the issue of undetected CPD, however, administration of oxytocin would have the singular disadvantage of subjecting some gravidas and their fetuses to uterotonic stimulation contrary to currently acceptable standards, which interdict such use in the presence of CPD because of the known hazards it may involve. In two studies in which active management was studied with these concerns in mind, there was indeed a reduction in the cesarean rate for dystocia among nulliparas, but concomitantly there was an almost equal and counterbalancing increase in the frequency of cesarean sections for fetal distress.[49,61]

References

1. U.S. says 349,000 cesareans in 1991 were not necessary. New York Times, April 23, 1993.
2. Notzon FC, Placek PJ, Taffel SM. Comparison of national cesarean section rates. N Engl J Med 1987;316:386–389.
3. Anderson GM, Lomas J. Recent trends in cesarean section in Ontario. Can Med Assoc J 1989;141:1049–1053.
4. Gleicher N. Cesarean section rates in the United States: the short-term failure of the National Consensus Development Conference in 1980. JAMA 1984;252:3273–3276.
5. Quilligan EJ. Cesarean section, 1988: To have or have not! West J Med 1988;149:700–703.
6. Marieskind HI. An evaluation of cesarean section in the United States: final report. Washington, DC: U.S. Department of Health and Human Services, Office of Assistant Secretary for Planning and Evaluation/ Health, 1979:16–18.
7. Hibbard LT. Changing trends in cesarean section. Am J Obstet Gynecol 1976;125:798–804.
8. Mann LI, Gallant J. Modern indications for cesarean section. Am J Obstet Gynecol 1979; 135:437–441.
9. National Center for Health Statistics. Detailed diagnosis and procedures, National Hospital Discharge Survey, 1987. Washington, DC: Vital Health Statistics 1989;13: 1–304.
10. Sokol RJ, Rosen MG, Bottoms SF, et al. Risks preceding increased primary cesarean birth rates. Obstet Gynecol 1982;59:340–346.
11. Boylan PC, Frankowski R. Dystocia, parity and the cesarean problem. Am J Obstet Gynecol 1986;155:455–456.
12. Iglesias S, Burn R, Saunders LD. Reducing the cesarean section rate in a rural community hospital. Can Med Assoc J 1991;145: 1459–1461.
13. NIH Consensus Development, 1980: statement on cesarean childbirth. NIH Publication No. 82-2067. Bethesda, Maryland: National Institutes of Health, 1980.
14. NIH Consensus Development statement on cesarean childbirth: the Cesarean Birth Task Force. Obstet Gynecol 1981;57:537–545.
15. Indications for cesarean section: final statement of the panel of the National Consensus Conference on Aspects of Cesarean Birth. Can Med Assoc J 1986;134:1348–1352.
16. Rosen MG, Alper MH, Bloomfield R, et al. NIH Consensus Development Task Force statement on cesarean childbirth. Am J Obstet Gynecol 1981;248:1082–1084.
17. Dillon WP, Choate JW, Nussbaum ML, et al. Obstetrical care and cesarean birth rates: a

program to monitor quality of care. Obstet Gynecol 1992;80:731–737.
18. Porreco RP. Meeting the challenge of the rising cesarean birth rate. Obstet Gynecol 1990;75:133–136.
19. Friedman EA. Failure of labor progression. In: Queenan JT, ed. Management of high risk pregnancy, 2nd Ed. Oradell, New Jersey: Medical Economics, 1985:584–594.
20. Friedman EA. Failure to progress. In: Friedman EA, Acker DB, Sachs BP, eds. Obstetrical decision making, 2nd Ed. Toronto: BC Decker/St Louis: CV Mosby, 1987:242–243.
21. Friedman EA. Failure to progress during labor. Contemp Ob/Gyn 1989;34(6):42–47.
22. Friedman EA. Labor: clinical evaluation and management, 2nd Ed. New York: Appleton-Century-Crofts, 1978:61–135.
23. Krapohl AJ, Myers GG, Caldeyro-Barcia R. Uterine contractions in spontaneous labor: a quantitative study. Am J Obstet Gynecol 1970;106:378–389.
24. Seitchik J, Chatkoff ML. Intrauterine pressure wave form characteristics in hypocontractile labor before and after oxytocin administration. Am J Obstet Gynecol 1975; 123:426–432.
25. Chelmow D, Kilpatrick SJ, Laros RK. Maternal and neonatal outcomes after prolonged latent phase. Obstet Gynecol 1993; 81:486–491.
26. Rosen MG, Debanne SM, Thompson K. Arrest disorders and infant brain damage. Obstet Gynecol 1989;74:321–324.
27. American College of Obstetricians and Gynecologists. Dystocia. ACOG Technical Bulletin 137. Washington, DC: American College of Obstetricians and Gynecologists, 1989.
28. Hauth JC, Hankins GDV, Gilstrap LC, et al. Uterine contraction pressures with oxytocin induction/augmentation. Obstet Gynecol 1986;68:305–309.
29. Hillis DS. Diagnosis of contracted pelvis. Ill Med J 1938;74:131–134.
30. Thorp JM. Study raises serious doubt about value of Mueller–Hillis method. Ob-Gyn News, April 1, 1993; Thorp JM, Pahel-Short L, Bowes WA. The Mueller-Hillis maneuver: can it be used to predict dystocia? Obstet Gynecol 1993;82:519–522.
31. Greenhill JP, Friedman EA. Biological principles and modern practice of obstetrics. Philadelphia: WB Saunders, 1974:651.
32. Friedman EA. Effects of labor and delivery on the fetus. In: Cohen MR, Acker DB, Friedman EA, eds. Management of labor, 2nd Ed. Rockville, Maryland: Aspen, 1989: 461–477.
33. Friedman EA, Neff RK. Labor and delivery: impact on offspring. Littleton, Massachusetts: PSG Publishing, 1987:130–145.
34. Cohen WR. The pelvic division of labor. In: Cohen WR, Acker DB, Friedman EA, eds. Management of labor, 2nd Ed. Rockville, Maryland: Aspen, 1989:19–60.
35. Hellman LM, Prystowsky H. The duration of the second stage of labor. Am J Obstet Gynecol 1952;63:1223–1233.
36. Jacobson L, Rooth G. Interpretative aspects on the acid-base composition and its variation in fetal scalp blood and maternal blood during labor. J Obstet Gynaecol Br Commonw 1971;78:971–980.
37. Cohen MR. Influence of the duration of second stage labor on perinatal outcome and puerperal morbidity. Obstet Gynecol 1977; 49:266–269.
38. Kadar N. The second stage. In: Studd J, ed. The management of labour. Oxford: Blackwell, 1985:271.
39. Saunders NSG, Paterson CM, Wadsworth J. Neonatal and maternal morbidity in relation to the length of second stage in labour. Br J Obstet Gynaecol 1992;99:381–385.
40. Paterson CM, Saunders NSG, Wadsworth J. The characteristics of the second stage of labour in 25,069 singleton deliveries in North West Thames Health Region, 1988. Br J Obstet Gynaecol 1992;99:377–380.
41. Kadar N, Cruddas M, Campbell S. Estimating the probability of spontaneous delivery conditional on time spent in the second stage. Br J Obstet Gynaecol 1986;93: 568–576.
42. Katz M, Lunenfeld E, Meizner I, et al. The effect of the duration of the second stage of labour on the acid-base state of the fetus. Br J Obstet Gynaecol 1987;94:425–430.
43. Nickelsen C, Thomsen SG, Keber T. Continuous acid-base assessment of the human fetus during labour by tissue pH and transient carbon dioxide monitoring. Br J Obstet Gynaecol 1985;92:220–225.
44. Moir JC, ed. Munro Kerr's operative obstetrics, 7th Ed. London: Balliere Tindall and Cox, 1964:3.

45. Seiler JS. The demise of vaginal operative obstetrics: a suggested plan for its revival. Obstet Gynecol 1990;75:710–715.
46. Friedman EA. Midforceps? No. Clin Obstet Gynecol 1987; 30:93.
47. O'Driscoll K, Meagher D. Active management of labor. London: WB Saunders, 1980: 23–31, 130–138.
48. O'Driscoll K, Foley M, MacDonald D. Active management of labor as an alternative to cesarean section for dystocia. Obstet Gynecol 1984;63:485–490.
49. Boylan P, Frankowski R, Rountree R, et al. Effect of active management of labor on the incidence of cesarean section for dystocia in nulliparas. Am J Perinatol 1991;8:373–379.
50. Goyert Gl, Bottoms SF, Treadwell MC, et al. The physician factor in cesarean birth rates. N Engl J Med 1989;320:706–709.
51. Gould JB, Davey B, Stafford RS. Socioeconomic differences in rates of cesarean section. N Engl J Med 198;321:233–239.
52. DeMott RG, Sandmire HF. The Green Bay cesarean section study: II. The physician factor as a determinant of cesarean birth for failed labor. Am J Obstet Gynecol 1992;166: 1799–1804.
53. Guillemette J, Fraser WD. Differences between obstetricians in cesarean section rates and the management of labour. Br J Obstet Gynaecol 1992;99:105–108.
54. Fraser W, Usher RH, McLean FH, et al. Temporal variation in rates of cesarean section for dystocia: does "convenience" play a role? Am J Obstet Gynecol 1987;156: 300–304.
55. Phillips RN, Thornton J, Gleicher N. Physician bias in cesarean sections. JAMA 1982; 248:1082–1084.
56. Evans MI, Richardson DA, Sholl JS, et al. Cesarean section assessment of the convenience factor. J Reprod Med 1984;29:670–675.
57. Taylor ES. Editorial comment. Obstet Gynecol Surv 1990;45:55–56.
58. Taylor ES. Editorial comment. Obstet Gynecol Surv 1990;45:379–380.
59. Seitchik J, Holden AEC, Castillo M. Amniotomy and oxytocin treatment of functional dystocia and route of delivery. Am J Obstet Gynecol 1986;155:585–592.
60. Seitchik J, Holden AEC, Castillo M. Spontaneous rupture of membranes, functional dystocia, oytocin treatment, and route of delivery. Am J Obstet Gynecol 1987;156: 125–130.
61. Satin AJ, Leveno KJ, Sherman ML, et al. High- versus low-dose oxytocin for labor stimulation. Obstet Gynecol 1992;80: 111–116.

4
Active Management of Labor

PETER BOYLAN

Active management of labor (AML) is an approach to the supervision of primigravid labor developed during the past 30 years at the National Maternity Hospital, Dublin, Ireland.[1] The approach represents a philosophy that is aimed at ensuring that as many women as possible achieve spontaneous vaginal birth with a minimum of appropriate intervention. The original intention of the approach was to decrease the incidence of prolonged labor.[2] In recent times the approach has become associated with low cesarean rates,[3,4] and this has stimulated considerable interest as a result of the rapid escalation in cesarean birth rates in developing countries over the past decade.[5,6]

The basic principles of AML may be divided into two broad categories, organizational and medical. The organizational category includes a regular series of antenatal education classes, the provision of a personal nurse for each woman in labor, the involvement of senior medical staff, responsibility delegated to midwifery staff, and strict audit of outcome. The specific medical components include early diagnosis of labor, early rupture of intact forewaters, and infusion of oxytocin under strict guidelines.

The organizational, or management, principles are applicable to all women in labor, but the medical interventions are applied *only* to primigravidae at *term* with a *single* fetus, *cephalic* presentation, and no fetal distress.

Background

In 1963 the originator of the concept, Kieran O'Driscoll, was elected Master at the National Maternity Hospital (NMH). He observed a worrying high incidence of prolonged labor, at that time defined as more than 36 h, and set about trying to decrease the incidence. In many cases the labors were associated with instrumental vaginal delivery with its associated trauma for both mother and infant. It was felt that this negative experience produced a permanent revulsion to childbirth in a substantial number of women. The development of the approach resulted in a rapid decline in the incidence of prolonged labor to the extent that a redefinition was possible, to 12 h, by the early 1970s.[7] Over the years there have been many refinements to the approach, and this chapter concerns the current implementation of AML.

Organizational Aspects

Antenatal Preparation

A series of six classes is held for all women expecting their first baby. Every primigravida is encouraged to attend these classes, with their partners if they so wish. The purpose of the classes is to educate the attending women about what is going to happen in labor so that anxiety levels can be reduced and nothing unexpected will happen. Every aspect of the forthcoming birth is discussed; a copy of the partogram is given to participants, and a tour of the labor ward is also arranged.

Personal Nurse

Each woman is assigned her own personal nurse when she is admitted to the labor floor, and this nurse stays with the woman throughout her labor, within the confines of duty rosters and standard breaks. The nurse assigned is a student midwife who has already completed her basic nursing training over a period of 3 years. These nurses are undergoing a 2-year course leading to a degree in Midwifery. Medical students attending the hospital function in the same way as the student nurses when involved on the labor floor. The function of the personal, professional attendant in labor is to provide emotional support at a professional level, thereby relieving the partner of any responsibility for the safe conduct of the labor. Partners are welcome throughout, but are not expected in any way to replace the professional support provided by the midwifery staff. Because the labors are relatively short and the only women in the labor ward are those who are in labor or having their labors induced, it is possible to assign an individual nurse to each woman in labor.

Senior Doctor Involvement

A senior obstetrician is responsible for the labor ward on a daily basis and is in attendance at regular intervals throughout the day. The function of the senior obstetrician is to provide support for the midwifery staff as well as being available for consultation on a constant basis. Regular rounds are done in the labor ward by the consultant. The senior resident staff function in a similar way and also perform regular rounds on the labor floor. Residents function at a much lower level and have no decision-making authority whatsoever. The senior midwife on duty communicates directly with the most senior resident on duty, equivalent in North American terms to a junior faculty professional. Junior doctors are regarded purely as postgraduate students and are there to learn but not to make decisions about individual cases, unless under the closest supervision by senior medical staff.

Midwifery Involvement

The labor ward is run on a day-to-day basis by the senior midwife on duty, and she has at least two other midwives reporting directly to her. There are in addition six student midwives on any one shift. The senior midwife is directly responsible for the running of the labor ward but liaises directly, and closely, with the senior medical staff. The approach is regarded as a team approach with everthing the midwife does underwritten by the Master of the Hospital at the time. Midwives' responsibilities are extensive but within clearly defined limits.

Audit of Results

All labor outcomes are reviewed by a senior obstetrician on a daily basis, and results are audited at a weekly labor ward audit meeting attended by the senior midwife on duty and by the senior resident staff. All operative deliveries and abnormal outcomes are reviewed with a view to training of the residency staff as well as reviewing procedures to fine-tune and maximize beneficial outcomes. The audit meeting is also attended by the antenatal education midwives so that their input is available from the classes.

Medical Components

Diagnosis of Labor

The diagnosis of labor depends, as all diagnoses in medicine depend, on a history and physical examination. Important points in the history are regular painful uterine contractions occurring at least once every 8 min. Strong supporting evidence in the history is a bloody show, the release of the mucus plug, or spontaneous rupture of membranes. Without regular painful contractions, the dignosis of labor is not entertained.

On physical examination, the most important feature is effacement of the cervix. The primigravid cervix is a different organ from the multigravid cervix, which has dilated on a previous occasion. The primigravid cervix tends to efface before it dilates, and in AML the diagnosis of labor is accepted with full cervical effacement, so long as it is accompanied by regular painful uterine contractions. In cases of doubt, the midwife has a grace period of 1 h after which the condition of the cervix is reassessed and a firm decision is made as to whether or not the woman is in labor. The diagnosis of labor is therefore accepted if the cervix is fully effaced and there are painful uterine contractions.

If the diagnosis of labor is rejected, the woman is discharged from the labor ward but retained in the hospital because of the possibility that labor may commence within the next several hours. If the woman expresses a firm preference for discharge home, her wishes are accepted. Before discharge from the labor unit, a nonstress test (NST) is performed. The woman is advised that labor may commence within the next 24 h or so and is asked to return should she believe that labor is commencing. Prior attendance at the antenatal classes facilitates acceptance of the rejection of the diagnosis of labor.

Artificial Rupture of Membranes

Artificial rupture of membranes is performed in that proportion of women in whom spontaneous rupture has not already occurred. In current practice, aproximately 30% of women have had spontaneous rupture of membranes before admission to the labor ward. The purpose of rupturing the membranes is twofold: (1) to ensure efficient uterine action, and (2) to observe the volume and color of amniotic fluid.

Inefficient uterine action expressed as failure of progressive cervical dilatation is always corrected in AML by infusion of oxytocin. This is not done, however, if the membranes are still intact. Rupture of intact membranes of itself will often be sufficient to ensure progressive cervical dilatation. The color and volume of amniotic fluid are observed because the fluid may be an indicator of prior "placental insufficiency." Absence of amniotic fluid, or thick meconium staining, suggests preexisting oligohydramnios and raises the specter of placental insufficiency. In these circumstances the fetus is regarded as being at increased risk of developing hypoxia in labor.

Oxytocin

Oxytocin is infused at a standard rate in cases in which progressive cervical dilatation does not occur. Cervical dilatation is assessed by vaginal examination 2 h after admission to the labor ward and rupture of membranes. If the cervix has not dilated by 2 cm in the intervening period, oxytocin is commenced. Approximately 50% of primigravidae require oxytocin to accelerate labor in the first stage. The proportion requiring oxytocin is largely determined by the cervical dilatation on admission. Oxytocin is infused at a standard rate commencing at 4 mu (milliunits) per minute and increased every 15 min, by 4 mu, to a maximum of 32 mu/min. Safety of oxytocin infusion is discussed later.

In the second stage of labor, oxytocin may be commenced for the first time if the head does not descend within 1 h following diagnosis of full dilatation. Management of second-stage labor is discussed separately here because of the influence of epidural anesthesia on progress in the second stage.

Progress in Labor

Progress in labor is assessed on a standard labor graph/partogram (Figure 4.1). The essence of the partogram is simplicity, and a minimum amount of relevant information is included. The graph represents progress judged by cervical dilatation against time. There is no reference to descent of the head

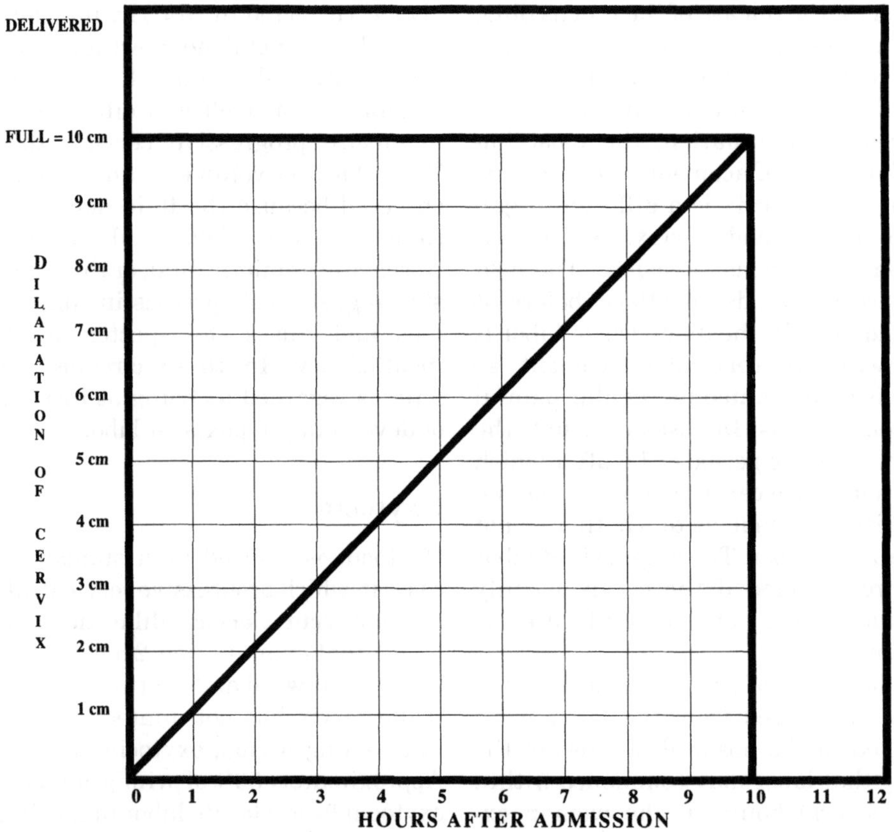

FIGURE 4.1. Standard labor graph, or partogram, used at the National Maternity Hospital, Dublin, Ireland.

in the first stage of labor. A conscious decision was made during the development of the partogram to keep it simple so that clarity would be enhanced. Progress in the first stage is judged by vaginal examination and the rate of cervical dilatation. The minimum accepted norm is 1 cm/h. The first vaginal examination takes place on admission to the labor ward, the second 2 h after admission, and generally speaking the third at the diagnosis of full dilatation of the cervix. The average number of vaginal examinations is three.

Progress in the second stage is judged by descent of the head. Failure of descent of the head in the second stage is treated by oxytocin infusion if this has not already commenced. The incidence of oxytocin infusion increases in the presence of epidural anesthesia because descent does not appear to occur with the same degree of efficiency as in the absence of epidural anaesthesia.

Safety

The central philosophy behind active management is to achieve spontaneous vaginal delivery in as high a proportion of cases as is compatible with good outcome. There is no question that, from the mother's point of view, spontaneous vaginal delivery is almost always the safer option. Cesarean section is of course preferable to a traumatic forceps delivery; this goes without saying. Safety issues in AML generally focus on the use of what appears to be high-dose oxytocin infusion. From the mother's point of view, the risk of oxytocin infusion is rupture of the uterus. In the world literature, however, there are few cases of rupture of the primigravid uterus.[8] In most of those cases reported, either the management was clearly out of line with standard practice or the birth occurred in circumstances in which there was minimal medical supervision. Often the latter occurrences have been in third world countries after extremely prolonged, obstructed labors. These circumstances have no relationship to standard practice in the current Western world. The major risk associated with oxytocin infusion from a mother's point of view therefore is that of water intoxication. This only occurs, however, when very large doses of oxytocin, far in excess of those recommended in AML, are infused in association with large volumes, in excess of 3 l, of salt-free fluid. The most likely occasion for this combination to occur is when induction of labor is accompanied by epidural anesthesia over a prolonged period of time. Water intoxication should not occur in standard practice.

The principal danger to the fetus associated with oxytocin infusion is the precipitation of fetal hypoxia through strengthening uterine contractions. There is no question that oxytocin infusion strengthens uterine contractions, as this is the very purpose for which it is given. Clearly the treatment of fetal hypoxia is delivery of the fetus and *not* infusion of oxytocin, which will only aggravate the situation in these circumstances.

There is no evidence that oxytocin infusion of itself increases the incidence of fetal hypoxia, provided it is given under adequate fetal supervision. Evidence from the National Maternity Hospital[9] and elsewhere[10] clearly indicates that oxytocin infusion does not increase the incidence of fetal death in labor, birth asphyxia, or hypoxic acid–base values at birth.

AML and Epidural Anesthesia

There is evidence that epidural anesthesia may increase the incidence of cesarean section and forceps delivery.[11] This has not been the experience at the National Maternity Hospital, where the epidural policy was liberalized very significantly at the beginning of 1991.[12] The reason why this increase has not occurred in the NMH is not entirely clear, but it may be that efficient uterine action is ensured early in labor by the appropriate exhibition of oxytocin and artificial rupture of intact membranes.

The major problem with epidural anesthesia in the management of primigravid labor arises in the second stage. Normally second-stage labor is diagnosed when the woman has an urge to push. At this stage the head is on the pelvic floor, and pushing, or the active phase of the second stage, may rightfully commence. With epidural anaesthesia, however, the second stage of labor may be diagnosed before the head has descended to a level where pushing is appropriate. This is a clinical judgment decision, which under AML is made by the senior midwife in attendance. If the head has not descended to a level where the midwife feels it is appropriate to commence pushing, oxytocin is infused if it is not already in place. One hour later, vaginal examination is repeated and pushing is commenced. After 1 h of pushing, if delivery is not imminent the clinical situation is reviewed by the senior doctor on duty.

Use of AML in Other Countries

Active management of labor has become associated with low rates of cesarean section, and this has generated much interest in recent years. Attempts have been made, with varying degrees of success, to implement various parts of AML in an attempt to lower the incidence of cesarean section. Turner et al.[13] and Robson et al.[14] successfully implemented AML in different units in British hospitals, resulting in significant declines in operative delivery and lower rates of cesarean section. Akoury et al.[15] implemented AML in Canada and had similarly successful results. Boylan et al.,[16] implementing AML in Texas, met with mixed success as a result of several factors including failure to comply with protocol. Outside

North America, AML has also been successfully implemented in Chile.[17]

The reasons why AML has not achieved the same results as in Dublin, at the NMH, may be misunderstanding of the importance of the organizational components of the approach and failure to implement the specifically medical components faithfully. A major difficulty with implementing AML has been fear of the effect of oxytocin infusion in the doses outlined. This fear is rational and is compounded by the medical-legal climate, particularly in North America. As pointed out, however, there is no evidence that oxytocin infusion of itself results in a poorer outcome, and indeed at the NMH it is associated with significantly better outcome than in other units.

Future Developments

Childbirth is unique in medicine insofar as it is so strongly influenced by the cultural environment in which it occurs. Women's expectations of childbirth vary enormously from country to country, and in such a multicultural environment as North America there is of course a wide variety of expectations. There is no "correct" incidence of cesarean section that is applicable to all societies. There are many subtle influences, which tend to increase the incidence or indeed keep it low. These factors are more related to differences in cultural background, not only of the laboring women but especially of the physicians attending those women. It is clear from attempts to implement AML in different racial groups that almost identical results can be achieved in a variety of different settings when the various components of AML are implemented most closely to those at the NMH.

It is also clear from experience both in Dublin and elsewhere that AML may have a significant role to play in reducing the incidence of cesarean section where this is believed to be inappropriately high. The question of what is an inappropriately high incidence of cesarean birth, however, is one that can only be answered on a local level.

References

1. O'Driscoll K, Meagher D, Boylan P. Active management of labor, 3rd Ed. London: Mosby Year Book Europe, 1993.
2. O'Driscoll K, Jackson RJA, Gallagher JT. The prevention of prolonged labor. Br Med J 1969;ii:477–480.
3. O'Driscoll K, Foley M, MacDonald D. Active management of labor as an alternative to cesarean section for dystocia. Obstet Gynecol 1984;63:485–490.
4. Socol ML, Garcia PM, Peaceman AM, Dooley SL. Reducing cesarean births at a primarily private university hospital. Am J Obstet Gynecol 1993;168:1748–1758.
5. Taffel DM, Placek PJ, Moien M. 1988 Cesarean section rate at 24.7 per 100 births—a plateau? N Engl J Med 1990;323:199–200.
6. Notzen FC, Cnattingius S, Bergsjo P, et al. Cesarean section delivery in the 1980's: international comparison by indication. Am J Obstet Gynecol 1994;170:495–504.
7. O'Driscoll K, Stronge JM, Minogue M. Active management of labour. Br Med J 1973;iii: 135–137.
8. Boylan P. Labor in the primigravid patient. Curr Probl Obstet Gynecol Fertil, 1991:14(1).
9. Cahill DJ, O'Herlihy C, Boylan P. Does oxytocin augmentation increase perinatal risk in primigravid labor? Am J Obstet Gynecol 1992;166:847–850.
10. Thorp J, Boylan P, Parisi VM. The influence of high dose oxytocin augmentation on umbilical acid base balance. Am J Obstet Gynecol 1988;159:670–675.
11. Thorp J, Hu DH, Albin RM, et al. The effect of intrapartum epidural analgesia on nulliparous labor: a randomized, controlled, prospective trial. Am J Obstet Gynecol 1993; 169:851–858.
12. Robson MS, Boylan P, McParland P, MacQuillan K, O'Neill M. Epidural analgesia need not influence the spontaneous vaginal delivery rate. In: Proceedings of the 13th Annual Meeting of the Society of Perinatal Obstetricians, San Francisco, 1993 (abstract & poster).
13. Turner MJ, Brassil M, Gordon H. Active management of labor associated with a

decrease in the caesarean section rate in nulliparae. Obstet Gynecol 1988;71:150–154.
14. Robson MS, Scudamore I, Walsh S. Can labour ward audit improve obstetric outcome? Oral presentation at XIII World Congress of Gynaecology and Obstetrics, 15–20 September, 1991, Singapore.
15. Akoury HA, Brodie G, Caddick R. Active management of labor and operative delivery in nulliparous women. Am J Obstet Gynecol 1988;158:255–258.
16. Boylan P, Frankowski R, Rountree R, Selywn B, Parrish K. Effect of active management of labor on the incidence of caesarean section for dystocia in nulliparas. Am J Perinatol 1991;8:373–379.
17. Masoli P, Pico V, Pellerano I. Manejo activo del parto. Experiencia en el Hospital Gustavo Fricke. Rev Chil Obstet Ginecol 1986;51: 223–230.

5
Vaginal Birth After Cesarean Section

BRUCE L. FLAMM

Vaginal Birth After Cesarean: The Doctor's Dilemma

Dr. Roy Pitkin, editor of *Obstetrics and Gynecology*, recently stated that "without question, *the* most remarkable change in obstetric practice over the last decade involves management of the woman with a prior cesarean delivery."[1] It would also probably be safe to say that few changes in the past decade have made obstetricians so angry. Why is this true?

Some consumer advocates have argued that it is a monetary issue and that doctors have resisted vaginal birth after cesarean (VBAC) because they are paid more for cesareans than for vaginal births. I do not believe this. Although the average cost for having a cesarean birth in the United States in 1991 was $3106 higher than for vaginal birth, only 20% of the difference ($611) was for physician fees.[2] In some areas, insurance companies are now paying doctors the same fee regardless of the mode of birth. I have talked to some of these doctors and many of them are still not excited about VBAC even though they are no longer paid more for cesareans.

These doctors point out several reasons why they prefer routine repeat cesareans. The first is convenience. A cesarean generally takes less than an hour and can be scheduled for a convenient time of day. A trial of labor will take many hours, and the birth may occur at any hour of the day or night. A second issue is the fear of litigation. Some doctors point out that, although the risk is low, if a uterine rupture or any significant problem does occur during a trial of labor an attorney will always be able to argue that the doctor should have "just" done an elective cesarean. The third common complaint that doctors voice about VBAC is that women often are not interested. Certainly many women who have had difficult first labors will not be excited about going through the whole process again. These three points are difficult to refute. Repeat cesareans are convenient, fairly low risk, and generally well accepted by patients. So why rock the boat? Twenty-five years ago that would have been a valid question. Only 5% of all births were by cesarean, and only a fraction of those were repeat operations. But today some hospitals in the United States are reporting cesarean rates greater than 50%.[3] Because increasing numbers of women presenting for prenatal care have already had at least one cesarean, any attempt at reducing the overall cesarean rate will be doomed to failure unless we can break the cycle of one cesarean leading to subsequent "routine" repeat operations.

The Impact of Repeat Cesareans on the Overall Cesarean Rate

In the United States, *the* single most common indication for performing a cesarean operation is "repeat cesarean."[4,5] It follows that if a physician or hospital is interested in reducing their cesarean rate, evaluating the repeat cesarean rate is a logical place to begin. Repeat cesarean has become so common that an individual physician could make a concerted effort to avoid cesarean section for breech or twin gestations and yet end up doing little to change his or her total cesarean rate. This is because only about 3% of infants are breech at term and some of these cases are not detected until the onset of labor. Even if detected early, the success rate for external cephalic version is only about 50%. Furthermore, many cases of failed external version are not suitable candidates for attempted vaginal delivery, and even a successful version does not guarantee a successful vaginal birth. Similarly, an effort to decrease the overall cesarean rate by attempting to avoid cesareans for twins and higher order multiple births will not be effective. Only about 2% of pregnancies involve multiple gestations and, depending on the fetal presentations, many of these cases are not suitable candidates for attempted vaginal delivery.

In contrast, about 11% of patients now presenting for prenatal care have a history of previous cesarean delivery. Data from the National Hospital Discharge Survey revealed that patients with prior cesareans accounted for approximately 441,000 of the 4,158,000 births that occurred in 1990. This represents almost 11% of all prenatal patients. Data from the same source also indicate that repeat operations accounted for 36% of all cesareans in 1990 (626,000 primary, 351,000 repeat, 977,000 total). Approximately 99% of these women have had prior low-transverse incisions and most are suitable candidates for trial of labor. Several studies have shown that about 75% of labors after previous cesarean will result in vaginal deliveries.[6–8]

The potential impact of VBAC on overall cesarean rates can be illustrated with the following example. Suppose doctors at a hospital doing 4000 deliveries per year decided to attempt to lower their 30% cesarean rate (currently 1200 cesareans per year). Assume they are currently delivering almost all breeches by cesarean and almost all patients with previous cesarean by repeat cesarean. The doctors might be able to utilize the technique of external cephalic version to avoid 36 cesarean operations (assuming a 3% rate of breech, 75% early detection rate, 50% version success rate, and that 90% of successful versions go on to deliver vaginally). If the same doctors decided to encourage VBAC, they might be able to avoid 250 cesareans (this assumes 11% prior cesarean rate, 75% trial of labor rate, and 75% VBAC rate among those who attempt trial of labor). The breech version program would result in a reduction of the overall cesarean rate from 30% to 29.1%. The VBAC program would result in a reduction of the overall cesarean rate from 30% to 23.8%.

This is a hypothetical example, but we have confirmed similar results with actual data at the hospital where I practice. At the Kaiser Riverside hospital (Riverside, California), the cesarean rate for 1992 was 12.7% (327 cesareans/2565 births). Some physicians find this number difficult to believe but it can be explained almost exclusively on the basis of an active VBAC program. There were 193 vaginal births after cesarean at our hospital in 1992. If we did not allow trial of labor, our 1992 overall cesarean rate would have been 20.3% (520 cesareans/2565 births), a figure very similar to the national average. As shown by the data in Table 5.1, this was not an unusual year. For each of these years the total cesarean rate would have been at least 60% higher if there had been no trial of labor (TOL) program (Table 5.2).

This discussion is not meant to discourage external cephalic version (or selective

TABLE 5.1. Cesarean statistics (actual), Kaiser Permanente Medical Center, Riverside, CA

	1990	1991	1992
Total births	2408	2548	2565
Primary cesareans	193	237	212
Repeat cesareans	108	107	115
Total cesareans	301	344	317
Total cesarean rate	12.5%	13.5%	12.7%
TOLs	247	280	230
Successful VBACs	196	236	193
Successful TOLs	79%	84%	84%

TOL, trial of labor; VBAC, vaginal birth after cesarean.

TABLE 5.2. Cesarean statistics (hypothetical), Kaiser Permanente Medical Center, Riverside, CA

	1990	1991	1992
Total births	2408	2548	2565
Primary cesareans	193	237	212
Repeat cesareans[a]	304	343	308
Total cesareans[a]	497	580	520
Total cesarean rate[a]	20.6%	22.8%	20.3%
Increase in cesearan rate[a]	65%	69%	60%

[a] If no trials of labor had been allowed.

vaginal breech delivery). The analogy is presented only to emphasize the relative importance of vaginal birth after cesarean in any program aimed at safely reducing the overall cesarean rate.

Uterine Rupture During Trial of Labor After Previous Cesarean Section

Trial of labor after previous cesarean has been a controversial subject for decades. No aspect of trial of labor has created more controversy than the subject of uterine rupture. More than 50 publications have addressed this issue. Many of the early publications on trial of labor after previous cesarean section did not differentiate between thin areas of the lower uterine segment found at the time of repeat cesarean (dehiscence or uterine window) and true uterine ruptures (complete or symptomatic). I will attempt to briefly summarize the vast amount of data on this subject in a few paragraphs.

In a review of 25 papers on trial of labor published between 1950 and 1980, Lavin et al.[9] found 5,325 VBACs and no maternal deaths from uterine rupture. The review listed 14 fetal deaths but 12 of these were caused by rupture of prior classical incisions. In 1985 I published a review of all 21 trial-of-labor papers published between 1980 and 1984.[10] The papers included more than 6000 trials of labor and more than 5000 VBACs without a single maternal death from uterine rupture. In 1987, a report from Dublin revealed a 0.45% uterine rupture rate in 1781 trials of labor.[11] There was one intrapartum fetal death in this series. Interestingly, 45% of these patients did not have electronic fetal monitoring during their trial of labor. In 1989, a group from Sweden reported 1006 trials of labor with a 0.6% rupture rate.[12] There were no serious complications in either the mothers or the infants. A recent review of more than 10,961 trials of labor at the Los Angeles County medical center found a 0.7% rate of uterine rupture.[13]

A meta-analysis published in 1991 that included many of the papers just listed found a combined rupture-dehiscence rate of 1.8% but did not differentiate between true uterine rupture and asymptomatic dehiscence.[14] At Kaiser Permanente hospitals in California, we found a 0.2% rate of uterine rupture (10 of 5733) during the years 1984–1988.[7] The incidence increased to 0.8% (39 of 5022) during 1990–1992. We recently reviewed our data on more than 11,000 trials of labor and found the overall incidence of uterine rupture during the past decade to be 0.5%.[15] With tens of thousands of patients now reported, it is seems safe to say that the incidence of true uterine rupture during planned trial of labor is less than 1%.

Although most reports have concluded that maternal and perinatal outcomes are often good even when uterine rupture does occur, two recent reports from Colorado and Utah emphasized that these ruptures can have very serious consequences.[16,17] These two reports include a total of 20 uterine ruptures from an unknown number of trials of labor. One of the papers estimates a uterine rupture incidence of 0.7%, which is consistent with published data. There were no maternal deaths in either of these reports, but there were four perinatal deaths and four cases of neurologic impairment. Fortunately, in 70% of their catastrophic uterine rupture cases (14/20), the infants had good outcomes. Interestingly, 3 of the 4 perinatal deaths in these reports apparently involved women laboring at home without fetal monitoring. These reports serve to point out that trial of labor is not risk free and that careful fetal monitoring is mandatory.

Intrauterine Pressure Catheters: Can They Diagnose Uterine Rupture?

Because an intrauterine pressure catheter (IUPC) rests inside the uterine cavity, one might expect the amplitude of contractions detected by internal monitoring to decrease with a uterine dehiscence and perhaps fall to zero with a complete uterine rupture. This would be a wonderful diagnostic aid, but unfortunately things are not so simple.

In 7 of the 8 complete uterine ruptures in the Colorado series previously described, IUPCs were used but apparently did not aid in the diagnosis.[16] Likewise, Rodriguez et al.[18] reviewed a series of 39 uterine ruptures (some patients had prior cesareans) in which IUPCs had been used and found no loss of uterine tone or cessation of labor. They concluded that the usefulness of the IUPC in making the diagnosis of uterine rupture was not supported.[18] A group in England reached the same conclusion.[19] In an interesting experiment (one which I wish I had thought of), a group in Georgia recently created "artificial catastrophic uterine ruptures" by simply leaving the IUPC in place while performing cesarean sections for CPD or fetal distress.[20] The monitoring record was kept on as the uterus was incised (simulated complete uterine rupture) and even as the baby was delivered through the incision (simulated complete fetal expulsion). Neither the uterine tone nor the peak uterine pressures (contraction amplitude) were changed by the simulated uterine ruptures.

All these studies seem to indicate that the IUPC is of little or no value in the detection of uterine rupture. Of course the IUPC has been shown to be of value during oxytocin augmentation of labor. But for women undergoing trial of labor without oxytocin, there seems to be no reason to insert an IUPC. Although it would have been useful if the IUPC had been a harbinger of uterine rupture, there may be a good side to this story as it means that there is no need to perform artificial rupture of membranes at the earliest possible moment just to insert a catheter. In fact, if the fetal heart is being monitored externally without difficulty, there may be no need to intervene at all.

Oxytocin Use During Trial of Labor After Previous Cesarean Section

I recently attended a malpractice trial to testify in defense of an obstetrician accused of negligence in a uterine rupture case. As I sat under oath on the witness stand, the plaintiff's attorney read a sentence aloud from the defendant doctor's pretrial deposition. When asked if he had ever heard of *Williams Obstetrics*, the doctor responded, "Oh yes, it's the bible of obstetrics." The attorney then placed a copy of the 18th edition of *Williams Obstetrics* in my hand and instructed me to "read aloud from the bible of obstetrics." The highlighted sentence on p. 345 said, "Oxytocin is a powerful drug, and it has killed or maimed mothers through rupture of the uterus and even more babies through hypoxia from markedly hypertonic uterine contractions." (If you doubt that such a statement is in Williams, you may wish to open a copy of the 1989 edition and turn to p. 345.[21]) As the jury mumbled in astonishment I tried to explain that the sentence was taken out of context and was apparently meant to apply to the outmoded oral (buccal) administration of the medication. However, the plaintiff's attorney shouted that I was not asked to give a medical lecture and cautioned me to just answer his questions. The judge sustained his admonishment and warned me to be silent. The plaintiff's attorney then instructed me to turn to p. 321 of the "bible of obstetrics" and read aloud the sentences highlighted in yellow. "Before delivery, the spontaneously laboring uterus is very likely to be exquisitely sensitive to oxytocin. Even with an intravenous dose of a few milliunits per minute, the pregnant uterus may contract so violently as to kill the fetus, rupture itself, or both."[21] The plaintiff's attorney then took the book from my hands and slammed it down on the table as if to say, "this case is closed." I started to explain that oxytocin is one of the most commonly used drugs in obstetrics and probably one of the safest but I was again silenced by the plaintiff's attorney and cautioned by the judge.

This scenario points out a couple of things. First, it is probably not wise for a defendant to identify any book as the "bible of obstetrics"! Second, the next edition of *Williams Obstetrics* might consider editing out these misleading and potentially damaging statements. The most recent edition of *Williams Obstetrics*, published in 1993, still presents oxytocin as a dangerous drug that has killed or maimed both mothers and babies. Third, it serves to remind obstetricians that even though oxytocin is a safe and efficacious medication, its use (even its proper use) does appear to result in a small but significant increase in the risk of uterine rupture.

In reality, oxytocin has been used to induce or augment labor in literally millions of women, and most of these cases have resulted in neither uterine rupture nor fetal damage. National statistics indicate that 21% of all labors were either induced or augmented in 1990 alone (863,000 cases).[22] Clearly, there are many times when the small potential risks of oxytocin are offset by the potential benefits. Indications for oxytocin use such as postdates pregnancy, premature rupture of the membranes, and slow progress in labor occur about as frequently in women planning trial of labor after previous cesarean as in the general obstetric population. If oxytocin could not be used during trial of labor, many (perhaps most) planned trials of labor would end up as repeat cesarean operations. Oxytocin is certainly not a panacea, and many readers are well aware of the fact that there are alternatives such as expectant management after spontaneous rupture of the membranes and ambulation during desultory labor. Nevertheless, there are many cases in which the decision boils down to whether to use oxytocin or to proceed with repeat cesarean operation. In this situation the key issue becomes "is oxytocin safe to use during labor after previous cesarean section?"

Oxytocin can cause uterine hyperstimulation, often accompanied by fetal bradycardia, regardless of whether a patient has had prior uterine surgery. In other words, oxytocin use carries some risk regardless of the presence or absence of a previous uterine scar. The real issue is whether oxytocin use *substantially* increases the risk of rupture of a previous uterine scar. I stress the word "substantially" because, as pointed out earlier, the use of oxytocin is almost certainly associated with some increased risk of uterine rupture during trial of labor. In multiparous patients, oxytocin utilization has been associated with an increased risk of rupture of the unscarred uterus. It would be surprising if the same relationship did not exist for patients with prior uterine scars.

Before 1980, almost no data existed on oxytocin utilization during trial of labor after previous cesarean section. In 1985 I reviewed 21 papers on trial of labor that had been published during 1980–1984 and found more than 600 cases in which oxytocin had been used.[10] Two cases involved uterine rupture, yielding a rupture rate of less than 0.5%, not significantly different than that expected without the use of oxytocin. In 1987 we reported on 485 patients treated with oxytocin during trial of labor after previous cesarean, and noted a small but not statistically significant increase in the incidence of uterine rupture.[23] Neilsen et al.[12] found no increased risk of uterine rupture in 406 trial-of-labor patients treated with oxytocin. Molloy et al.[11] found no increased risk of uterine rupture in 271 trial-of-labor patients treated with oxytocin but noted an apparently increased risk in 30 patients who also had epidural analgesia. A meta-analysis of the papers reporting on oxytocin use during trial of labor noted that the numbers were still small but that use of oxytocin appeared to be unrelated to perinatal mortality rates.[14]

Trial of Labor with More Than One Previous Cesarean Operation

A review of trial-of-labor papers published from 1950 to 1980 only identified about 100 cases involving more than one previous cesarean operation.[9] I reviewed the literature from 1980 to 1984 and was only able to find another 100 cases.[10] More recently, Farmakides and associates[24] reported on 57 cases (39 two-prior, 18 three-prior) and noted no uterine ruptures. Pruett et al.[25] reported on 55 cases (51 two-prior, 3 three-prior) and noted no complete uterine ruptures. Interestingly, 42 of these patients had previous uterine incisions of unknown type and two had known low vertical incisions. Novas and Gleicher (Novas et al.[26]) reported 1 uterine rupture of 36 trials of labor (27 two-prior, 9 three-prior) but later learned that this patient had two prior classical cesareans. Phelan et al.[27] reported data on trial of labor in 501 patients with two prior cesareans. Sixty-nine percent of the patients in this large series delivered vaginally and there were no true uterine ruptures; 248 (57%) of these women were treated with oxytocin. Between 1982 and 1984 in Kaiser Permanente medical centers in California, we collected data on 245 patients undergoing trial of labor with two *or more* prior cesareans. Sixty-eight percent of these women delivered vaginally, and the incidence of uterine rupture did not differ significantly from that found in women with one prior cesarean section.[7] Since that paper was published our group has collected data on more than 500 additional cases involving two prior cesareans, and preliminary analysis indicates no increased risk of uterine rupture.

With more than 1000 cases reported in the literature it seems quite reasonable to allow trial of labor in the patient with two previous cesarean operations. However, there are very little data available on the outcome of trial of labor with three or more previous cesareans. The 1988 revision of the *ACOG Guidelines for Vaginal Delivery After a Previous Cesarean Birth* recommends that, "A woman with two or more previous cesarean deliveries with low transverse incisions who wishes to attempt vaginal birth should not be discouraged from doing so in the absence of contraindiations." Because data to support the "or more" portion of that statement are minimal, it would probably be prudent to inform women with *more than two* previous cesarean deliveries that they may be at greater risk for uterine rupture if they opt for trial of labor.

Suspected Fetal Macrosomia

One problem that often develops before a planned trial of labor is "suspected fetal macrosomia." This is a great concern to both the patient and her physician. No woman wants to go through many hours of labor and end up with a repeat cesarean because of failure to progress. This is a particular concern for women whose initial cesarean had been performed following failure to progress. Doctors share this concern. Perhaps this explains why (with the exception of diabetic pregnancies) suspected fetal macrosomia is rarely listed as an indication for primary cesarean yet is frequently given as an indication for elective repeat cesarean. The issues of shoulder dystocia and perineal trauma must also be addressed in any discussion of macrosomia.

The original edition of the *ACOG Guidelines for Vaginal Delivery After a Previous Cesarean Birth* and both revisions of these guidelines have included precautionary statements regarding estimated fetal weight. The 1984 revision stated that "the estimated fetal weight should be less than 4000 grams." The 1988 revision stated that "the effects of labor on the patient with an estimated fetal weight of more than 4000 grams have not been substantiated."

There are two important questions relating to suspected fetal macrosomia and trial of labor; the first concerns the difficulty in predicting fetal weight and the second the risk of trial of labor, assuming that the fetal weight could be predicted with certainty.

Can fetal macrosomia be accurately diagnosed before birth? A study conducted in the 1970s found that only 35% of macrosomic infants could be correctly diagnosed within 1 pound of actual birth weight using abdominal palpation.[28] This was the famous study which revealed that estimates made by experienced obstetricians were no better than those made by nurses and medical students. Although dozens of recent studies have reported on sonographic estimation of fetal weight, many have cautioned that the predictive accuracy declines as fetal weight increases toward 4000 g.[29–31] In some cases these errors approached 450 g (1 pound), and large numbers of infants were incorrectly labeled as macrosomic. A review of the literature concluded that "Clearly, the current ultrasonic weight estimation procedures are not accurate enough to be used effectively in the detection of macrosomia defined by weight criteria."[32]

If technology advances to the point that fetal macrosomia can be accurately diagnosed, should trial of labor (TOL) be avoided with fetal weight over 4000 grams? To answer this question we investigated the outcomes of hundreds of trials of labor where the birthweight was more than 4000 g.[33] There was no diagnostic dilemma here because we considered actual birth weights, not at estimates made before birth. Of 1776 patients who underwent TOL during 1984–1985 at eight California hospitals, 301 delivered infants weighing 4000 g or more (8 pounds, 13 ounces). In the birthweight range of 4000–4500 grams, 58% delivered vaginally. In the birthweight range more than 4500 grams, (4500 g = 9 pounds, 15 ounces) 43% delivered vaginally. When compared to the 1475 TOLs with actual birthweights less than 4000 grams, there were no significant differences in perinatal or maternal morbidity or mortality. Other investigators have reported vaginal birthrates greater than 65% in TOLs with birthweights more than 4000 g.[11,34]

The patient with an apparently large fetus is a problem that obstetricians face almost daily. It is tempting to recommend cesareans in all such cases and avoid the potential risks. This is especially true if the patient has had a prior cesarean. Unfortunately, this policy would result in many operations with little or no improvement in outcomes. Abdominal delivery of all infants with suspected fetal macrosomia would necessitate huge numbers of cesarean operations, many of which would yield normal-sized infants. Even if the fetal weight could be predicted with no doubt, the foregoing data show that most macrosomic infants will deliver vaginally. It is also well known that many cases of shoulder dystocia occur with normal-sized infants.

The reassuring data presented here will not completely solve the dilemma presented by the patient who has the combined problems of prior cesarean delivery and current suspected fetal macrosomia. Because the rate of vaginal birth is similar to that of patients with nonmacrosomic infants, it seems that the real issue pertains to the risk of shoulder dystocia (and perineal trauma) rather than the risk of failed trial of labor. Although our study was reassuring in this regard, the numbers may not have been large enough to detect small differences in outcome. This is particularly true in the birthweight range of 4500 g and more, where we had only 61 cases.

One thing does seem clear with regard to suspected fetal macrosomia: It must be understood that shoulder dystocia and its potential sequelae (both maternal and perinatal) are often not preventable and will continue to occur in spite of the very best obstetrical care. Tort reform, which clearly acknowledges this simple fact, would go a long way toward eliminating many frivolous lawsuits and many unnecessary cesarean operations.

Unknown Scar Type

The inability to determine what type of incision was made in the uterus at the time of a previous cesarean should be differentiated from the failure to make any attempt at obtaining prior records. The importance of documenting that an attempt was made to obtain this information is discussed next (see VBAC: Avoiding the Pitfalls, later in this chapter). However, in many cases it is simply not possible to get this information. The records of the previous operation may have been destroyed, or the operation may have been performed in another part of the world. In other cases the patient may present for prenatal care late in the pregnancy or even in labor, allowing little or no time to confirm the previous scar type. This was the case in the studies described next.

A study from Baylor College of Medicine compared outcomes of 300 patients with unknown uterine scar type to 88 patients with documented low transverse scars.[35] They found that the incidence of uterine rupture did not differ statistically between the two groups and concluded that "allowing a trial of labor in a patient with an unknown uterine scar does not place the mother or fetus at greater risk than the risk for a patient with a known low cervical transverse incision." An even larger amount of information on this subject comes from studies at the University of Southern California/Los Angeles County Medical Center; of 751 patients having a trial of labor, 592 (79%) had "unknown" scar types and 82% of these patients were delivered vaginally.[8] A more recent paper from the same institution reports several thousand trials of labor and indicates that the previous uterine scar type was unknown in 80%–90% of cases.[13]

Several studies have reported preliminary results with the use of ultrasound in an attempt to detect potential scar defects before labor or to determine whether the previous scar is vertical or low transverse. The results have been been disappointing. One study of 56 patients preceding repeat cesareans found what appeared to be some type of "defect" of the scar in 21% of cases.[36] The authors noted that their incidence of defects was 4- to 10 fold that of previous reports. Because the risk of uterine rupture with trial of labor is of the order of 1% or less, they correctly concluded that "It is possible that ultrasound technology overdiagnoses anatomic variations that would not cause abnormal outcomes."

When prior operative records cannot be obtained, it is reassuring to recall that approximately 99% of cesareans performed in the past decade involve a low transverse uterine incision. Thus only about 1% of "unknown" incisions will actually be vertical or classical. We can probably reduce the risk even further by thinking about *why* these types of incisions are made. If records are unavailable and the patient explains that her first cesarean was performed because her baby was transverse or because the placenta was blocking the birth canal, a classical incision may have been performed. Likewise, if the patient explains that her baby was breech and weighed 4 pounds, there is reason for concern. In most cases, however, the patient will explain that her first cesarean was performed because of slow progress in labor, suspicious fetal monitoring, term breech, genital herpes, or other common conditions for which classical or vertical uterine incisions would rarely be used.

Vaginal Delivery of Twins After Previous Cesarean Section

Although most obstetricians recommend elective repeat cesarean for women who present with the combination of prior cesarean delivery and twin gestation in the current pregnancy, trial of labor appears to be a reasonable alternative. A review of the literature found 30 cases of trial of labor with twins, all of which resulted in vaginal deliveries.[37] The review mentioned 1 maternal death from elective repeat cesarean section for twins. The same group also reported 15 of their own cases, all of whom delivered vaginally. Another study reported 25 trials of labor, and in 18 of these cases both infants were delivered vaginally.[38] There were no uterine ruptures. They pointed out that concerns about rupture caused by uterine overdistension with twins did not seem to be substantiated in the literature. With fewer than 100 cases reported in the literature, this question has not been answered definitively but for the woman who strongly desires vaginal birth, trial of labor in a twin gestation seems to be a reasonable option.

Vaginal Breech Delivery After Previous Cesarean Section

Many physicians in the United States no longer perform any vaginal breech deliveries. However, there is ample evidence in the literature to support selected vaginal breech delivery. This is discussed in detail in Chapter 6 of this book. Even for those physicians who are comfortable with vaginal breech delivery, the patient who presents with a prior breech presentation and a history of a previous cesarean presents a dilemma. There have been a few reports on such patients in the medical literature. In 1963 a group in Ireland reported 22 breech deliveries after previous cesarean and noted no complications.[39] Two more recent studies also addressed this issue. A group in Israel allowed trial of labor in 47 patients with breech presentation, and 37 delivered vaginally.[40] A group in the United States recently reported 27 cases with 13 vaginal breech deliveries.[41]

Yet another option exists for this group of patients. We recently reported on 56 attempts at external cephalic version in patient with prior low transverse cesareans; 46 of these attempts were successful and there was no morbidity.[42] When I presented the initial results of this study at the ACOG combined district VIII/IX meeting in Seattle in 1990, I asked for a show of hands to see if anyone in the audience had any experience with external cephalic version in patients with prior cesareans. Much to my surprise, about half the attendees were already doing the procedure in these patients. Since that paper was published we have attempted version in at least 50 more patients without complications. Although the numbers are still relatively low, it appears that the risk of uterine rupture during external cephalic version is low.

In summary, few experts would argue with a doctor who recommends elective repeat cesarean for a patient with breech presentation and a history of a previous cesarean. However, for patients who strongly wish to avoid repetitive operations, both the options discussed here appear to be reasonable alternatives.

Trial of Labor After a Single-Layer Closure of a Low Transverse Uterine Incision

In the past several years the technique of single-layer closure of the uterine incision has been rapidly gaining in popularity. For many physicians it has become routine. As early as 1986 an obstetric textbook was recommending the single-layer closure.[43] The authors pointed out that

In view of the increasing acceptance of a trial of labor following cesarean section, the uterine incision should be closed in a fashion calculated to result in optimal healing. The popular method of repairing a transverse incision with a continuous locking stitch reinforced by a continuous inverting layer produces good hemostasis, but can also leave a poorly vascularized mass of tissue that interferes with healing. A second inverted layer created by using a continuous Lembert's or Cushing's stitch is customary but is really needed only when opposition is unsatisfactory after application of the first layer.

This is an interesting and reasonable point of view. Two layers do not necessarily heal better or stronger than one. In fact, the opposite may be true. I attempted to study this at our hospitals but found that most cesarean operative reports contained the statement, "the uterine incision was closed in the usual fashion."

Other investigators have had more luck. Pruett et al. at Baylor College of Medicine studied 57 trials of labor after documented single-layer uterine closure and found no complete uterine ruptures.[35] They did note a 2.8% incidence of dehiscence (incomplete, asymptomatic separation) in this group and cautioned that more patients would have to be studied before a definitive statement could be made. More recently, researchers at the University of Alabama compared 149 women with single-layer closures to 143 women with the traditional two-layer closure and found no complete uterine ruptures in any of the 292 trials of labor.[44] An incidence of partial scar separation of approximately 2% was found in each group.

Prostaglandin E_2 Gel Induction of Patients with a Prior Cesarean Section

The recent obstetric literature has been filled with reports on the use of prostaglandin E_2 (PGE_2) for cervical ripening. Although most of these studies have excluded patients with prior cesareans, some data are available. Prostaglandin was used for cervical ripening in 143 trials of labor in an English study. Of these women, 108 went on to deliver vaginally, and there were no complications.[45] In another study, 16 of 19 patients treated with 1.5 mg vaginal PGE_2 pessaries went on to deliver vaginally without complications.[46] The most recent study reported 25 patients with unfavorable cervix who were treated with 1 mg PGE_2 gel intracervically at the University of Texas.[47] 18 patients went on to deliver vaginally and there no major complications. However, uterine scar rupture has been reported with the use of PGE_2 and hence it must be used with caution. We are now evaluating the use of PGE_2 in patients with prior cesarean delivery and will soon be reporting the results of more than 300 cases.

Elective Repeat Cesarean Versus Trial of Labor

Most studies on trial of labor have focused on the results of women who actually attempted vaginal birth, not on a comparison to women having elective repeat cesareans. The best way to compare the two alternatives would of course be a randomized controlled trial. However, such a trial would be a logistic nightmare. Once randomized, women would have to open a sealed envelope to determine whether or not they would proceed to the operating room. Many women would be required to proceed with an operation that they really did not want and that dozens of studies have shown to be unnecessary. Other women would be required to have a trial of labor that they did not desire and some of these women would eventually demand a cesarean, thus creating postrandomization selection bias. Furthermore, because of the uncommon nature of major complications in either group, such a randomized trial would have to be very large. Also, because of the presence of the telltale abdominal scar, it would be difficult if not impossible to blind attending obstetricians to the fact that they were dealing with a patient with a scarred uterus.

Having given up on the idea of conducting a randomized controlled trial, we decided to do the next best thing. We had already published the results of a multicenter study involving more than 5000 trials of labor.[6,7] Patients who opted for elective cesarean were not evaluated in that study. In 1990 we began a prospective study of *all* pregnant women at all 10 Southern California Kaiser Hospitals. This study would include all patients who opted for elective repeat cesarean and would thus allow for a comparison to those who had trial of labor. Because this study was prospective but not randomized there was a possibility that we would be comparing apples to oranges. For example, it would be likely that the elective repeat cesarean group would include significantly more women with multiple prior cesareans and other risk factors. On the other hand, certain predetermined bad outcomes would be preferentially shifted to the trial-of-labor group. For example, patients with unexplained intrauterine fetal deaths or preterm labor involving an extremely premature infant are generally allowed to deliver vaginally rather than by repeat cesarean operation. Having recognized the possibility of nonequivalence of the two groups, we used multivariate regression methods to adjust for differences in baseline factors to evaluate the independent contribution of trial of labor to perinatal outcome.

During a 3-year interval, we studied 7229 patients; 5022 (70%) had a trial of labor and 2207 had elective repeat cesarean operations.[15] Of those patients opting for labor, 3747 (75%) went on to deliver vaginally. When the two groups were compared, the length of hospital stay, the incidence of postpartum transfusion, and the incidence of postpartum fever were all significantly higher ($p \leqslant .0001$) in the elective repeat cesarean group. The incidence of uterine rupture was less than 1%. This study did not prove that trial of labor is safer or better than elective repeat cesarean. The study did demonstrate, on a large scale, that equivalent outcomes could be achieved without the need for repeat cesarean operations in the majority of cases. This had been predicted by using hypothetical data and a decision analysis process a decade before the study began.[48]

VBAC: Avoiding the Pitfalls

During the past 10 years, I have reviewed the medical records of several cases in which trial of labor resulted in a bad outcome. Some of these cases involved superior physician and nursing care and therefore reflect the fact that childbirth, regardless of mode of delivery, is not risk free. However, other cases offered information from which we can all learn.[49] Some of these cases resulted in lawsuits. Analysis of the records has allowed me to come up with a few clinical "pearls of wisdom" that may make trial of labor a safer experience.

Attempt to Confirm the Type of Uterine Incision Made at the Time of the Initial Cesarean

Although this may seem to be a trivial statement, it is not. As trial of labor becomes more common, doctors and nurses seem to worry less about previous incisions. This can be a serious mistake. Many obstetricians who cover the labor and delivery area for colleagues may have had the experience, as I have, of trying to confirm a prior incision type in the middle of the night. On many occasions I found myself reviewing a clinic chart that documented that the patient had 10 or 12 prenatal visits with absolutely no mention of any attempt to obtain the old records. This is not a pleasant experience at 3 A.M. I recently heard of a malpractice case that illustrates the hazards of this type of oversight. A catastrophic uterine rupture occurred during a trial of labor. No attempt had been made to document to prior incision type. After the poor outcome, a simple call to the hospital where the primary cesarean had been performed confirmed that the initial operation had been a classical cesarean. Earlier knowledge of this fact could have saved the patient and her doctor a lot of grief.

From a medical-legal perspective it is important to document in the patient's chart that an attempt has been made to try to determine the previous uterine incision type. However, there is a big difference between failure to attempt to get the records and not being able to do so after a reasonable effort. The inability to obtain prior medical records is certainly not a contraindication to trial of labor. As was discussed earlier in this chapter, several papers have shown that trial of labor with "unknown" scar type is relatively safe.

Act Quickly for Suspected Fetal Distress

Malpractice cases involving the management of patients with prior uterine incisions generally do *not* focus on whether the patient was a reasonable VBAC candidate. ACOG guidelines and the medical literature make it clear that, with the exception of known prior classical or vertical uterine incisions, there are few contraindications to trial of labor. The focus of litigation, as is true in most obstetric cases, will be on the fetal monitor recordings. Although proponents of universal fetal monitoring have long held that the mountains of fetal monitor strips accumulating across America will be used to defend doctors, I am not so sure. I have met "expert" plaintiff's witnesses who can find subtle "late" decelerations at some time during the course of almost any labor. But years of experience with trial of labor leads to the conclusion that in many cases you do not need a magnifying glass to detect signs of complete uterine rupture. The rupture often presents with the sudden onset of severe variable decelerations or a prolonged bradycardia. Such findings during the course of a trial of labor should lead to an immediate trip to the operating room. There is no point in wasting time performing ultrasound examinations or fetal scalp samples. These cases should be treated with the same urgency as suspected abruption or umbilical cord prolapse. By following this policy we have delivered many vigorous infants with excellent outcomes in spite of complete uterine ruptures. Of course, the policy has also led to emergency operations in which the uterus was found to be totally intact. We believe that this is not an error in management. Even with this aggressive policy, only 1% or 2% of all trials of labor will involve a "stat" cesarean.[50]

Use Oxytocin with Care

A malpractice attorney once told me that oxytocin put all his children through college. I do not think he was joking. Although the studies listed earlier reveal that it is reasonable to use oxytocin during a trial of labor, the admonition for caution is worth heeding. If the cervix is not closed, consideration should be given to the use of an intrauterine pressure catheter. This will not necessarily yield any valid information about uterine rupture (as has been discussed), but it will allow a better analysis of the contractions that are being generated. Because oxytocin has been associated with rupture of even the *unscarred* uterus, we must not be cavalier with its use during trial of labor. A recent malpractice case involved uterine rupture during a trial of labor in which oxytocin was being administered at almost 60 mu/min. Although oxytocin protocols differ from one hospital to the next, it might be wise to limit oxytocin dosages in patients with prior cesareans to 20 or 30 mu/min. A recent review pointed out that, despite widespread use and many research investigations, no consensus exists with respect to the appropriate dosing of oxytocin.[51] However, active management regimens that use a maximum dose of 40 mu/min or more have often[51] although not always[52] involved nulliparous patients who are at lower risk for uterine rupture.

References

1. Pitkin R. Once a cesarean? Obstet Gynecol 1991;77(6):939.
2. 1992 Sourcebook of health insurance data. Washington, DC: Health Insurance Association of America, 1992.

3. Steinbrook R. Half the cesarean operations in U.S. called unnecessary. Los Angeles Times, January 27, 1989 (Part 1, p 14).
4. Taffel S, Placek P, Liss T. Trends in the United States cesarean section rate and reasons for the 1980–85 rise. Am J Public Health 1987;77:955–959.
5. Shiono P, McNellis D, Rhoads G. Reasons for the rising cesarean delivery rates: 1978–1984. Obstet Gynecol 1987;69:696–700.
6. Flamm B, Lim O, Jones C, Fallon D, Newman L, Mantis J. Vaginal birth after cesarean section: results of a multicenter study. Am J Obstet Gynecol 1988;158:1079–1084.
7. Flamm B, Neuman L, Fallon D, Thomas S, Yoshida M. Vaginal birth after cesarean section: results of a five-year multicenter study. Obstet Gynecol 1990;76:750–754.
8. Paul R, Phelan J, Yeh S. Trial of labor in the patient with a prior cesarean birth. Am J Obstet Gynecol 1985;151:297–303.
9. Lavin J, Stephens R, Miodovnik M, Barden T. Vaginal delivery in patients with a prior cesarean section. Am J Obstet Gynecol 1982; 59:135–148.
10. Flamm B. Vaginal birth after cesarean section: controversies old and new. Clin Obstet Gynecol 1985;28:735–745.
11. Molloy B, Sheil O, Duignan N. Delivery after cesarean section: review of 2176 consecutive cases. Br Med J 1987;294:1645–1647.
12. Nielsen T, Ljungblad U, Hagberg H. Rupture and dehiscence of cesarean section scar during pregnancy and delivery. Am J Obstet Gynecol 1989;160(3):569–573.
13. Miller D, Paul R, Diaz F. Vaginal birth after cesarean delivery: a ten-year experience. Obstet Gynecol 1994;84:255–258.
14. Rosen M, Dickinson J, Westhoff C. Vaginal birth after cesarean: a meta-analysis of morbidity and mortality. Obstet Gynecol 1991; 77(3):465–470.
15. Flamm B, Goings J, Yunbao L, Wolde-Tsadik G. Elective repeat cesarean delivery versus trial of labor: a prospective multicenter study. Obstet Gynecol 1994;83:927–932.
16. Jones R, Nagashima A, Hartnett-Goodman M, Goodlin R. Rupture of low transverse cesarean scars during trial of labor. Obstet Gynecol 1991;77(6):815–817.
17. Scott J. Mandatory trial of labor after cesarean delivery: an alternative viewpoint. Obstet Gynecol 1991;77(6):811–814.
18. Rodriguez M, Masaki D, Phelan J, Diaz F. Uterine rupture: are intrauterine pressure catheters useful in the diagnosis? Am J Obstet Gynecol 1989;161:666–669.
19. Beckley S, Gee H, Newton JR. Scar rupture in labor after previous cesarean section: the role of uterine activity measurement. Br J Obstet Gynaecol 1991;98:265–269.
20. Devoe L, Croom C, Youssef A, Murray C. The prediction of "controlled" uterine rupture by the use of intrauterine pressure catheters. Obtet Gynecol 1992;80:626–629.
21. Conningham FG, MacDonald PC, Gant NE, eds. Williams Obstetrics, 18th Ed. East Norwalk, Connection: Appleton Lange, 1989: 321, 345.
22. Ventura S, Taffel S, Mathews T. Advance report of maternal and infant health data from the birth certificate, 1990. Month Vital Stat Rep 1993;42(2):1–31.
23. Flamm B, Goings J, Fuelberth N, Fischermann E, Jones C, Hersh E. Oxytocin during labor after previous cesarean section: results of a multicenter study. Obstet Gynecol 1987; 70:709–712.
24. Farmakides G, Duvivier R, Schulman H, Schneider E, Biordi J. Vaginal birth after two or more previous cesarean sections. Am J Obstet Gynecol 1987;156:565–566.
25. Pruett K, Kirshon B, Cotton D, Poindexter A. Is vaginal birth after two or more cesarean sections safe? Obstet Gynecol 1988;72:163–165.
26. Novas J, Myers S, Gleicher N. Obstetric outcome of patients with more than one previous cesarean section. Am J Obstet Gynecol 1989; 160:364–367.
27. Phelan J, Ahn M, Diaz F, Brar H, Rodriguez M. Twice a cesarean, always a cesarean? Obstet Gynecol 1989;73(2):161–165.
28. Ong HC, Sen DK. Clinical estimation of fetal weight. Am J Obstet Gynecol 1972;112:877–880.
29. Miller JM, Brown HL, Khawli OF, Pastorek JG, Galbert HA. Ultrasonic identification of the macrosomic fetus. Am J Ostet Gynecol 1988;159:1110–1114.
30. Miller JM, Kissling GA, Brown HL, Gabert HA. Estimated fetal weight: applicability to small and large-for-gestational-age fetus. JCU 1988;16:95–97.
31. Benacerraf BR, Gelman R, Frigoletto FD. Sonographically estimated fetal weights: Accuracy and limitation. Am J Obstet Gynecol 1988;159:1118–1121.

32. Deter RL, Hadlock FP. Use of ultrasound in the detection of macrosomia: A review. J Clin Ultrasound 1985;13:519–524.
33. Flamm B, Goings J. Vaginal birth after cesarean section: is suspected fetal macrosomia a contraindication? Obstet Gynecol 1989;74:694–697.
34. Phelan J. Trial of labor in women with macrosomic infants. J Reprod Med 1984;29: 36–40.
35. Pruett K, Kirshon B, Cotton D. Unknown uterine scar and trial of labor. Am J Obstet Gynecol 1988;159:807–810.
36. Michaels W, Thompson H, Boutt A, Schreibner B, Michaels S, Karo J. Ultrasound diagnosis of defects in the scarred lower uterine segment during pregnancy. Obstet Gynecol 1988;71:112–120.
37. Gilbert L, Saunders N, Sharp F. The management of multiple pregnancy in women with a lower-segment cesarean scar. Is repeat cesarean section really the safe option? Br J Obstet Gynecol 1988;95:1312–1316.
38. Strong T, Phelan J, Ahn M, Sarno A. Vaginal birth after cesarean delivery in the twin gestation. Am J Obstet Gynecol 1989;161: 29–32.
39. Allahbadia N. Vaginal delivery following cesarean section. Am J Obstet Gynecol 1963; 85(2):241–249.
40. Ophir EOM, Yagoda A, Markovits Y, Rojansky N, Shapiro H. Breech presentation after cesarean section: always a section? Am J Obstet Gynecol 1989;161:25–28.
41. Sarno AP, Phelan JP, Ahn MO, Strong TH. Vaginal birth after cesaran delivery: trial of labor in women with breech presentation. J Reprod Med 1989;34:831–833.
42. Flamm B, Fried M, Lonky N, Saurenman-Giles W. External cephalic version after previous cesarean section. Am J Obstet Gynecol 1991;165:370–372.
43. Hibbard L. Cesarean section and other surgical procedures. In: Gabbe S, Niebyl J, Simpson J, eds. Obstetrics, normal and problem pregnancies, New York: Churchill Livingstone, 1986:522–523.
44. Tucker M, Hauth J, Hodgkins P, Owen J, Winkler C. Trial of labor after a one-layer or two-layer closure of a low transverse uterine incision. Obstet Gynecol 1993;168:545–546.
45. MacKenzie I, Bradley S. Vaginal prostaglandins and labor induction for patients previously delivered by cesarean section. Br J Obstet Gynecol 1984;91:7.
46. Goldberger S, Rosen D, Michaeli G, Markov S, Ben-Nun I, Fejgin M. The use of PGE_2 for induction of labor in parturients with a previous cesarean section scar. Acta Obstet Gynecol Scand 1989;68(2):523–526.
47. Blanco J, Collins M, Willis D, Prien S. Prostaglandin E_2 gel induction of patients with a prior low transverse cesarean section. Am J Perinatol 1992;9(2):80–83.
48. Shy K, LoGerfo J, Karp L. Evaluation of elective repeat cesarean section as a standard of care: an application of decision analysis. Am J Obstet Gynecol 1981;139: 123–129.
49. Flamm B. VBAC: low risk, not no risk. Contemp Ob/Gyn 1991;October(10):24–32.
50. Gibbs C. Planned vaginal delivery following cesarean section. Clin Obstet Gynecol 1980; 23(2):508–515.
51. Baker ER, D'Alton ME. Management of labor in the nullipara. Clin Consult Ob/Gyn 1992; 4:218–230.
52. O'Driscoll K, Foley M, MacDonald D. Active management of labor as an alternative to cesarean section for dystocia. Obstet Gynecol 1984;63:485–490.

6
Breech Presentation

LUIS A. CIBILS

It is generally accepted that the normal *fetal lie* at term is longitudinal, which indicates that the maternal and fetal spines are parallel. Furthermore, the normal presentation at term is cephalic, meaning that the cephalic pole of the fetus presents to the pelvic inlet and the birth canal. The physical relationship between fetus and birth canal is the most favorable for delivery when these two conditions are present. It follows that any deviation from this physical relationship is an abnormality of variable importance. *Breech* presentation is such an abnormality, one in which the lie is longitudinal but the presenting part is the lower pole of the fetus. There are several variants in attitude of breech presentation, depending on the relationship of the fetal body and the lower limbs, and also the lower limbs position: they are known as complete, frank, footling (single or double), and kneeling. Each one of these attitudes has specific clinical characteristics and types of complications which make the importance of accurate diagnosis of more than academic interest.

Breech presentation in any attitude is associated with a number of complications that contribute heavily to increase the perinatal mortality (PNM) and perinatal morbidity of infants born by the breech. More detailed discussion follows in various sections of this chapter; suffice to say here that among the more prevalent conditions associated with breech presentation are premature labor and delivery, congenital malformations, intrauterine growth retardation, abnormal implantation of the placenta, prolapse of the umbilical cord, and uterine malformations.

Incidence

The relatively excessive volume of amniotic fluid in proportion to fetal volume, normally occurring in mid- and preterm gestation, facilitates the instability of the fetal lie and presentation. As a consequence the prevalence of abnormal lie and presentation is as high as 45% in late second trimester.[1] Likewise, the prevalence of breech presentation ranges from 24% to 33% in late second and early third trimester.[1,2] With the advance of pregnancy, these numbers decrease steadily because there is proportionally less amniotic fluid, and by 33 weeks gestation only 10% to 14% are still in breech presentation. About one-half of these fetuses will convert spontaneously to cephalic presentation by the time they reach full term,[1–3] when the prevalence of breech presentation ranges from 3% to 7% according to standard textbooks. It is therefore understandable that when labor ensues prematurely the incidence of breech presentation is inversely related to the age of gestation. In other words, premature labors carry a high incidence of breech presentations, and the earlier the labor, the higher the incidence of breech presentations.

Importance

Many very interesting studies have been carried out in the past several years regarding the morbidity and mortality of infants presenting by the breech. Those observations were essentially promoted by the need to determine the potential causes of their substantially greater PNM and morbidity.[4–8]

Breech presentations seem to have intrinsic characteristics, such as relatively lower placental weights,[5] an extremely high incidence of congenital anomalies,[5–9] and premature deliveries.[5–8,10] Breech-delivered infants appear to have an inherently lower weight at birth even when corrections are made for gestational age, and this deficit seems to persist in early infancy.[11] On the other hand, the observation that fetuses with a severe congenital handicap appear to present by the breech in an inordinately high proportion prompted Braun to ask "why the fetus failed to assume the vertex position."[12] The average length of the umbilical cord has been shown to be longer in cephalic presentations, the observers suggesting that this may be caused by increased fetal activity.[13] This postulated variation in activity may be related to other anatomic differences such as congenital hip dislocation, found by Clausen to be nearly four times more frequent among infants born in breech presentation.[14] Whether this abnormality is the cause or consequence of the fetal position in utero is not clear; however, Luterkort found that there is an extremely high proportion of hip joint instability among breech deliveries, particularly prevalent among those in frank attitude.[15] The probability that inherently anatomic or physiologic anomalies predispose to breech presentation has been reinforced by the observation by Nelson and Ellenberg[16] that about one-third of breech presentations with cerebral palsy had major malformations unrelated to the central nervous system (CNS).[16] These observations confirmed a prior report by Kauppila[6] about CNS and hip malformations.

This partial review of the numerous abnormal conditions associated with breech presentation and delivery should make us pause and be very cautious when confronted with one such case. Likewise, one must take these factors into consideration when reviewing published material, or when considering management recommendations based on dramatic differences with cephalic presentations. Some long-term follow-up studies of survivors delivered at various gestational ages and according to route of delivery are reviewed later in this chapter.

One of the aspects of breech presentation receiving even more attention than the problems mentioned in the previous paragraphs is the mode of delivery advised for those fetuses. In the United States, about 85% of all breech presentations are currently delivered by cesarean section, regardless of attitude or gestational age.[17] This represents around 10% of all Cesarean sections, close to 4% of all deliveries or approximately 100,000 operations done every year. The importance of these numbers with regard to morbidity, mortality, and health care expenditures are thoroughly discussed in other chapters of this book. The perusal of those chapters would suggest that a decision to perform a cesarean section should not be taken lightly and without considering every aspect of obstetric management, including maternal morbidity and mortality.

Perinatal Mortality

Perinatal mortality traditionally has been quoted as being 5- to 10-fold higher for breech deliveries compared to cephalic presentations. This fact is still currently observed. However, it is important to attempt to study separately the various pathologic factors that cluster around breech presentations. The PNM ranges from nearly 10%[8] to 25%,[5] depending on the population studied or the years when the material was collected. In our material of nearly 1150 consecutive cases observed over a 16-year period, there was a 24% uncorrected PNM, which is severalfold higher than for cephalic presentation.

Prematurity occurs with extreme frequency among all series, probably for the reasons just discussed, and ranges from 22% to 35%,[5–7] which is three- to fourfold higher than the 6%–8% usually quoted as the prevalence for the general population in this country. In our material the prematurity rate was 53% by weight, that is, fetuses weighing less than 2500 g, and 51% if classified by gestational age (36 weeks or less).[18] This extremely high rate of prematurity is the direct cause of nearly one-half of all perinatal (PN) deaths (11% of all births in our population) and almost two-thirds of neonatal deaths (NND). There is an intrinsic trap with this condition because the more premature a labor, the higher the chances of a breech presentation,[1,2] and thus more cases of extreme prematurity are born in breech.

Premature rupture of membranes is, of course, the triggering of many premature labors but at the same time predisposes to prolapse of the umbilical cord, which may be a life-threatening complication when producing cord compression. It occurs with variable frequency depending on the attitude of the presentation or age of gestation. It is variably quoted as about 5% for term-size infants,[6] while for others of the order of 3%,[19] or even less than 1%.[20] It is fortunate that the majority of these fetuses are delivered in excellent condition, perhaps because the mechanical condition of the cord dropping between the legs does not predispose to significant cord compression and also because of the usual rapid termination of pregnancy when the condition is recognized. In our total population (term and preterm), it occurred in 7% of cases.

Uterine malformations are classically mentioned as causing breech presentation, but actually are mentioned as contributing to only a small percentage of cases in the publications discussing the point.

Tables 6.1, 6.2, and 6.3 summarize the perinatal results observed by authors who published the results of materials studied with different methods and collected under various circumstances. There is an improvement as the years have passed, but the mortality and morbidity are still very high compared to known figures for cephalic presentations.

Mode of Delivery

Neonatal Effects

From the foregoing discussion it is clear that breech presentation at the time of delivery is associated with a number of high-risk conditions. The prevailing concept, during the early to middle part of this century, was that much of the PNM and morbidity could be

TABLE 6.1. Perinatal results (all ages)[a]

Reference	Number of cases	Mortality	Prematurity	Malformations	Corrected perinatal mortality (PNM)
Hall et al., 1965[21]	6,044	12.3%	24.4%	4.7%	6.5%
Brenner et al., 1974[5]	1,016	25.4%	36.0%	6.0%	11.0%
Kauppila, 1975[6]	2,227	12.5%	20.0%	10.0%	
Fianu, 1976[7]	1,878	7.2%	15.1%		
Lewis et al., 1979[22]	579	10.4%	24.2%		
Mann and Gallant, 1979[23]	457	19.9%	>24.0%	>1.5%	5.9%
Sachs et al., 1983[24]	9,626				2.5%
White and Cibils, 1984[10]	302	7.9%	32.1%	12.0%	6.9%
Schutte et al., 1985[8]	4,628	9.7%	17.1%	4.4%	3.5%
Fortney et al., 1986[25]	10,749				5.2%
Kiely, 1991[26]	17,587	9.7%			4.9%
Brown et al., 1994[18]	843	24.0%	52.7%	9.4%	14.9%

[a] The cases published until 1983 include only infants weighing more than 1000 g. Only the publications after 1983 include fetuses weighing 500 g and more. The calculations were made from information and figures given in the respective publications.

TABLE 6.2. Perinatal results (term, 2500 g or more)[a]

	Number of cases	Mortality	Malformations	Corrected perinatal mortality (PNM)
Hall et al., 1965[21]	4,568	4.0%	1.1%	2.9%
Rovinsky et al., 1973[4]	2,145	3.2%	2.1%	0.6%
Kauppila, 1975[6]	1,779	2.8%	9.0%	
Fianu, 1976[7]	1,595	2.8%		
Sachs et al., 1983[24]	7,028			0.4%
White and Cibils, 1984[10]	205	1.5%	6.0%	1.5%
Tatum et al., 1985[20]	580	1.1%	1.1%	0%
Schutte et al., 1985[8]	3,837	2.5%	3.5%	1.4%
Fortney et al., 1986[25]	8,546			1.6%
Kiely, 1991[26]	13,694	2.5%		1.2%
Brown et al., 1994[18]	309	1.8%		0%

[a] The cases published until 1983 include only infants weighing more than 1000 g. Only the publications after 1983 include fetuses weighing 500 g and more. The calculations were made from information and figures given in the respective publications.

TABLE 6.3. Perinatal results (prematures, less than 2500 g)[a]

	Number of cases	PNM	NNM	Corrected PNM	Malformations	PNM, VLBW	Corrected PNM, VLBW
Hall et al., 1965[21]	1476	34.3%	18.6%	22.3%	6.4%	70.7%	50.0%
Kauppila, 1975[6]	448	51.1%	32.8%	32.4%	10.9%	90.1%	41.3%
Fianu, 1976[7]	283	32.2%			17.0%	66.2%	
Sachs et al., 1983[24]	1264			11.7%			38.5%
White and Cibils, 1984[10]	97		21.6%	18.6%	25.0%	57.0%	52.0%
Schutte et al., 1985[8]	791	44.6%	20.5%	18.9%	6.2%	67.5%	54.0%
Fortny et al., 1986[25]	2203			19.2%			56.7%
Kiely, 1991[26]	3893			17.8%			40.5%
Cibils et al., 1994[27]	444	42.5%	28.3%	22.5%		64.5%	53.3%

PNM, perinatal mortality; NNM, neonatal mortality; VLBW, very low birthweight.
[a] The cases published until 1983 include only infants weighing more than 1000 g. Only the publications after 1983 include fetuses weighing 500 g and more. The calculations were made from information and figures given in the respective publications.

prevented by improvements in conducting the delivery and strict adherence to appropriate techniques.[28] However, significantly higher complication rates than in cephalic presentations were still observed. As a consequence, in 1956 Goethals[29] suggested that cesarean section should be liberally indicated to deliver breech presentations at term when occurring in primiparas.[29] Shortly thereafter, in 1959, R.C. Wright[30] postulated that "any patient over 35 weeks gestation who enters labor with a living baby in breech presentation should be delivered by Cesarean section, providing there is no maternal disease that contraindicates abdominal delivery." This, at the time radical, approach did not gain immediate wide acceptance but it was slowly introduced in the obstetric practice, first in this country and then abroad. By the early 1980s in the city of New York nearly 65% of all breeches were delivered abdominally[26] while in some hospitals in Montreal (Canada) this incidence was only 22% in the early 1970s but rapidly reached 94% by 1979.[31] This dramatic change in practice adopted in the North American continent spilled over other areas of the world with more or less massive effect. In some South American countries the figures are similar to these for Canada, while in Europe the incidence of cesareans in breech presentation was much lower, reaching as much as 30% on the average for a large number of departments surveyed by Kubli et al. in the mid-1970s,[32] or as in the Netherlands where in 1982 it was "only" 26.5%.[8]

After several years of this practice some authors decided to review the outcomes, and compared cephalic and breech deliveries as well as abdominal and vaginal deliveries in both groups. For some it was not very convincing that morbidity and mortality were better among cesarean-delivered fetuses, in particular for term-size fetuses.[23,25] Others strongly urged the systematic abdominal delivery of fetuses less than 1500 g, even acknowledging that there were no data to substantiate that recommendation.[33] The same advice was given by others, who collected their material several years later, because their PNM was significantly better for those prematures delivered by Cesarean section.[34–36] However, a number of authors have been unable to observe a difference in PNM ascribable to mode of delivery for these small infants.[37–41] From a recent large multicenter study, Malloy et al.[42] concluded that "there were no statistical significant associations in any of the birth weight strata between neonatal death or IVH and the method of delivery."

It is extremely important to very carefully evaluate the potential confounding factors in those reports. Two review publications analyzing the same materials arrived at completely different conclusions for a variety of reasons. Westgren and Paul[43] tabulated 14 publications ostensibly comparing outcomes of vaginal and cesarean deliveries in very low birthweight (VLBW) breech presentations. However, they included a study in which breech deliveries were compared to cephalic vaginal deliveries and undetermined presentations born by cesarean sections.[33] They tabulated more than 8000 breech presentations when, in fact, in the quoted publications only 823 and 523 cases were VLBW breeches, respectively delivered vaginally and abdominally. Furthermore, the conclusions reached by these reviewers were contrary to that of some of the authors they quote in the table,[38,44] most likely because they disregarded the extremely important variable of changing weight-specific mortality from 500 to 1500 g birthweight by pooling the material in a single group. Myers and Gleicher[45] reviewed and reanalyzed six of those publications, in addition to two others, and reached the opposite conclusion, that the route of delivery does not influence neonatal outcome in these small infants, as they divided the cases into birthweights of less than 1000 g and more than 1000 g. In our own institution, the population analyzed when adjusted for fetal weights and gestational age revealed no difference in neonatal outcome among those VLBW infants delivered vaginally or by cesarean section.[27] This observation corroborates what was found in 1987 in a small European collaborative study reported by Thiery, who concluded that "we are no longer convinced . . . nor does our study support Cesarean delivery of all tiny neonates."[46]

Some authors have evaluated the long-term association of mode of delivery with neurologic or intelligence quotient (IQ) development in prematures. Effer and coworkers[38] found no difference up to 3 years of age, an observation replicated by Faber-Nijholt et al.[47] among either "optimally matched" at 3–10 years of age or "paired" at 18 months. From the foregoing it appears that those who advocate routine cesarean section for the VLBW fetus presenting by the breech have not been able to objectively demonstrate the validity of their recommendation on the basis of medical facts. In spite of this, a current textbook states that "the evidence presented above" (meaning references 37–39, 43, 46) "is sufficient to support this practice: to perform a Cesarean delivery for any live fetus presenting breech who weighs less than 2000 grams and more than 700 grams (26 weeks' gestation)."[48] It seems that there is a misinterpretation of "the evidence."

The management of the term breech fetus does not significantly vary from that observed for the prematures or VLBW fetuses. The recommendation made by Goethals in 1956[29] for primiparas and later extended by Wright[30] that all breech presentations should be delivered by cesarean section became the standard of practice in this country in the late 1970s and early 1980s. This dictum spread through Canada[31] and Europe.[46] In view of the alarming rise of cesarean section rates in this country, which reached nearly 25% in 1988,[17] efforts were made to evaluate some of the indications with the aim of controlling its progressive escalation. Among the contributing conditions, breech presentation appeared to deserve an evaluation. Two randomized studies were conducted in this country for frank[49] and nonfrank[19] presentations, and both failed to demonstrate any immediate perinatal benefit for the fetus but showed a significantly increased morbidity for the mother among those who had an abdominal delivery.

Numerous observational studies made in various areas of the world were also aimed at assessing the advantages, if any, of abdominal delivery over the vaginal route for these infants. Barlow and Larsson[50] identified patients "who could give birth vaginally without any mortality or persistent morbidity." Oian and coworkers[51] did not observe any change in perinatal outcome when the cesarean section rate increased fourfold among their breech presentations, thus corroborating what had been seen by Green et al.[31] a few years earlier in Montreal when their cesarean section rate for breeches increased from 22% to 94% without any appreciable change in neonatal asphyxia, trauma, or death. Several other relatively small series, gathered in this country and abroad, likewise failed to report any immediate beneficial effect for the fetuses delivered by cesarean section.[10,46,52–60] Two recent large reviews of material collected in single institutions in this country by Flanagan et al.[61] and by Croughan-Minihane and coworkers[62] found no effect of route of delivery on neonatal morbidity or mortality in their populations. It appears that when strict technical criteria are enforced in single institutions, these results can be achieved. The latter authors concluded "the adjusted relative risk estimate for the combined outcome category of head trauma, neonatal seizures, cerebral palsy, mental retardation, or spasticity was 0.5 in vaginal delivered infants (95% CI 0.1–3.2)."[62] Likewise, the observations carried out in our hospital failed to show any difference whatsoever among fetuses delivered either way; in fact, all the deaths among the cases with fetuses weighing more than 1500 g were in the group receiving an elective cesarean section, while none occurred among vaginal deliveries;[18] the difference was not significant.

Relatively few studies have focused on the long-term developmental differences among these neonates. Luterkort and collaborators followed prospectively more than 130 breech presentations in late gestation from which slightly more than one-half turned to cephalic. When discussing mode of delivery they concluded that "according to multiple linear regression analysis the breech presentation and the vaginal delivery per se were not factors in influencing the neurological score."[3] Westgren and Ingemarsson[57] stated that "vaginal delivery for full term breech presentations show no significant differences in outcome brought about by birth route." Otamiri et al.[63] later confirmed those observations when, from evaluation of their own material, they wrote "it may be concluded that infants born by elective Cesarean section and by vaginal breech delivery showed deviations in neurological adaptation during the first 5 days of age ... the neurologic dysfunction was transient ... the infants showed normal growth, neurology and psychomotor development at 6 months follow-up."[63] Nevertheless, several authors have been able to observe at the time of delivery among infants in breech presentation (regardless of route) what appears to be defective development when compared to their cephalic-presenting peers. Luterkort et al.[3] observed, when comparing cephalic with breech presentations who were small for gestational age (SGA) that the latter had lower neurologic scores; however, "children in the breech group born vaginally or by elective Cesarean section had similar neurological scores." Likewise, Westgren and Ingemarsson[57] concluded that "prospective follow-up studies and carefully matched controlled studies with sophisticated neurological evaluations indicate that breech infants, regardless of mode of delivery, will score slightly less favorably that infants born in vertex presentation". From similar observations, Hytten[64] thought that "we have come to assume that the baby may be brain damaged by the trauma of breech delivery, but it may often be that the baby is presenting by the breech because it is already brain damaged." After reviewing their own material, Cox and collaborators[37] concluded that "the value of routine Cesarean section in the absence of other obstetric pathology for low birth weight breeches in unproven and may be deleterious to mother and baby."

Maternal Effects

It is likely, because almost all normal mothers will accept any sacrifice for the sake of their infants, that little mention is made in publications and discussions about maternal morbidity related to mode of delivery of the breech-presenting fetus. The few publications in which this point is reviewed uniformly report a significant incidence of maternal febrile morbidity among the patients receiving a cesarean section.* The need of blood transfusion because of excessive bleeding[19,20,49] or even the need to perform hysterectomy as the result of intraoperatory complications is periodically reported.[20,49] The occurrence of postoperatory pelvic abscesses or evisceration following cesarean section has been observed by several authors.[20,49,65,66] Maternal mortality caused by postoperatory misadventures is more commonly observed among cesarean deliveries.[46]

The commonly quoted incidence of postoperative complications, ranging from 20% to 89%, is much higher than that observed (1%–6%) among vaginal deliveries.[20,49,65,66] Clearly, vaginal delivery offers more immediate safety to the mothers. This is true even when considering the frequent need to do a mediolateral episiotomy[48] that may extend to third- or fourth-degree laceration.[20] Appropriate repair should avoid further problems.

An important aspect to remember about cesarean delivery is the frequent necessity of a repeat procedure in a subsequent pregnancy, in spite of the current vogue and relative success in attempting vaginal delivery under that circumstance.[17] Many times, in particular when the operation is performed for VLBW fetuses requiring a vertical incision with extension into the corpus, the subsequent cesarean is inevitable because of the risk of dehiscence or rupture with those scars.

* References 19, 20, 49, 53, 65.

Characteristics of Breech Labor

From the foregoing it appears that, barring medical or obstetric contraindications to labor, a trial of labor should be permissible in most cases of breech presentation. However, to properly conduct such labor the obstetrician must be well informed of the particular aspects affecting the fetus and mother during labor.

The well-known differences due to parity that are observed in the cervix apply equally to breech presentation. Likewise, the uterine contractility pattern is indistinguishable from that observed in cephalic presentations (Fig. 6.1), and its effectiveness to dilate the cervix is the same.[67] Also, the labor curve pattern against time is not different from cephalic presentations.[68] Monitoring of the fetus, either biochemical or electronic, has been applied to the breech-presenting fetus. In this case, it has a characteristic pattern different in some aspects from the average cephalic presentation. Few authors studied these points specifically. The first publication discussing specifically the breech-presenting fetus was by Teteris et al.,[69] but it described

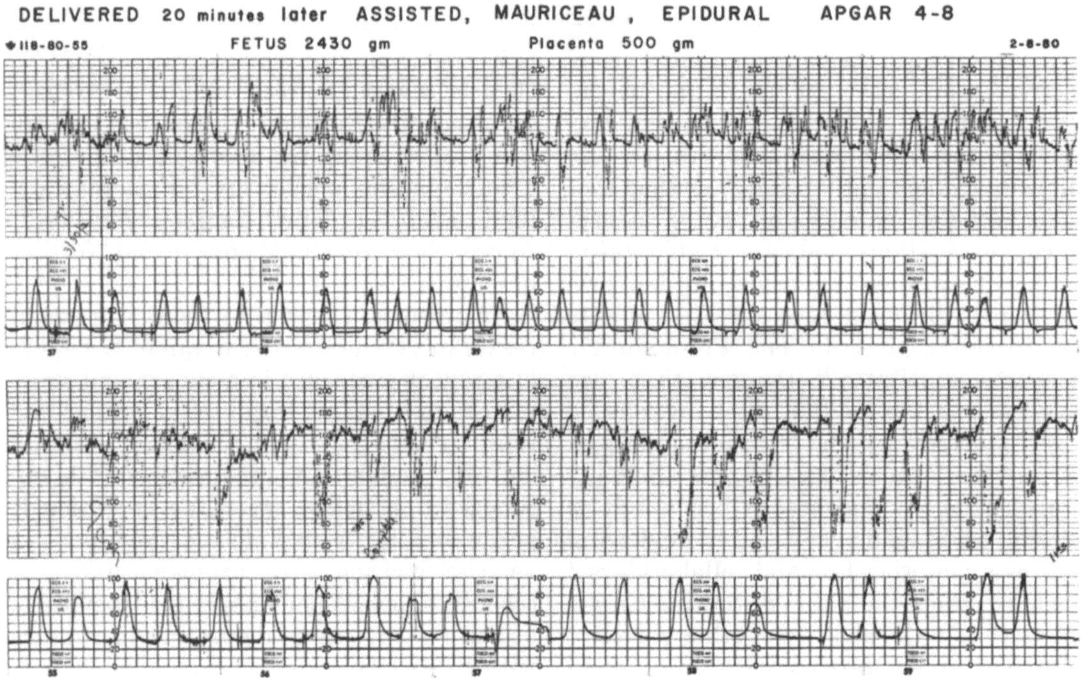

FIGURE 6.1. Premature rupture of membranes (PRM). Induced labor, primipara, 36 weeks gestation. *Top half* of chart, 50 min of direct monitoring show oxytocin-driven early first stage (3 cm dilatation) with normal contraction pattern. The fetal heart rate (FHR) shows large accelerations blunted by decelerations (cord compression) on a baseline of 135 beats per minute. *Bottom part,* 140 min after the end of top part of recording, under epidural anesthesia, shows advanced first-stage uterine contractility (UC) pattern (9 cm to complete). The FHR baseline has risen to about 160; the accelerations are blunted earlier by deep variable decelerations. Twenty minutes after the end of the tracing, a fetus in good condition was delivered by assisted technique with Mauriceau maneuver. Recording paper, 1 cm/min.

only the characteristics and differences from cephalics in the second stage. They emphasized the predictable sustained deceleration during breech delivery and implicated as a possible mechanism the "chest compression" during the process of delivery. Wheeler and Greene[70] described the frequent occurrence of variable decelerations and baseline changes, and correlated them with neonatal depression and pH changes. They, rightfully, attributed the fetal heartrate (FHR) alterations to cord compressions, and ascribed the accelerations observed to "abdominal compression of the fetus during contractions." Hill and collaborators[71] confirmed the observations made by Wheeler but reported much difficulty in obtaining meaningful tracings in second stage and thus advised the concomitant use of biochemical monitoring.

White and Cibils[10] studied specifically FHR alterations in this presentation and reported a 63% incidence of variable decelerations ("cord compression pattern") with one-third of them having the associated "late" component indicating transient hypoxemia. They further reported the importance of assessing changes of the baseline (tachycardia, fixed or saltatory) occurring in association with the decelerations as means of assessing fetal well-being. In their experience these associations predicted a high incidence of depressed 1-min Apgar scores (Figures 6.1 and 6.2) compared to those who

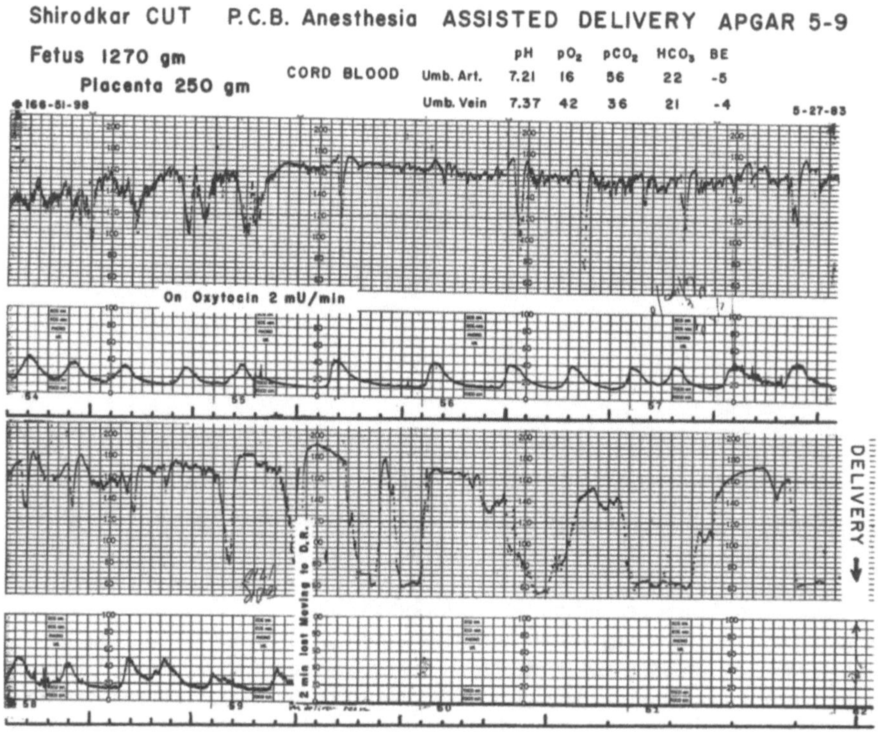

FIGURE 6.2. Induction of labor, multipara PRM, 27 weeks. Frank breech. *Top* tracing and left third of *lower* tracing show 54 min of advanced first-stage UC and FHR pattern under paracervical block anesthesia. Note mild tachycardia and variable decelerations. *Lower right* two-thirds shows the second-stage FHR tracing with deep sustained variable decelerations followed by tachycardia and fixed baseline. Assisted breech delivery occurred at the end of the tracing. Shown on top Apgar scores, fetal weight, and umbilical cord gases: mild respiratory acidosis. In this and subsequent figures the time marks below the tracings indicate minutes; note in the last 6 min the paper speed was increased to 3 cm/min. (From Cibils LA,[72] by permission of Parthenon Publishing.)

had only simple variable decelerations with a normal baseline (Figure 6.3). On the other hand, they observed accelerations in a very high percentages of cases (73%), either isolated or preceding decelerations, and correlated them with good outcome (Figures 6.3 and 6.4). Their presence was attributed to partial compression of the cord (umbilical vein) and, when observed in late first stage or second stage, predicted a neonate in very good condition. In a later report, Cibils[72] observed the almost invariable occurrence the significant sustained decelerations (cord compression pattern) during late second stage and the process of delivery, as illustrated in Figures 6.2 through 6.4. This probably represents almost complete interruption of umbilical cord circulation and sets the stage for early respiratory acidosis.

Repetitive decelerations caused by cord compression of course *may* induce alterations in fetal homeostasis when they either occur too frequently, are of very long duration, or when the fetal condition is already borderline compromised by high-risk factors or insufficient intervillous space circulation. The understanding of this pathophysiology[73] is essential for proper management of those cases; the management could be either continued observation or rapid intervention (see technique of vaginal delivery) depending on the characteristics of the tracing. The

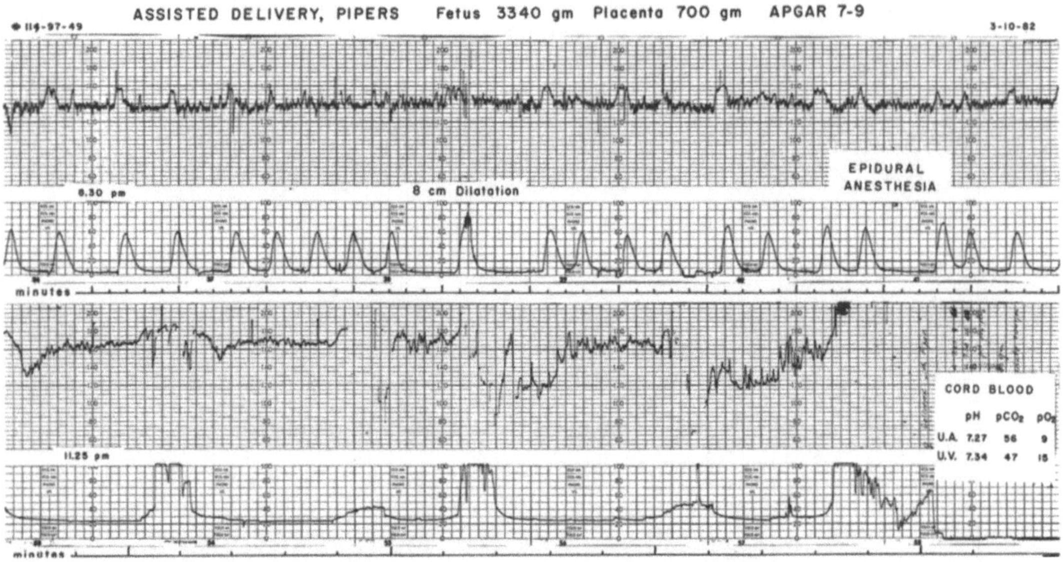

FIGURE 6.3. Induction of labor, secundipara, 41 weeks. Frank breech. *Top* tracing shows 60 min of advanced first-stage direct tracings under effective epidural. Normal oscillating baseline shows accelerations triggered by almost every contraction. *Lower* tracing shows 17.5 min (paper speed, 3 cm/min) of last part of second stage: accelerations were blunted by variable decelerations when the patient pushed. Assisted delivery with Piper forceps occurred at the last contraction. Fetal data and cord gases are shown. (From Cibils LA,[72] by permission of Parthenon Publishing.)

changes in fetal blood homeostasis, when they occur, will be manifested by the forementioned baseline alterations, and deterioration of aicd–base balance. Thus the fetal blood pH will fall, first because of accumulation of CO_2 (respiratory acidosis) (Figure 6.2) and, if the situation continues, the added loss of buffer bases (metabolic acidosis) (Figure 6.5).

The rapid induction of respiratory acidosis with falling blood pH was shown in 1972 by Eliot and Hill in an excellent study that also documented the rapid recovery in the first 5 min of life.[74] Kubli and coworkers[32] reported the high proportion of umbilical artery acidosis, which they attributed to cord compression, and elaborated on the unpredictability of that outcome. At about the same time, Hill and collaborators[71] suggested the routine use of fetal blood sampling in the management of second-stage breech deliveries in view of their difficulties to obtain good FHR tracings. The fact that there is an excellent correlation between buttocks blood samples and umbilical arterial blood pH was documented by Brady and coworkers.[75] Subsequent observations by others compared vaginal and abdominal breech deliveries and found no differences in umbilical artery pHs.[76,77] However, others found lower umbilical artery pHs among breech vaginal deliveries when compared to cesarean breech deliveries,[78] but did not find to be of clinical significance.[60,77] This observation prompted Socol et al.[77] to conclude that "the liberal use of Cesarean section may improve Apgar scores but will probably not appreciably improve the acid-base status of the breech neonate." This finding should, in fact, be expected if one remembers that during delivery there is a sustained cord compression, the duration of which dictates the degree of respiratory acidosis. The difference in pH and base buffers between umbilical arterial and venous blood documents this point,[72] and should therefore be taken into

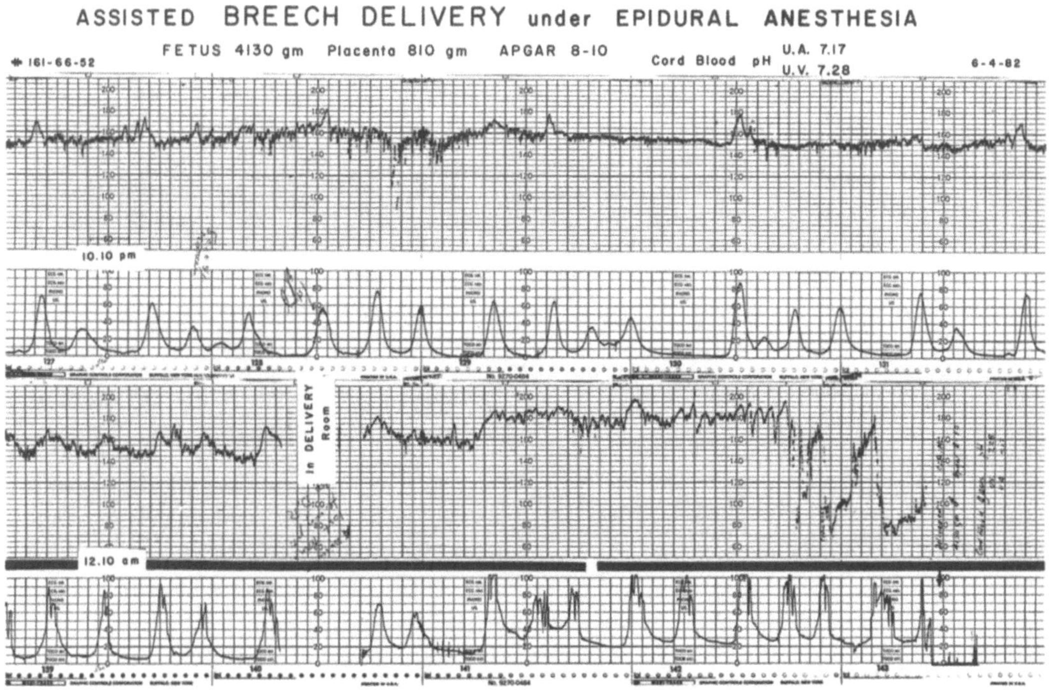

FIGURE 6.4. Enhanced labor, multipara, at term. Frank breech. *Top* tracing shows 50 min of advanced first stage. Normal UC pattern triggered some FHR accelerations. *Bottom* tracing, 80 min later, shows the last 45 min of second stage: pushing with contractions and accelerations, tachycardia in latter part. Two sustained variable decelerations preceded the assisted delivery of good neonate with cord around neck. Fetal information and cord gases are shown on top. (From Cibils LA,[72] by permission of Parthenon Publishing.)

serious consideration for the technique of delivery as illustrated in Figures 6.2, 6.4, and 6.5.

From the foregoing it seems clear that labor in breech presentations presents characteristics predisposing, if unchecked, to fetal deterioration on short notice; thus, continuous observation and adequate monitoring are essential for correct management. Another well-known intrapartum risk is the possibility of cord prolapse, which occurs in 1% to 20% of cases depending on the attitude of the breech presentation and the length of the pregnancy. Fortunately, severe cord compression is not frequently observed, and there is almost always time for successful intervention. In our material there was a 7% cord prolapse.[18] This accident may occur in a quite unexpected way even when the FHR tracing is completely normal (Figure 6.6).

One of the most frequently mentioned indications for cesarean sections in cases of trial of labor is the so-called dystocia or "poor progress," the implication being that the fetal head will be too large for the pelvis and thus that either the uterine contractility is abnormal or the cervix does not dilate as a result of the disproportion. It is hard to understand how this theoretical mechanical maladaptation can play any part in the process of the first stage of labor, but it is dutifully mentioned in every textbook available.

The Use of Oxytocin

As a corollary of the concept just discussed, it is also stated that oxytocin should not be used to attempt to correct the labor abnormality and that instead a cesarean section should be carried out.[19,48,79] The intravenous

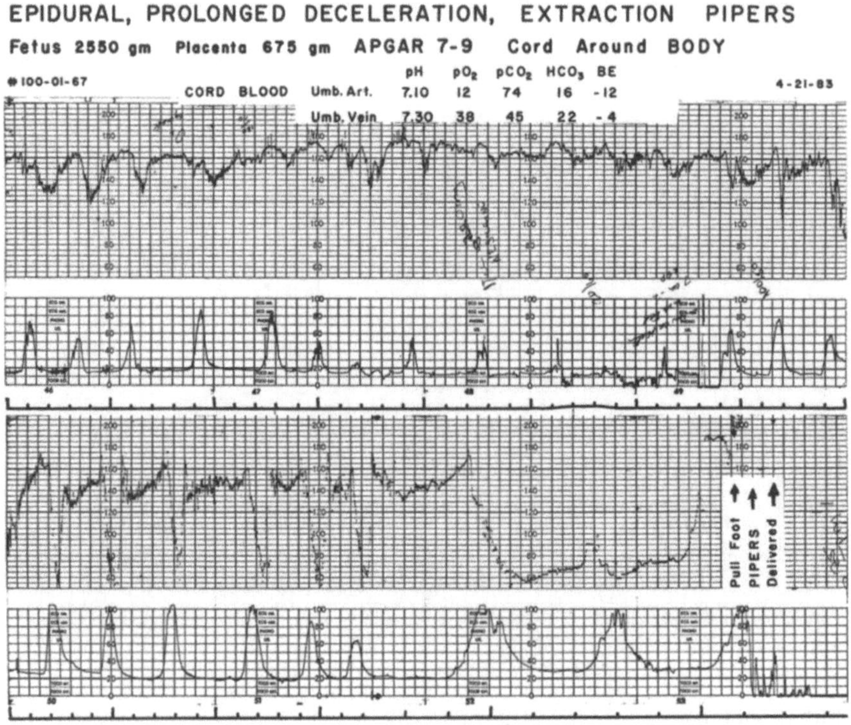

FIGURE 6.5. Induced labor, grand multipara, at term for PRM. Complete breech. Continuous 72 min of late first stage and second stage under effective epidural anesthesia. *Top*: average variable decelerations and mild tachycardia, occasional accelerations. *Bottom*: deep variable decelerations blunt the accelerations on an oscillating baseline. As descent progressed with pushes, the decelerations did not recover (the last 10 min of recording at 2 cm/min) and prompted extraction of the fetus with the cord around the body, thus explaining the sustained deceleration as this part of the fetus passed through the birth canal. Likewise, this mechanical condition explains the large arteriovenous blood gases differences and the mixed acidosis. (From Cibils LA,[72] by permission of Parthenon Publishing.)

infusion of oxytocin in physiologic doses has been, for the past 30 years, the standard therapeutic means to attempt to correct labor abnormalities or induce labor when indicated. Strangely, even when some clinicians would allow spontaneous labor to proceed with a breech presentation, the physiologic stimulation of uterine contractions (enhancement) would be absolutely contraindicated in the same patients.[48,79] This widespread practice, which inflates the number of unnecessary cesarean sections, has never been shown to have a rational practical basis. On the contrary, the majority of publications in which this subject is discussed clearly indicate that the use of oxytocin does not increase the incidence of undesirable outcomes.*

Similarly, the use of physiologic doses of oxytocin to induce labor is taught to be contraindicated[48] while the review of published series indicate that this technique may benefit the patient without increasing the risk for the fetus.† The rate of induced labors ranges in those series from 4% to 60%. Most interesting, in the majority of those series of induced labors there is an important component of primiparas, an absolute contrain-

---

*References 4–6, 10, 18, 20, 27, 49, 61, 70, 80.
†References 10, 18, 19, 27, 46, 49, 61, 70, 80.

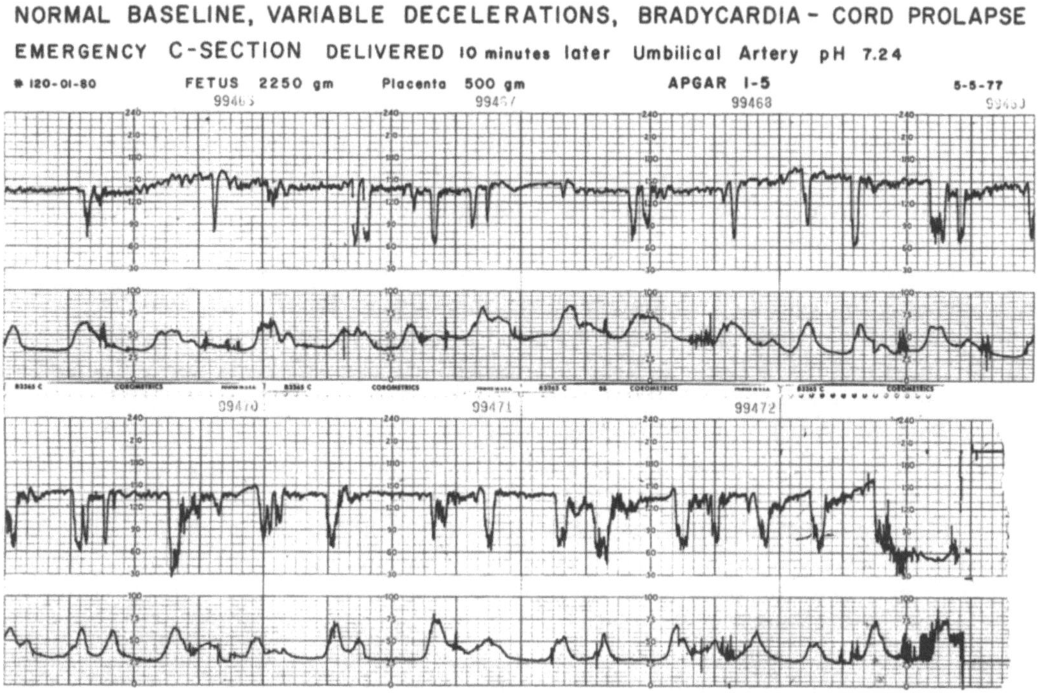

FIGURE 6.6. Spontaneous labor, secundipara, 34 weeks gestation. Frank breech. Continuous 78 min of mid first stage: normal UC and FHR baseline, variable decelerations triggered by most contractions. The last deceleration did not recover; an exam documented prolapse of the cord. This was followed by delivery in 10 min, by cesarean section under general anesthesia, of good fetus with Apgars 1 and 5 and umbilical artery pH of 7.24.

dication in the lore of breech presentation. It seems clear from the published evidence that the appropriate use of oxytocin to enhance labor may safely avoid many unnecessary operations, and that induction of labor, within the realm of correct indications, is as safe as for cephalic presentations. In our material, enhancement was given in 14% of cases (Figure 6.4), while induction of labor was started in 10% of the patients[18] (Figures 6.1, 6.2, 6.3, and 6.5).

Risks During Delivery

Vaginal

There are very characteristic risks during the delivery process of the breech presentation because of the mechanical relationships of the fetus and birth canal.

The Fetus

The risk of prolapse of the umbilical cord during labor has already been mentioned; of course, this is an unavoidable accident that has been observed as occurring many times during second stage and requiring prompt intervention.[4,5] Most of the other potential mechanical problems should be avoidable when the obstetrician follows closely the various technical steps during the delivery process. These are described in the standard obstetrics textbooks,[48] but some personal variations by the author are based on the observations about FHR changes previously described. The variations in technique practiced in our institution consist of continuous FHR monitoring during second stage. If the decelerations trigger baseline abnormalities and the delivery does not occur promptly (see Figures 6.2 and 6.4), or the decelerations are very prolonged, extraction should be accomplished (see Figure 6.5).

The fundamental principle, to avoid iatrogenic complications, is that once the delivery process starts (whether assisted or by extraction), it has to be a continuous, gentle, smooth traction following all the mechanical adaptations of the mechanism of breech delivery. First and foremost, it has to be started during a contraction and maternal bearing-down effort, preferably aided by gentle fundal pressure by an assistant. What appears to be a similar technique has been described by Arulkumaran and coworkers,[81] who with it obtained neonates in better condition than with the classical "assisted" technique of waiting to assist until the scapulas are visible.

A further variation in technique from classical textbook teaching,[48] at our institution, is that the anterior shoulder is delivered first, followed by gentle rotation toward the fetal back to bring the posterior shoulder below the pubis and deliver it as anterior. Piper's forceps are applied when the shoulders have been delivered unless spontaneous delivery of the head occurred while getting ready for the application. It has been shown by Milner[82] that application of forceps to the after-coming head is associated with significant reduction in neonatal mortality among fetuses weighing between 1 and 3 kg. Pulling from the limbs or body without a contraction or appropriately applied gentle fundal pressure facilitates the extension of the fetal arms, and thus this becomes an iatrogenic complication. A further complication of the extension of the arms occurs when the extended arm slides behind the fetal neck, thus creating the "nuchal arm," a more complicated and mechanical dangerous problem requiring skill and gentleness for an atraumatic solution. Sherer and collaborators[83] reported a 4% incidence of nuchal arm during first stage, a figure much too high for our experience. As stated, this is essentially an iatrogenic and thus avoidable complication.[49]

It was mentioned earlier that compression of the umbilical cord is a mechanical, inevitable occurrence during the passage of the fetus through the birth canal. However, the other difficulties or traumas, as described in textbooks and publications, are more likely preventable with adequate assessment of fetus–pelvis relationship and skillful handling of the delivery process. In breech presentation there is *no room for a trial of labor*, as is permissible in cephalic presentations with "borderline" pelvises. When there is the slightest doubt that the fetus may be too large or "tight" for a given pelvic canal, a cesarean section should be carried out.

The classic method to evaluate the pelvis has been x-ray pelvimetry.[19,20,48,79] This test has the added advantage of defining the fetal attitude and demonstrating the relationship of fetal head and spine (ruling out extension or hyperextension of the head), as well as excluding malformations, all very important conditions that should be weighed when deciding about route of delivery. Others have proposed the use of computed tomography, which would give information similar to x-ray pelvimetry but with lesser amount of irradiation for the fetus and mother.[61,84] The concept currently more widely accepted is that only frank attitudes should be considered for vaginal delivery. However, in a randomized study Gimovsky and coworkers[19] reported no difference in neonatal outcomes among nonfrank breeches delivered either way. Likewise, others have had no complications[20,61] when allowing vaginal deliveries in these cases, an experience confirmed in our own institution.[18,27]

Hyperextension of the head, on the other hand, is a rather dangerous condition with varied frequency, as high as 7%–14%,[20,35] which predisposes to severe spinal cord injuries when vaginal delivery is accomplished.[86] In an excellent review Abroms and coworkers[87] reported the extremely high incidence of spinal cord transection (21%) among those delivered vaginally, and no severe sequelae among those delivered abdominally. It seems that the safest route of delivery is the latter.[20] On occasions the hyperextended position of the fetus near term may already represent a permanent scar of earlier damage to the spinal cord.[88] The very high incidence of congenital malformations is shown in Tables 6.1 through 6.3; many of these are incompatible with life.[85] This incidence is even higher among VLBW fetuses (see Table 6.3), making a diagnosis very desirable before deciding on route of delivery. Less severe congenital malformations are also very common, in particular among fetuses of higher weight.[6–8,14,15,49]

There are numerous reports of various types of mechanical difficulties or neonatal trauma following breech vaginal deliveries. One of the mechanical accidents most frequently mentioned as occurring during vaginal delivery is the so-called head entrapment,[48,49] either resulting from a small pelvis,[85] or, more often, described in small fetuses and ascribed to "cervical dystocia."[38,85,89] This is a mislabeled diagnosis, because the cervix is made of connective tissue and thus is always completely thinned out, having minimal resistance even when dilating to allow passage of a premature fetus. What "traps" the fetal head is the lowermost part of the corpus where it joins with the lower segment (the same area that makes the Bandl's ring), when it is allowed to clamp on the fetal neck. When it happens, because the corpus is a powerful, thick muscle contracting, only good general anesthesia can relax it quickly to facilitate an atraumatic delivery. There was only one "trapped head" in our material of more than 370 cases of vaginal deliveries; that was a VLBW fetus admitted already dead antepartum.[18] It is probable that the combination of good anesthesia and the coordinated continuous action of obstetrician and patient (or assistant) contributed to the prevention of this dreaded complication.

Injury of the brachial plexus is usually the aftermath of difficulties to deliver the head and subsequent strong pulling. It is reported to have occurred in the majority of series discussing complications.* Fractures of clavicles and humerus are also occasionally observed.[31,90]

Intracranial hemorrhage is particularly damaging because its sequelae are often incapacitating. It occurs more often in smaller fetuses,[85,91] but larger ones are not exempt from this risk.[31] The worst possible complication from mechanical difficulties with the head is severe injury or severance of the spinal cord[92] with quadriplegia. Skull fractures have also been observed.[31]

All these traumatic complications are probably the result either of hasty maneuvers to deliver the infant or of misjudgment of fetal size in allowing too large a fetus to go through the birth canal. On rare occasions the progress of labor may be so advanced that vaginal delivery may be the only feasible alternative.[20] From the foregoing one may surmise that the mechanics of breech delivery are more complicated than for cephalic, at least for the obstetrician. Furthermore, it seems reasonable to state that the great majority of traumatic complications should be preventable if the fetopelvic relationship is adequately diagnosed and the delivery process conducted with the right technique. There are four cases reported unofficially,[93] and one completely described,[94] of abdominal deliveries of breech-presentation infants partially exteriorized through the vagina with minimal deleterious fetal effects. A good estimation of fetal size and pelvic capacity should make unnecessary this very dramatic situation.

The Mother

The risks for the mother are those related to potential lacerations and infections of the birth canal and uterus. For term-sized fetuses it is usually advised to perform a mediolateral episiotomy which, if improperly done, may extend through the sphincter and rectum.[20] Others report only on overall morbidity of about 7%.[49] This would include postpartum endometritis, which occurred in 4% of patients in our obstetric population. Occasionally the episiotomy wounds may

---

* References 20, 31, 49, 85, 89, 90.

break down. Uterine atony has also been described.[49]

Cesarean Section

The understanding of the mechanism of breech delivery through an abdominal incision is a mandatory requirement for any obstetrician doing a cesarean section, as it is if the birth would be accomplished by the vaginal route. This is necessary because, to accomplish the delivery as atraumatically as possible, the fetus has to be pulled out following the appropriate maneuvers through a tight uterine incision, often made through a thick muscular uterine wall.

The Fetus

The indication for an abdominal delivery is that, ostensibly, it will be much less traumatic and potentially less damaging for the fetus than vaginal delivery. Unfortunately that is not always the case, as periodically there are reports of accidental lacerations made to the fetus during incision of the uterus.[61] Likewise, during the extraction of the fetus, if the extraction maneuvers are not gentle and smooth fractures of lower limbs may occur,[4] and have been reported to be either metaphyseal of femur and tibia near the joint[95] or in the middle of the thigh[20,96,97] (Figure 6.7). One such fracture occurred in our series during an emergency operation of a term fetus admitted with a prolapsed cord. The humerus may be fractured when the proper maneuvers to deliver the arms are not made, that is, flexion of the forearm at the elbow and not in the middle of the arm.[90] Difficulties to deliver the head create the condition for brachial plexus injuries because of uncontrolled pull from the body.[20,90] This may be particularly likely to occur during delivery of small fetses through transverse incisions on poorly developed lower segments.

The Mother

Unlike vaginal deliveries, with cesarian section there are greater chances of complications for the mother. Although conduction anesthesia is the method of choice for abdominal deliveries, general anesthesia often may be necessary because of either the urgency of the case, the patient's request, or the necessity of uterine relaxation for an atraumatic delivery. The relative risk of this type of anesthesia in obstetrics, which may range from "severe difficulties"[20,49] and aspiration[85] to maternal death,[46] is well known. Among our cases, 54% received general anesthesia and the remainder received conduction. The most common complication intrinsic to the technique of cesarean section is the extension of the incision to the uterine pedicles[49] when the transverse incision is used. This is the most popular type of incision for term or near-term pregnancies, and when it extends triggers important blood loss that may require transfusion.[20,49,61]

Among all cesarean sections, in our material, 55% were given a transverse incision, nearly all for fetuses weighing more than 1500 g; there were 15 extensions but none required transfusion. In certain circumstances the tears may be so large or the bleeding so difficult to control that hysterectomy may be necessary.[49] It has also been reported that while attempting hemostasis ureteral injuries occurred.[49] Wound infections are a known complication of cesarean section[65] and have been reported in the majority of breech series.[20,49,61] The high incidence of endomyometritis and parametritis is a well-recognized post-cesarean complication, ranging from 15% to 70% even when using prophylactic antibiotics.[20,49,61,79] A more dramatic infectious complication is dehiscence of the wound.[20,79] The severity or extension of some infections may require a hysterectomy for adequate treatment.[20,49] Secondary morbidity following blood transfusion has been observed,[49] a complication much feared now because of the potential transmission of viral infections by blood products.

An often-overlooked long-term risk of cesarean section for the mother is the necessity of a repeat operation in a subsequent pregnancy. It is true that the recommendation of a trial of labor is more and more accepted if breech presentation has been the

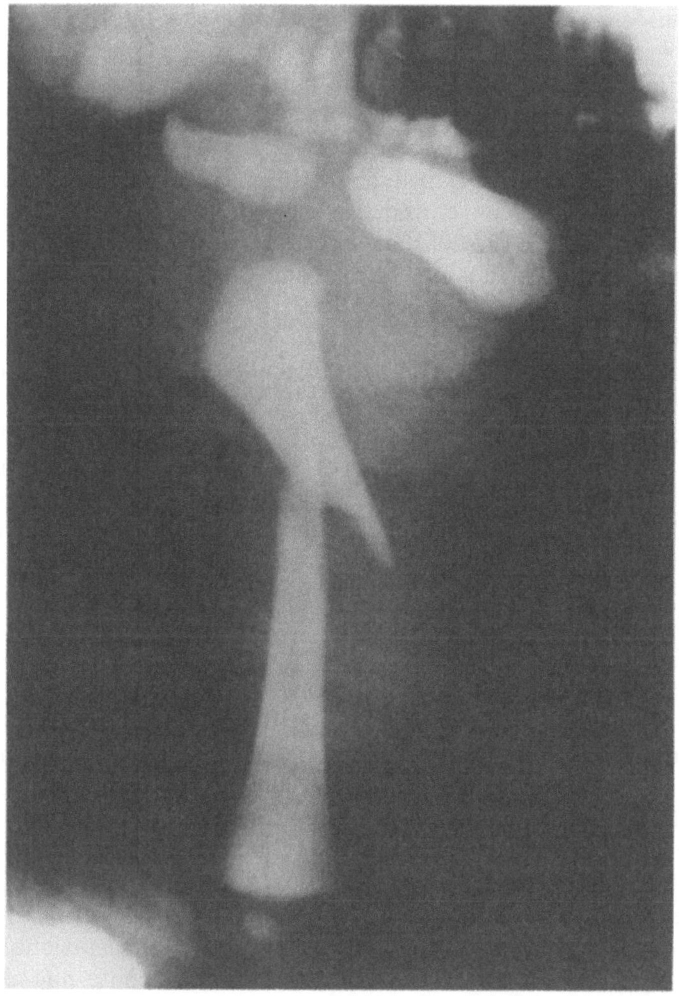

FIGURE 6.7. Flat plate of newborn at term, 3700 g, delivered by elective cesarean section. Midshaft femoral fracture. Not evident, but later observed on healing, was another fracture, of distal metaphyseal femur. Good healing and function were noted at 9 months. (From Vasa R and Kim MR,[96] reprinted with permission from The American Journal of Perinatology 7:46–48, 1990, Thieme Medical Publishers, Inc.)

indication. However, because of the trend to operate on almost all breeches of less than 2500 g, which constitute a very high percentage of them (53% in our population)[18] and the requirement of a vertical incision for safe delivery, the need of a repeat operation in the future is already determined. Thus, the potential morbidity of one or more future pregnancies with a scarred uterus and subsequent cesarean section should be considered in the equation.

Management

From the foregoing review of published experience it seems clear that the generally applied approach of cesarean delivery to breech presentation[17] is not based on clinical facts. It is, rather, the conclusion based on biased or incomplete analyses of information passed from previous generations and the refusal to objectively review more current observations. A further, widely stated, reason given for recommending cesarean section for almost all breech presentations[48] is that not enough cases occur to give adequate training to house officers.[48,79] The same is said about forceps deliveries but both are in reality fallacious arguments because the cases that can be used to transmit knowledge are, instead, managed with a rapid cesarean section. At least, the impressively higher morbidity of this route of delivery compared to vaginal is not denied but, in spite of that, is rarely taken into consideration when undertaking the operation.

The obstetrician managing a patient with a fetus in breech presentation must consider several options. Can the abnormal presentation be safely corrected? If the answer is in the affirmative, and the maneuver is successful, that part of the obstetric problem was solved. When the correction is unsuccessful, then the decision must be made between abdominal and vaginal delivery. These various options are now discussed.

Prophylaxis

The ideal approach to offset the effect of abnormal presentation on perinatal morbidity (mother and fetus) is its prevention. That premise being an unattainable goal, the next best thing to try is to minimize the number of abnormal presentations. In reference to breech presentation, attempts have been made for many years utilizing an "external cephalic version."

This maneuver, well described in all obstetric textbooks, fell into disrepute in the mid-1960s,[98,99] when the popularity of elective cesarean section was on the rise, and continued in disfavor until relatively recently. A few isolated crusaders maintained the art and continued to practice it with great benefit to their patients. Among them, Ranney[99] reported his experience of several years with excellent results. Its popularity was slowly regained after a report in Europe, where the use of uterine relaxants was predicated, by Saling and Mueller-Holve.[100] By the middle of the 1980s the version became a popular maneuver in the major perinatal centers but was seldom used in so-called private practice settings. Actually, it is a relatively simple procedure requiring gentleness and patience from the obstetrician and a number of clinical prerequisites. If successful, the chances of a return to breech presentation decrease with the fetus closer to term. For the best chance of success, the following are required: the membranes must be intact with normal amniotic fluid volume; the breech must not be engaged; and the patient must be cooperative and able to tolerate a β-mimetic substance to relax the uterus. In terms of equipment, it is extremely useful to have available a real-time ultrasound scanner to follow the fetal position (and its heartrate) during the maneuvers. In addition, continuous FHR monitoring before and after the procedure should provide timely information to either abort the version, if abnormal before it is completed, or to begin prompt intervention if the fetal heart rate becomes and stays abnormal after completion. Often transient abnormalities will be observed.[101,102] Success has been reported variously as between 48% and nearly 100% for those attempting the version.* Others report lowering of cesarean section rate either among their total population, or among those who were successfully converted to cephalic as compared to those not converted.[101,103–105,107–109] In any case, a substantial number of breech presentations could be made cephalic if the procedure is done at the right moment as correlating with the appropriate technique.[99] Some factors have been mentioned as correlating with the rate of success, such as parity and amniotic fluid volume[105,106,110,111] or earlier gestational age.[99,103] Conversely, failure rate has also been reported to be promoted by placental location,[103] specifically the cornual location for some.[110,112] In our experience relative oligohydramnios has been a frequent reason to discontinue attempts at version.

There are some general requirements, as have been indicated. However, there are few reports of series in which the version was safely and successfully accomplished despite relative contraindications. Some have done versions with a high rate of success on uteri with cesarean section scars,[113] which is generally considered a contraindication. Others have accomplished versions without the use of tocolytics (uterine relaxant drugs),[99,114,115] an important element in the currently recommended technique.

---

*References 61, 98, 99, 102–107.

Complications have been observed following external version and adscribed to its execution. Compound presentation and dystocia[105] could be considered relatively benign. Others of more serious consequence have been observed in moderate-sized series or as isolated cases. Fetomaternal passage of significant amounts of blood has been documented[102,116,117] and seems to be likely in view of the manipulations necessary over the placental implantation. In our institution a placental implantation over a large area of the anterior aspect of the uterus is a contraindication to version. Unexplained fetal deaths[98,102] or placental abruptio with subsequent perinatal deaths have been observed shortly after external versions.[98,118,119] Direct physical trauma to the fetus has also been reported, this in spite of the "protection" provided by the many layers of tissues and the amniotic fluid separating it from the hands of the operator. Bruising and brachial palsy have been produced[120] as well as severe spinal cord injury.[121] It is hard to escape the conclusion that in these cases the overzealous determination of the operator must have played a major role in the complication. A maternal fatality from amniotic fluid embolism has been reported.[102]

To recapitulate, the safe prophylaxis of breech presentation is possible in nearly two-thirds of cases provided the operator respects the general contraindications, follows the proper technique, and acts with gentleness.[99] Furthermore, it is important to observe the patients with FHR monitoring to detect alterations, and to obtain Kleihauer–Betke tests to investigate possible fetomaternal bleeds and act accordingly. In the case of an Rh-negative mother, prophylactic anti-D gamma globulin should be administered.

Choice of Route of Delivery

It has been accepted by the majority of authors who reviewed their own material that there is no difference in the corrected mortality or morbidity for fetuses delivered vaginally or abdominally if they weigh more than 2000 g, and that maternal morbidity is much higher following cesarean section. In spite of that, about 85% of all breeches in this country are still delivered by cesarean section.[17]

From the perusal of the most recent publications comparing vaginal and abdominal deliveries referred in the preceding pages, it seems that vaginal delivery of the breech-presenting fetus, regardless of weight, should be the first contemplated option. However, there are criticisms that attempt to discredit the validity of those reports. In particular, it is universally stated that those studies are not randomized and prospective. That is quite correct, and the option should be to undertake such an epidemiologically more refined study. The problem arises when one tries to execute such undertaking because most obstetricians in this country would change their current approach of systematic cesarean section for fetuses of less than 2000 g[122] if others do a satisfactory study. Likewise, in the United Kingdom the conduct is very similar.[123] Numerous reasons (excuses) are given to justify the continuous practice of cesarean sections in breech presentations and, as seen in the previous pages, none of these are of very solid medical value. The so-called medical-legal argument is discussed in Chapter 12 of this book. Interestingly, the conduct to subject the mothers to a cesarean section has never been proven to be justified, but the majority of obstetricians responsible for making the decision are reluctant to carry out the prospective study they claim should clarify the problem.[124] Penn[125] attempted to carry out a multicenter randomized prospective study of VLBW breech fetuses (25 large separate departments agreed with the protocol and were enrolled) but had to discontinue that effort after 15 months because only 13 cases had been recruited in the study.

Vaginal Delivery

When one evaluates a breech-presenting pregnancy in labor, or near labor, it is essential to remember that there are absolute contraindications to vaginal delivery and when they are present it should not be tried. In previous pages it was discussed that suspected or clear cephalopelvic disproportion is one such contraindication. Many authors continue to rely on x-ray pelvimetry, as part of their protocols, to permit or contraindicate a trial of labor,* while others suggest the use of computed tomography to minimize irradiation exposure.[59,126] Magnetic resonance imaging (MRI) gives comparable measurements without ionizing radiation.[127] In our institution x-ray pelvimetry is not routinely done. Instead, a thorough assessment of fetal size and clinical pelvimetry are the deciding factors. Hyperextension of the fetal head should preclude labor and vaginal delivery. An abdominal radiograph, which at the same time may rule out severe congenital malformations, is necessary to correctly make this diagnosis. A prolapsed cord with incompletely dilated cervix cannot be safely delivered vaginally, and as in the other situations mentioned, mandates a cesarean section. Insufficient skills of the person in charge to conduct labor and delivery is also an absolute contraindication to attempt a vaginal delivery. The first three are not controllable by the obstetrician and therefore not remediable, and about the last one there are some further comments.

There are also relative contraindications to vaginal breech delivery. The presence of a uterine scar has been one of them, but with the current trend of encouraging vaginal deliveries after previous cesarean section this is a dwindling contraindication, and it has been shown to be safe for breech presentations.[53,113,128,129] The nonfrank breech presentation has been considered as a mandatory indication for cesarean section even in centers where vaginal birth is attempted for breech presentations. The ostensible reason for this conduct is the known higher incidence of cord prolapse among them (4%–12% compared to 1%–2%). However, in the only published prospective randomized study of nonfrank breech presentation, prolapse of the cord occurred in 4% of trials of labor, and all cases were delivered by cesarean section in good condition.[19] In our own material, of more than 1140 cases, prolapse occurred in 7% of cases while frank presentations occurred in 47% of that total. Among the last 840 cases there were 48% nonfrank presentations delivered vaginally with corrected perinatal mortality and morbidity not different from those delivered by cesarean section,[18,27] even though there were 6% prolapses among completes and 14% among footlings.

The use of physiologic doses of oxytocin to correct desultory labor resulting from abnormal uterine contractility was discussed previously. Suffice here to say that there is no evidence that oxytocin has a deleterious effect on the fetus, and clearly it is the method of choice to improve unsatisfactory contractions pattern (see Figure 6.4). Premature rupture of the membranes with unripe (unfavorable) cervix is also often mentioned as a contraindication to labor. If one uses oxytocin to induce or enhance labor in cephalic presentation with this syndrome there is no rational basis for not doing the same with breech presentation (see Figures 6.1 and 6.2). This should be understandable because the uterine contractility pattern and efficiency are the same as in cephalic presentations.[67,68]

As discussed previously, labor in breech presentation presents special characteristics with regard to the fetus. Nearly 70% of these fetuses will develop variable decelerations during first stage,[10,72] and almost all during second stage (Figures 6.1 through 6.5). It is critical to understand and distinguish correctly the decelerations that are benign from those which represent progressive alteration of fetal homeostasis. Those will be, essentially, associated changes of the baseline expressed as tachycardia and fixed baseline (diminished or absent variability).[73] When these alterations are observed, an assessment of the stage of labor should be undertaken by the clinician to evaluate the likelihood of rapid progress toward complete

* References 19, 20, 49, 50, 61.

dilatation and vaginal delivery (see Figure 6.2). If this progress is not prognosticated as occurring before fetal deterioration, an abdominal delivery should be indicated at once and carried out expeditiously.

Intrinsic to the mechanics of delivery, which entails compression of the umbilical cord during passage of the body and head, is the transient curtailment of oxygen supply to the fetus for a variable period of time. There is a concurrent accumulation of carbon dioxide and fall of pH. In our material, 49.5% of breech cases showed an umbilical artery pH of 7.20 or less compared to 36.5% among cesarean deliveries, but at 5 min there were no differences in Apgar scores.[18] This particular characteristic of the breech delivery requires that continuous FHR monitoring be maintained until delivery is complete because unexpected sustained compression of the cord may occur at any time during second stage, as the fetus descends in the birth canal (see Figure 6.5). This situation constitutes an indication to proceed with extraction which, if properly done, should not adversely affect the fetus. An extraction was carried out in 15% of vaginal deliveries in our cases.

A very important factor contributing to good labor and smooth delivery is adequate control of pain. Classically, pain control of labor in breech was carried out by means of systemic analgesia (natural or synthetic opiates), in first stage, and by local anesthesia, preferably pudendal block, for second stage.

With the widespread use of conduction analgesia-anesthesia in obstetrics in the past 20 years, it was tried with excellent results for breech presentation. It was shown that, although second stage may be moderately lengthened, the expulsive forces were not obliterated,[129,130] and that the delivery maneuvers could be executed without haste and with the patient's cooperation under controlled conditions. Furthermore, when there is good anesthesia during first stage one can proceed almost immediately with an abdominal delivery if the need arises because of an unanticipated emergency. The technique of delivery has been briefly described under the heading "Risks During Delivery, Vaginal." Crawford[129] observed that there were fewer cases of depression among the neonates born from labors under epidural anesthesia, and Darby and coworkers[130] observed better 5-min Apgar scores among the same group. In our population of vaginal deliveries, conduction anesthesia was usd only in 35%, but it is important to clarify that 18% of these deliveries were stillborns, the majority of the latter done under systemic analgesia and local anesthesia.

Cesarean Section

Only a few additional pertinent points are needed to be made to complete the discussion given in previous pages under the heading "Risks During delivery. Cesarean Section." It was stated there that an atraumatic delivery is the objective, and to accomplish that it is essential to use a combination of appropriate anesthesia and gentle extraction. The age of gestation and the size of the fetus are important determinants of how these are conducted.

The method of anesthesia controls the presence or absence of contraction during the process of extracting the fetus. On the other hand, the presence and size, or absence, of a lower segment determines the type of incision more favorable for an easy delivery, and this is directly related to gestational age as well as prior labor. In view of these different situations it is necessary to discuss the technique to apply under the circumstances.

Term Fetus

When one makes the decision to proceed with a cesarean section on a fetus at term, but the mother is not in labor, it is necessary to evaluate the development of the lower segment by a thorough pelvic examination. Deep fornices and a nearly effaced cervix indicate a good lower segment and the possibility of an easy delivery under conduction anesthesia, with a low transverse incision. The same could be said for patients in labor operated for fetal distress. However, it may be necessary to do a vertical incision to avoid fetal trauma to limbs if the presenting part is too deeply engaged.[96,97] The abdominal incision is part of the route to be negotiated by the fetus and therefore it has as much importance as the uterine incision; its inadequacy may contribute to neonatal trauma.[85] In our total population, transverse incision of the uterus was used only in 54% of cases, but among the fetuses weighing more than 1500 g this incision was done in 65% of cases. Conversely, a classical incision or a vertical with extension into the corpus was done in only 12% of the group of fetuses more than 1500 g, of which 29% (of the 554 cases) were premature by weight.[18] It is highly likely that all these incisions were made on the cases weighing less than 2000 g. Low vertical incisions were made on the remainder, 23%.

The Very Low Birthweight Fetus

These fetuses, according to the widely held view, are those that most benefit from an abdominal "atraumatic" delivery. As shown elsewhere in this chapter, from the review of more recently published series,[27,37–42,46] that concept has not proven to be correct. Every step should be taken to fulfil the premise of an atraumatic delivery. These are very fragile fetuses and therefore the anesthesia as well as the uterine incision must be appropriate. When labor is conducted under regional anesthesia and there is a reasonably developed lower segment, it is safe to proceed with cesarean section, if indicated, under the same type of anesthesia. However, for emergency situations or an inadequately formed lower segment it is necessary to operate under general anesthesia with good uterine relaxation. In other words, the obstetrician must wait until the uterus is well relaxed before proceeding with the maneuvers to extract the fetus, as it is mandatory when conducting any intrauterine manipulation.

Conduction anesthesia is not satisfactory under these circumstances because the uterus continues to contract at its own rhythm with caudal, spinal, or epidural blockade.[131–133] The substances used to induce general anesthesia by "rapid sequence," that is, thiopental, succinylcholine, or drugs of similar effect, do not relax the human uterus at doses utilized for cesarean section.[134] Only the volatile anesthetics (nitrous oxide, halothane, fluothane, penthrane, etc.) will trigger uterine relaxation after they reach the appropriate blood levels in about 3–5 min after their administration starts.[134] The surgeon should wait until these drugs produce their relaxing action on the uterus before attempting to extract the fetus. In 54% of our cesarean sections (all fetal weights), general anesthesia was used but it was given to 70% of the very low birth weight (VLBW) group. The remaining 30% of these received conduction anesthesia. It appears that there is an intrinsic higher risk of neonatal depression among breech-presenting fetuses to have lower Apgar scores and more often need of intubation than among cephalics, as it ws shown by Calvert for healthy term-size neonates.[135]

In addition to the method of anesthesia used one must decide on the best type of uterine incision to accomplish an easy and smooth extraction. As was stated, this step is crucial in handling these small infants, particularly if the presenting part is deep in the pelvis or the uterus is not well relaxed. Westgren and Paul[43] reported that for their VLBW population of less than 30 weeks gestation about 60% were low vertical, and the remaining 40% equally divided between classical and low transverse; for the same weights but 31 weeks gestation and older about 50% were low transverse, 30% low vertical, and 20% still classical. In our material of VLBW infants, regardless of gestational age, only 18% were delivered by a transverse incision, 42% a low vertical incision, and 40% by classical or low vertical with extension into the corpus.

The combination of a relaxed uterus and an appropriate vertical incision is, in our judgment, the only technique that permits a truly atraumatic delivery. A contracting uterus ("tight") and a low transverse incision are an invitation to a "head entrapment" as has been described by some to occur even with term-sized breeches.[85] "Entrapment" of the whole fetus has been described to occur in VLBW infants delivered by low (vertical and transverse) incisions, necessitating extension into the corpus.[43] For this to occur, the mothers must not have been given anesthesia with uterine relaxant substances as is our recommendation. Likewise, a tight uterus (contracting) or an insufficient transverse incision are direct contributors to fractures observed in low birthweight (LBW) as well as term-size neonates.[95–97] It is important to reemphasize that the extraction maneuvers must be carried out with great gentleness and following the mechanics of breech delivery. The described traumatic complications to the fetus may occur unless the proper technique is followed.

As far as intraoperative maternal complications are concerned, they may be related to the type of incision used on the uterus. Extension and lacerations of the uterine artery or vein and their branches, causing severe blood loss and often the need of red cell transfusion, have been reported with transverse incisions.[49,136] Ureteral, bladder, and small bowel wounds have also been observed.[49,136] Complications labeled "severe" have been observed significantly more often when the operation is performed on pregnancies of less than 32 weeks.[136] This is the group of patients for whom the current trend recommends systematic cesarean section. Significant blood loss, even without extensions of the incision or lacerations, but requiring blood replacement have also been observed with regularity.[19,20,49,136] Nonfatal severe complications of anesthesia[49] or even maternal death from complications of general anesthesia have been also reported.[46] The postoperative complications, particularly infections, have already been mentioned (see pages 69 and 80). There is no need to repeat the risks involved in this area.

One of the long-term risks for the mother who undergoes a cesarean section, particularly those receiving a classical or a low vertical incision with extension into the corpus, is the necessity to be subjected to repeated cesarean sections in subsequent pregnancies. Of course this entails facing again the same intraoperatory risks just described. In addition, scar dehiscences occur in a number of post-cesarean pregnancies, some of which may become complete ruptures if not operated by chance or diagnosed by very astute clinicians. Another also very serious complication observed in pregnancies following cesarean section, only recently recognized, is the significantly increased risk of placenta previa, proportionally growing very rapidly with each subsequent operation.[137–139] A relative rarity in obstetrics, placenta accreta (even increta or percreta), occurs with alarming frequency in pregnancies evolving in uteri with a scar from previous cesarean section.[138–142] The association of placenta previa with a scar from a previous cesarean section carries a risk as great as 67% of being a placenta accreta. Of course, this entity almost always requires emergent hysterectomy.

Conclusions

The detailed review of the available literature based on comparable cases of breech presentations managed either by cesarean section or vaginal delivery indicates that there is no clear benefit for the fetus to be delivered abdominally. However, the morbidity facing the mother is manyfold higher and often of significant severity. These two statements appear to be valid for all breech presentations, regardless of gestational age. Thus labor, either spontaneous, enhanced, or induced, appears to offer the most sensible medical approach to deliver a breech presentation that could not be verted to cephalic.

Of course there should be stringent requirements if the obstetrician wants to avoid the potenial complications of an operation for the mother and, at the same time, offer the infant a safe delivery. The contraindications must be observed. Continuous FHR and uterine contractions (UC) monitoring should be applied when labor is permitted. The tracings should be artifact free and continued until delivery of the head. Furthermore, an obstetrician competent in the interpretation of the tracings should follow the case. Finally, the person who will conduct the delivery should be skilled in the conduction of breech deliveries (either assisted or complete extraction). Adequate pain relief by a competent anesthesiologist (or anesthetist) is very important for a smooth second stage. Only with these prerequisites should labor and delivery be planned.

The argument that not enough trained obstetricians are available nowadays is inconsistent with the conduct applied to other high-risk conditions. These cases (hypertensives, diabetics, premature labors, placenta previas, intrauterine growth retardations, etc.) are referred to Level III centers staffed, ostensibly, by obstetricians "possessed of special knowledge and . . . special competence in maternal-fetal medicine," who manage them. It seems uncongruous that these specialists will not have the necessary training to conduct a safe vaginal breech delivery when they have been certified to possess such special knowledge. Persistent breech presentation should, therefore, be referred to these centers as is any other high-risk pregnancy. If these obstetricians have not acquired the necessary skills to conduct a breech delivery, they owe to their patients to undertake the extratraining needed to become proficient in it and thus avoid for them the immediate and long-term risks inherent to cesarean section, without benefit to their infants and done solely because of their insufficient skills.

Not all breech presentations are amenable to vaginal delivery, as are not all cephalic presentations. Cesarean section should be done only for the standard obstetric (fetal or maternal) indications. As for vaginal deliveries, skill in the mechanics of smooth atraumatic extraction is required, and the judgment to give the most convenient incision is essential.

References

1. Scheer K, Nubar J. Variation of fetal presentation with gestational age. Am J Obstet Gynecol 1976;125:269–270.
2. Hill LM. Prevalence of breech presentation by gestational age. Am J Perinatol 1990;7: 92–93.
3. Luterkort M, Polberger S, Persson PH, et al. Role of asphyxia and slow intrauterine growth in morbidity among breech delivered infants. Early Hum Dev 1986;14:19–31.
4. Rovinsky J, Miller JA, Kaplan S. Management of breech presentation at term. Am J Obstet Gynecol 1973;115:497–513.
5. Brenner WE, Bruce RD, Hendricks CH. The characteristics and perils of breech presentation. Am J Obstet Gynecol 1974;118:700–712.
6. Kauppila O. The perinatal mortality in breech deliveries and observations on affecting factors. Acta Obstet Gynaecol Scand [suppl 39], 1975:1–78.
7. Fianu S. Fetal mortality and morbidity following breech delivery. Acta Obstet Gynaecol Scand [suppl 56], 1976:1–85.
8. Schutte MF, van Hemel OJS, van de Berg C, et al. Perinatal mortality in breech presentations as compared to vertex presentations in singleton pregnancies: an analysis based upon 57,819 computer-registered pregnancies in the Netherlands. Eur J Obstet Gynecol Reprod Biol 1985;19:391–400.
9. Nissell P, Bistoletti P, Palme C. Preterm breech delivery. Acta Obstet Gynaecol Scand 1981;60:(363–366.
10. White PC, Cibils LA. Clinical significance of fetal heart rate patterns during labor: VIII. Breech presentations. J Reprod Med 1984;29:45–51.
11. Luterkort M, Polberger S, Weldner BM, et al. Growth in breech presentation. Ultrasound and post-partal assessment of growth in 225 fetuses presenting by the breech in the 33rd gestational week. Acta Obstet Gynecol Scand 1986;65:157–160.
12. Braun FHT, Jones KL, Smith DW. Breech presentation as an indicator of fetal abnormality. J Pediatr 1975;86:419–421.
13. Soernes T, Bakke T. The length of the human umbilical cord in vertex and breech presentations. Am J Obstet Gynecol 1986; 154:1086–1087.

14. Clausen I, Nielsen TK. Breech position, delivery route, and congenital hip dislocation. Acta Obstet Gynaecol Scand 1988;67:595–597.
15. Luterkort M, Persson PH, Polberger S, et al. Hip joint instability in breech pregnancy. Acta Paediatr Scand 1986;75:860–863.
16. Nelson KB, Ellenberg JH. Antecedents of cerebral palsy: multivariate analysis of risk. N Engl J Med 1986;315:81–86.
17. Taffel SM, Placek PJ, Moien M, et al. 1989 U.S. cesarean section rate studies—VBAC rate rises to nearly one in five. Birth 1991;18:73–77.
18. Brown L, Karrison T, Cibils LA. Mode of delivery and perinatal results in breech presentation. Am J Obstet Gynecol 1994;171:28–34.
19. Gimovsky MI, Wallace RL, Schifrin BS, Paul RH. Randomized management of the nonfrank breech presentation at term: a preliminary report. Am J Obstet Gynecol 1983;146:34–40.
20. Tatum RK, Orr JW, Soong SJ, et al. Vaginal breech delivery of selected infants weighing more than 2000 grams. Am J Obstet Gynecol 1985;152:145–155.
21. Hall JE, Kohl SG, O'Brien F, et al. Breech presentation and perinatal mortality. Am J Obstet Gynecol 1965;91:665–677.
22. Lewis BV, Seneviratne HR. Vaginal breech delivery or cesarean section? Am J Obstet Gynecol 1979;134:615–618.
23. Mann LI, Gallant JM. Modern management of the breech delivery. Am J Obstet Gynecol 1979;134:611–614.
24. Sachs BP, McCarthy BJ, Rubin G, et al. Cesarean section: risks and benefits for mother and fetus. JAMA 1983;250:2157–2159.
25. Fortney JA, Higgins JA, Kennedy KI, et al. Delivery type and neonatal mortality among 10,749 breeches. Am J Public Health 1986;76:980–986.
26. Kiely JL. Mode of delivery and neonatal death in 17,587 infants presenting by the breech. Br J Obstet Gynaecol 1991;98:898–904.
27. Cibils LA, Karrison T, Brown L. Factors influencing neonatal outcomes in the very-low-birth-weight fetus (<1500 grams) with a breech presentation. Am J Obstet Gynecol 1994;171:35–42.
28. Piper EB, Bachman C. The prevention of fetal injuries in breech delivery. JAMA 1929;92:217–221.
29. Goethals TR. Cesarean section as the method of choice in management of breech delivery. Am J Obstet Gynecol 1956;71:536–549.
30. Wright RC. Reduction of perinatal mortality and morbidity in breech delivery through routine use of Cesarean section. Obstet Gynecol 1959;14:758–763.
31. Green JE, McLean F, Smith LP, et al. Has an increased cesarean section rate for term breech delivery reduced the incidence of birth asphyxia, trauma and death? Am J Obstet Gynecol 1982;142:643–648.
32. Kubli F, Boos W, Ruuttgers H. Cesarean section in the management of singleton breech presentation. 5th Eur Congr Perinat Med Uppsala (Sweden), 1976:69–75.
33. Goldenberg RL, Nelson KG. The premature breech. Am J Obstet Gynecol 1977;127:240–244.
34. Main DM, Main EK, Maurer MM. Cesarean section versus vaginal delivery for the breech fetus weighing less than 1500 grams. Am J Obstet Gynecol 1983;146:580–584.
35. Tejani N, Verma U, Shiffman R, Chayen B. Effect of route of delivery on periventricular/intraventricular hemorrhage in the low-birth-weight fetus with a breech presentation. J Reprod Med 1987;32:911–914.
36. Weissman A, Blazer S, Zimmer EZ, et al. Low birth weight breech infant: short term and long term outcome by method of delivery. Am J Perinatol 1988;5:289–292.
37. Cox C, Kendall AC, Hommers M. Changed prognosis of breech—presenting low birth weight infants. Br J Obstet Gynaecol 1982;89:881–886.
38. Effer SB, Saigal S, Rand C, et al. Effect of delivery method on outcomes in the VLBW breech infant: is the improved survival related to C-section or other perinatal care maneuvers? Am J Obstet Gynecol 1983;145:123–128.
39. Bodmer B, Benjamin A, McLean FH, Usher RH. Has use of Cesarean section reduced the risks of delivery in the preterm breech presentation? Am J Obstet Gynecol 1986;154:244–250.
40. Davidson S, Cohen WR. Influence of presentation on neonatal outcome of vaginally delivered low birth weight infants. A

matched pair analysis. J Perinatol 1990;10: 38–42.
41. Westgren LMZ, Songster G, Paul RH. Preterm breech delivery: another retrospective study. Obstet Gynecol 1985;66:481–484.
42. Malloy MH, Onstad L, Wright E, et al. The effect of Cesarean delivery on birth outcome in very low birth weight infants. Obstet Gynecol 1991;77:498–503.
43. Westgren M, Paul RH. Delivery of the low birth weight infant by cesarean section. Clin Obstet Gynecol 1985;28:752–762.
44. Rosen M, Chik L. The effect of delivery route on outcome in breech presentation. Am J Obstet Gynecol 1984;148:909–913.
45. Myers SA, Gleicher N. Breech delivery: why the dilemma? Am J Obstet Gynecol 1987; 156:6–10.
46. Thiery M. Management of breech delivery. Eur J Obstet Gynecol Reprod Biol 1987;24: 93–103.
47. Faber-Nijholt R, Huisjes HJ, Towen BCL, Fidler VJ. Neurological followup of 281 children born in breech presentation: a controlled study. Br Med J 1983;286:9–12.
48. Cunningham FG, MacDonald PC, Gant NF, Leveno KJ, Gilstrap LC III. Williams obstetrics, 19th Ed. Norwalk: Appleton and Lange, 1993:493–520.
49. Collea JV, Chein C, Quilligan EJ. The randomized management of term frank breech presentation. Am J Obstet Gynecol 1980;137:235–244.
50. Barlow K, Larsson G. Results of a five-year prospective study using a fetopelvic scoring system for term singleton breech delivery after uncomplicated pregnancy. Acta Obstet Gynecol Scand 1986;65:315–319.
51. Oian P, Skramm I, Hannisdal E, Bjoro K. Breech delivery. An obstetrical analysis. Acta Obstet Gynecol Scand 1988;67:75–79.
52. Jaffa AJ, Peyser MR, Ballas S, Toaff R. Management of term breech presentation in primigravidae. Br J Obstet Gynaecol 1981; 88:721–724.
53. Ophir E, Oettinger M, Yagoda A, et al. Breech presentation after cesarean section: always a section? Am J Obstet Gynecol 1989;161:25–28.
54. Watson WJ, Benson WL. Vaginal delivery for the selected frank breech infant at term. Obstet Gynecol 1984;64:638–640.
55. Mecke H, Weisner D, Freys I, Semm K. Delivery of breech presentation infant at term. An analysis of 304 breech deliveries. J Perinat Med 1989;17:121–126.
56. Bingham P, Hird V, Lilford RJ. Management of the mature selected breech presentation: an analysis based on the intended method of delivery. Br J Obstet Gynaecol 1987;94:746–752.
57. Westgren LM, Ingemarsson I. Breech delivery and mental handicap. Bailliere's Clin Obstet Gynaecol 1988;2:187–194.
58. Wisestanakora W, Herabutya Y, Prasertsawat P, Thanantaseth C. Fetal outcome in term frank breech primapara delivered vaginally and by elective cesarean section. J Med Assoc Thail 1990;73[suppl 1]:47–51.
59. Christian SS, Brady K, Read JA, Kopelman JN. Vaginal breech delivery: a five-year prospective evaluation of a protocol using computer tomographic pelvimetry. Am J Obstet Gynecol 1990;163:848–855.
60. Roumen FJ, Luyben AG. Safety of term vaginal breech delivery. Eur J Obstet Gynecol Reprod Biol 1991;40:171–177.
61. Flanagan TA, Mulchahey KM, Korenbrot CC, et al. Management of term breech presentation. Am J Obstet Gynecol 1987;156: 1492–1502.
62. Croughan-Minihane MS, Petitti DB, Gordis L, Golditch I. Morbidity among breech infants according to method of delivery. Obstet Gynecol 1990;75:821–825.
63. Otamiri G, Berg G, Ledin T, et al. Influence of elective cesarean section and breech delivery on neonatal neurological condition. Early Hum Dev 1990;23:53–66.
64. Hytten FE. Breech presentation: is it a bad omen? Br J Obstet Gynaecol 1982;89:879–880.
65. DePalma RT, Cunningham FG, Leveno KJ, Roark ML. Continuing investigation of women at high risk for infection following cesarean delivery. Obstet Gynecol 1982;60: 53–59.
66. Duff P. Pathophysiology and management of post-cesarean endometritis. Obstet Gynecol 1986;67:269–276.
67. Arulkumaran S, Ingemarsson I, Gibb DM, Ratman SS. Uterine activity in spontaneous labor with breech presentation. Aust NZ J Obstet Gynaecol 1988;28:275–278.
68. Friedman EA. Clinical management of labor, 2nd Ed. New York: Appleton-Century-Crofts, 1978:168–175.

69. Teteris NJ, Botschner AW, Ullery JC, Essig GF. Fetal heart rate during breech delivery. Am J Obstet Gynecol 1970;107: 762–765.
70. Wheeler T, Greene K. Fetal heart rate monitoring during breech labour. Br J Obstet Gynaecol 1975;82:208–214.
71. Hill JG, Eliot BW, Campbell AJ, Pickett-Heaps AA. Intensive care of the fetus in breech labour. Br J Obstet Gynaecol 1976; 83:271–275.
72. Cibils LA. Fetal heart rate patterns during labor in breech presentation. In: Proceedings of the XII World Congress of Gynecology and Obstetrics (Rio de Janeiro), Vol. 5. London: Parthenon, 1989:Chap. 26.
73. Cibils LA. Electronic fetal-maternal monitoring. Antepartum, Intrapartum. Boston: Wright-PSG, 1981:341–368.
74. Eliot BW, Hill JG. Method of breech management incorporating use of fetal blood sampling. Br Med J 1972;4:703–706.
75. Brady K, Duff P, Read AJ, Harlass FE. Reliability of fetal buttock blood sampling in assessing the acid-base balance of the breech fetus. Obstet Gynecol 1989;74:886–888.
76. Luterkort M, Marsal K. Umbilical cord acid-base state and Apgar score in term breech neonates. Acta Obstet Gynecol Scand 1987;66:57–60.
77. Socol ML, Cohen L, Depp R, et al. Apgar scores and umbilical cord arterial pH in the breech neonate. Int J Gynaecol Obstet 1988; 27:37–43.
78. Christian SS, Brady K. Cord-blood acid base values in breech presenting infants born vaginally. Obstet Gynecol 1991;78:778–781.
79. Pauerstein CJ. Clinical obstetrics. New York: Wiley, 1987:511–524.
80. Gimovsky M, Petrie RH, Todd WD. Neonatal performance of the select term vaginal breech delivery. Obstet Gynecol 1980;56: 687–691.
81. Arulkumaran S, Thavarasah HS, Ingemarsson I, Ratnam SS. An alternative approach to assisted vaginal breech delivery. Asia Oceania J Obstet Gynaecol 1989;15:47–51.
82. Milner RDG. Neonatal mortality of breech deliveries with and without forceps to the aftercoming head. Br J Obstet Gynaecol 1975;82:783–785.
83. Sherer DM, Menashe M, Palti Z, et al. Radiologic evidence of nucal arm in the breech-presenting fetus at the onset of labor: an indication for abdominal delivery. Am J Perinatol 1989;6:353–355.
84. Kopelman JN, Duff P, Karl RT, et al. Computed tomographic pelvimetry in the evaluation of breech presentation. Obstet Gynecol 1986;68:455–458.
85. Gimovsky ML, Paul RH. Singleton breech presentation in labor: experience in 1980. Am J Obstet Gynecol 1982;143:733–739.
86. Bhagwanani SG, Price HV, Laurence KM, Giuz B. Risks and prevention of cervical cord injury in the management of breech presentation with hyperextension of the fetal head. Am J Obstet Gynecol 1973;115: 1159–1161.
87. Abroms IF, Bresnan MJ, Zuckerman JE, et al. Cervical cord injuries secondary to hyperextension of the head in breech presentation. Obstet Gynecol 1973;41:369–378.
88. Cattamanchi GR, Tamaskar V, Egel RT, et al. Intrauterine quadriplegia associated with breech presentation and hyperextension of fetal head: a case report. Am J Obstet Gynecol 1981;140:831–833.
89. Gimovsky ML, Petrie RH. The intrapartum and neonatal performance of the low birth weight vaginal breech delivery. J Reprod Med 1982;27:451–454.
90. Bistoletti P, Nisell H, Palme C, Lagercrantz H. Term breech delivery. Acta Obstet Gynecol Scand 1981;60:165–171.
91. Cahill DJ, Turner MJ, Stronge JM. Breech presentation: is a reduction in traumatic intracranial hemorrhage feasible? J Obstet Gynaecol 1991;11:417–419.
92. Castillo M, Quencer RM, Green BA. Cervical spinal cord injury after traumatic breech delivery. AJNR 1989;10:S99.
93. Sandberg EC. The Zavanelli maneuver extended: progression of a revolutionary concept. Am J Obstet Gynecol 1988;158: 1347–1352.
94. Iffy L, Apuzzio JJ, Cohen-Addad N, et al. Abdominal rescue after entrapment of the aftercoming head. Am J Obstet Gynecol 1986;154:623–624.
95. Alexander J, Gregg JEM, Quinn MW. Femoral fractures at cesarean section. Case reports. Br J Obstet Gynaecol 1987;94: 273.

96. Vasa R, Kim MR. Fracture of the femur at cesarean section: case report and review of literature. Am J Perinatol 1990;7:46–48.
97. Barnes AD, Van Geem TA. Fractured femur of the newborn at Cesarean section. A case report. J Reprod Med 1985; 30:203–205.
98. Bradley-Watson PJ. The decreasing value of external cephalic version in modern obstetric practice. Am J Obstet Gynecol 1975;123:237–240.
99. Ranney B. The gentle art of external cephalic version. Am J Obstet Gynecol 1973;116:239–251.
100. Saling E, Mueller-Holve W. External cephalic version under tocolysis. J Perinat Med 1975;3:115–122.
101. Rabinovici J, Barkai G, Shalev J, Majshiach S. Fetal heart rate changes following external cephalic version under tocolysis near term. Int J Gynaecol Obstet 1987;25:277–281.
102. Stine LE, Phelan JP, Wallace R, et al. Update on external cephalic version performed at term. Obstet Gynecol 1985;65:642–646.
103. Morrison JC, Myatt RE, Martin JN, et al. External cephalic version of the breech presentation under tocolysis. Am J Obstet Gynecol 1986;154:900–903.
104. Dyson DC, Ferguson JE, Hensleigh P. Antepartum external cephalic version under tocolysis. Obstet Gynecol 1986;67: 63–68.
105. O'Grady JP, Veille JC, Holland RL, Burry KA. External cephalic version: a clinical experience. J Perinat Med 1986;14:189–196.
106. Hellstrom AC, Nilsson B, Stange L, Nylund L. When does external cephalic version succeed? Acta Obstet Gynecol Scand 1990; 69:281–285.
107. Hanss JW. The efficacy of external cephalic version and its impact on the breech experience. Am J Obstet Gynecol 1990;162: 1459–1463.
108. Donald WL, Barton JJ. Ultrasonography and external cephalic version at term. Am J Obstet Gynecol 1990;162:1542–1545.
109. Mohamed K, Seeras R, Coulson R. External cephalic version at term. A randomized controlled trial using tocolysis. Br J Obstet Gynaecol 1991;67:63–68.
110. Ferguson JE, Armstrong MA, Dyson DC. Maternal and fetal factors affecting success of antepartum external cephalic versions. Obstet Gynecol 1987;70:722–725.
111. Tan GW, Jen SW, Tan SL, Salmon YM. A prospective randomized controlled trial of external cephalic version comparing two methods of uterine tocolysis with a non-tocolytic group. Singapore Med J 1989;30: 155–158.
112. Hofmeyr GJ, Sadan O, Myer IG, et al. External cephalic version and spontaneous version rates: ethnic and other determinants. Br J Obstet Gynaecol 1986;93:13–16.
113. Flamm BL, Fried MW, Lonky NM, Giles WS. External cephalic version after previous cesarean section. Am J Obstet Gynecol 1991;165:370–372.
114. Robertson AW, Kopelman JN, Read JA, et al. External cephalic version at term: is a tocolytic necessary? Obstet Gynecol 1987; 70:896–899.
115. Van-Veelen AJ, Van-Cappellen HW, Flu PK, et al. Effect of external cephalic version in late pregnancy on presentation at delivery: a randomized controlled trial. Br J Obstet Gynaecol 1989;96:916–921.
116. Marcus RG, Crewe-Brown H, Kawitz S, Katz J. Feto-maternal hemorrhage following successful and unsuccessful attempts at external cephalic version. Br J Obstet Gynaecol 1975;82: 578–580.
117. Gjode P, Rasmussen K, Jorgensen J. Feto-maternal bleeding during attempts at external version. Br J Obstet Gynaecol 1980; 87:571–573.
118. Kasule J, Chimbira THK, Brown I McL. Controlled trial of external cephalic version. Br J Obstet Gynaecol 1985;92:14–18.
119. DeRosa J, Anderle LJ. External cephalic version of term singleton breech presentation with tocolysis: a retrospective study in a community hospital. J Am Osteopath Assoc 1991;91:351–357.
120. Petrikovsky BM, DeSilva HN, Fumia FD. Erb's palsy and fetal bruising after external cephalic version: Case report. Am J Obstet Gynecol 1987;157:258.
121. Chapman GP, Weller RO, Normand ICS, Gibbens D. Spinal cord transection in utero. Br Med J 1978;2:398.
122. Amon E, Sabai BM, Anderson GD. How perinatologists manage the problem of the presenting breech. Am J Perinatol 1988;5: 247–250.
123. Penn ZJ, Steer PJ. How obstetricians manage the problem of preterm delivery with special reference to the preterm

breech. Br J Obstet Gynaecol 1991;98: 531–534.
124. Penn ZJ, Steer PJ. Reasons for declining participation in a prospective randomized trial to determine the optimum mode of delivery of the preterm breech. Controlled Clin Trials 1990;11:226–231.
125. Penn ZJ. Personal communication, 1993.
126. Aronson D, Kier R. CT pelvimetry: the foveae are not an accurate landmark for the level of the ischial spines. AJR 1991;156: 527–530.
127. van Loon AJ, Mantingh A, Thijn CJ, Mooyaart EL. Pelvimetry by magnetic resonance imaging in breech presentation. Am J Obstet Gynecol 1990;163:1256–1260.
128. Sarno AP, Phelan JP, Ahn MD, Strong TH. Vaginal birth after cesarean delivery. Trial of labor in women with breech presentation. J Reprod Med 1989;34:831–833.
129. Crawford JS. An appraisal of lumbar epidural blockade in patients with a singleton fetus presenting by the breech. J Obstet Gynaecol Br Commonw 1974;81:867–872.
130. Darby S, Thornton CA, Hunter DJ. Extradural analgesia in labour when the breech presents. Br J Obstet Gynaecol 1976;83:35–38.
131. Cibils LA, Spackman TJ. Caudal analgesia in first stage of labor: effect on uterine activity and cardiovascular system. Am J Obstet Gynecol 1962;84:1042–1051.
132. Hopkins EL, Hendricks CH, Cibils LA. Cerebrospinal fluid pressure in labor. Am J Obstet Gynecol 1965;93:907–916.
133. Matadial L, Cibils LA. The effect of epidural anesthesia on uterine activity and blood pressure. Am J Obstet Gynecol 1976; 125: 846–854.
134. Cibils LA. Electronic fetal-maternal monitoring—antepartum, intrapartum. Boston: Wright-PSG, 1981:411–471.
135. Calvert JP. Intrinsic hazard of breech presentation. Br Med J 1980;281:1319–1320.
136. Nielsen TF, Hokegard KH. Cesarean section and intraoperative surgical complications. Acta Obstet Gynecol Scand 1984;63:103–108.
137. Singh PM, Rodrigues C, Gupta AN. Placenta previa and previous cesarean section. Acta Obstet Gynecol Scand 1981; 60:367–368.
138. Clark SL, Koonings PP, Phelan JP. Placenta previa/accreta and prior cesarean section. Obstet Gynecol 1985;66:89–92.
139. Nielsen TF, Hagberg H, Ljunblad U. Placenta previa and antepartum hemorrhage after previous cesarean section. Gynecol Obstet Invest 1989;27:88–90.
140. Read JA, Cotton DB, Miller FC. Placenta accreta: changing clinical aspects and outcome. Obstet Gynecol 1980;56:31–34.
141. Fox H. Placenta accreta, 1945–1969. Obstet Gynecol Surv 1972;27:475–490.
142. Breen JL, Neubecker R, Gregori CA, Franklin JE. Placenta accreta, increta, and percreta. Obstet Gynecol 1977;49: 43–47.

7 Cesarean Section for Fetal Distress

DAVID A. MILLER AND RICHARD H. PAUL

Cesarean section is the major surgical procedure most frequently performed in the United States, accounting for 966,000 (23.5%) of the estimated 4,111,000 live births in 1991.[1] Nearly 89,000 (9.2%) of these operations were performed for the indication of fetal distress. Thus, cesarean deliveries for fetal distress represented 2.16% of all live births in 1991. In the history of cesarean delivery, performance of the procedure in a living patient solely in the interest of fetal safety is a relatively recent phenomenon. Delivery of a fetus through an incision in the maternal abdominal wall has been reported throughout history. As early as 3000 B.C., Egyptian law required that a postmortem cesarean be performed in all cases of maternal death late in pregnancy so that the stillborn infant might be buried separately.[2] Rarely, the operation was performed in an effort to salvage a fetus following the death of the mother.

Before the late nineteenth century, staggering maternal mortality rates of 75%–100% limited the use of cesarean in living patients to rare cases in which all other means of delivery had failed. Although isolated cases of maternal and infant survival were reported as early as 1500 A.D.,[3] it was not until the late 1800s that advances in aseptic technique, routine employment of uterine sutures, and transition from the classical to the low transverse uterine incision led to significant reduction in maternal mortality. By 1900, the maternal mortality rate of the procedure had fallen to 10%.[4] Throughout this century, improvements in anesthetic and blood transfusion techniques, the development of antibiotics, and advances in medical therapy for maternal illnesses have further contributed to reduction in maternal mortality. By 1970, the maternal mortality rate for cesarean birth was 113.8 per 100,000 deliveries,[2] and estimates place the current cesarean-related maternal mortality rate as low as 22.3 per 100,000.[5]

The dramatic decline in the maternal risk of cesarean delivery has led to more liberal use of the procedure, including increased acceptance of operative intervention for purely fetal indications. In fact, as maternal mortality has continued to fall, the percentage of cesareans performed for the diagnosis of fetal distress has risen more than 10 fold, from 0.8% in the mid-1930s[6] to 9.2% in 1991.[1] Despite significant advances in fetal assessment, however, the ability to accurately diagnose fetal distress remains controversial.

Fetal Heart Rate Auscultation

Auscultation of the fetal heart, described as early as the seventeenth century in Le Goust's "Humani Foetus Historia,"[7] was first reported in Western medical literature by Mayor[8] in 1818. In 1822, Jean Alexandre Le Jumeau, Vicomte de Kergaradec,[9] reported his observations of fetal heart sounds using Laennec's stethoscope, and proposed that auscultation of the fetal heart could be useful in confirming pregnancy, diagnosing multiple pregnancy, determining fetal position, and judging the state of fetal health or disease by changes in strength and frequency of the heart tones. Later, Kennedy,[10] Schwartz,[11] Winckel,[12] and others used auscultation to describe fetal heartrate (FHR) changes associated with umbilical cord compression, head compression, and fetal distress. Kilian,[13] in 1849, and Winckel,[12] in 1893, proposed indications for forceps delivery based upon FHR abnormalities such as tachycardia, bradycardia, "irregularity," and "impurity of tone." Schwartz[11] and Seitz[14] speculated that FHR changes were associated with changes in fetal oxygenation. Remarkably, these observations were made using only the stethoscope (mediate auscultation), or the ear of the examiner placed directly upon the maternal abdomen (immediate auscultation). It was not until 1917 that Hillis described the modified stethoscope known today as the DeLee–Hillis fetoscope.[15,16]

Electronic Fetal Monitoring

In 1906, Cremer[17] recorded the first fetal electrocardiograph (ECG), marking the beginning of the era of electronic fetal monitoring (EFM). By placing one electrode on the maternal abdomen above the fundus and another in the vagina, he was able to demonstrate small fetal electrical impulses among the higher voltage maternal signals. Despite technologic improvements, the quality of abdominal fetal ECG tracings remains unreliable, and the clinical usefulness of the technique is limited.

The concept of direct application of the ECG electrode to the fetus in-utero was introduced in the 1950s,[18–20] with results clearly superior to those obtained abdominally. The first practical clinical electronic fetal monitor became available in the United States in 1968 and, throughout the 1970s, fetal monitoring became increasingly incorporated into obstetric management. By 1978, more than half of all labors were electronically monitored.[21] The National Center for Health Statistics reported that, in 1990, EFM was used in 73% of the 4,110,563 total births in the United States.[22]

Fetal Asphyxia

The objective of EFM is to identify the fetus in distress so that measures might be taken in time to avert permanent damage or death. The term fetal distress, although commonly used in the modern practice of obstetrics, remains poorly defined. It has been described as "a condition in which fetal physiology is so altered as to make death or permanent injury a probability within a relatively short period of time."[23] Some authors base the definition upon FHR abnormalities,[24,25] while others require abnormal fetal blood gas values or low Apgar scores. Nevertheless, the term is generally understood to denote disruption of normal fetal oxygenation, ranging from mild hypoxia to profound asphyxia. Hypoxia refers to the reduction of oxygen supply to tissues below physiologic levels. Asphyxia, derived from the Greek word *asphuxia* meaning "a stopping of the pulse," is defined as "a condition in which an extreme decrease in the amount of oxygen in the body accompanied by an increase of carbon dioxide leads to loss of consciousness or death,"[26] or as a "lack of oxygen in respired air, resulting in impending or actual cessation of apparent life."[27] Parer and Livingston[28] defined fetal asphyxia as the "insufficiency or absence of exchange of the respiratory gases." The term is usually considered to imply the presence of metabolic acidosis as well.

Historically, the clinical diagnosis of birth asphyxia has been based upon a variety of findings, including meconium passage, abnormal FHR patterns, low Apgar scores, abnormal blood gases, and neonatal neurologic abnormalities. When present together these findings are highly suggestive of a birth-related asphyxial insult. Isolated abnormal findings, however, correlate poorly with birth asphyxia and subsequent neurologic impairment. In 1989, Gilstrap et al.[29] recommended that the diagnosis of birth asphyxia be reserved for infants who are severely depressed (5 min Apgar $\leqslant 3$) and acidotic (pH < 7.00) at birth, require resuscitation, and have seizures in the first day of life.

In 1991, the American College of Obstetricians and Gynecologists (ACOG) Committee on Obstetrics, Maternal and Fetal Medicine stated that "a neonate who has had severe hypoxia close to delivery that is severe enough to result in hypoxic encephalopathy will show other evidence of hypoxic damage including all of the following: (1) a profound umbilical artery metabolic or mixed acidemia (pH < 7.00), (2) persistence of an Apgar score of 0–3 for longer than 5 min, (3) neonatal neurologic sequelae, e.g., seizures, coma, hypotonia, and (4) multiorgan system dysfunction, e.g., cardiovascular, gastrointestinal, hematologic, pulmonary, or renal." They further declared that "the term birth asphyxia is imprecise and should not be used."[30] However the old terminology is well entrenched, and continues to generate confusion and debate.

Asphyxia at the Cellular Level

At the cellular level, asphyxia is thought to trigger a cascade of events, including membrane depolarization, disruption of energy metabolism, altered release and reuptake of neurotransmitters, ion shifts, protease activation, phospholipid degradation, and free radical production.[31] It is clear that profound and prolonged asphyxia may lead to cell death and, eventually, death of the organism. Levels of asphyxia that are sublethal to the organism, however, may result in clinical evidence of cellular dysfunction. Complications of asphyxial myocardial injury include cardiac conduction defects, myocardial dysfunction, and congestive heart failure. Manifestations in the gastro intestmal (GI) tract include hypoxic-ischemic mucosal injury, stress ulcers, hepatic injury, and necrotizing enterocolitis. In the lungs, asphyxial sequelae include meconium aspiration, persistent pulmonary hypertension, impaired surfactant production, and respiratory distress syndrome. Asphyxial renal injury may lead to renal insufficiency or acute renal failure. Hematologic manifestations include thrombocytopenia, neutropenia, and disseminated intravascular coagulation. Clinical manifestations of asphyxia are frequently seen in the central nervous system. Impaired neuronal water-regulatory mechanisms and disruption of the blood–brain barrier may cause cerebral edema and neuronal necrosis. Resultant disruption of normal membrane depolarization, neurotransmission, and receptor stimulation may lead to seizures and respiratory depression. Finally, the most widely publicized and unequivocally feared consequence of fetal asphyxia is cerebral palsy.

Asphyxia and Cerebral Palsy

Cerebral palsy (CP) is "a chronic disability, characterized by aberrant control of movement and posture, appearing early in life and not the result of recognized progressive disease."[32] It may be accompanied by mental retardation (41%) and seizures (23%)[33] as well as cortical visual impairment. Three forms of the disorder are recognized: the spastic syndromes (quadriplegia, hemiplegia, and diplegia), ataxic syndromes (simple ataxia and ataxic diplegia), and the dyskinetic syndromes (choreoathetosis, dystonic/tonus changing). Spastic diplegia and quadriplegia are associated with periventricular leukomalacia, seen most commonly in premature, low birthweight infants Choreoathetoid cerebral palsy is associated with neonatal hyperbilirubinemia and kernicterus as well as hypoxic-ischemic injury to the basal ganglia. The insult responsible for the later development of CP may occur at any time during the prenatal, perinatal, or postnatal periods.[34]

Cerebral palsy is classified, along with mental retardation, learning disorders, autism, and epilepsy, as a major disorder of neurodevelopment. Unlike most other developmental disorders, the relationship between CP and abnormal or difficult birth has long been recognized. The degree of correlation, however, is controversial and is the subject of evolving understanding.

In 1862, William John Little,[35] an English orthopedic surgeon, presented to the Obstetrical Society of London his treatise "On the influence of abnormal parturition, difficult labours, premature birth and asphyxia neonatorum, on the mental and physical condition of the child, especially in relation to deformities." The report reviewed the birth histories of children with spastic rigidity, and found a high incidence of preterm delivery, breech presentation, prolonged labor, late onset of crying and respiration, and neonatal convulsions and stupor. Little concluded that infantile spastic palsies could be caused by virtually nothing other than abnormalities of the birth process.

Schreiber,[36] in 1938, reviewed the birth records of 500 patients with cerebral symptoms, and noted a 70% incidence of birth apnea. In 1951, Lilienfeld et al.[37,38] reported higher incidences of placenta previa, malpresentation, prematurity, and abruptio placenta in children with CP than in controls. In 1955, Eastman and DeLeon[39] reviewed the obstetric records of 96 patients with CP and noted that the immediate neonatal condition was described as "poor" (abnormal respiratory behavior, flaccidity, cyanosis) in 41% of the cases compared to only 2% of controls. In addition, they reported higher incidences of third-trimester bleeding, prematurity, breech delivery, midforceps delivery, shoulder dystocia, prolonged second stage, fetal distress, prolonged neonatal apnea, intrapartum maternal fever, and prolonged neonatal fever in infants later diagnosed with CP. Of note, there were also significantly more congenital anomalies (polydactyly, facial clefts) in the CP group.

In 1962, Eastman et al.[40] reviewed 753 cases of CP and found high rates of prematurity, multiple gestation, malpresentation, resuscitation requirement, hemolytic disease, and hypoxia caused by cord prolapse, abruption, and preeclampsia. In the same year, Steer and Bonney[41] studied the histories of 317 patients with CP and found that 41 (13%) were attributable to kernicterus and other neurologic disease. Of the remaining 276 (87%) cases, there were 116 (43%) with no historical findings suggestive of anoxia, and 160 (57%) in whom the possibility of anoxia existed. In 92 of the latter, anoxia was presumed to have been present solely on the basis of neonatal incubator requirement. "Possible anoxia" was diagnosed in the remaining 68 cases by a range of criteria including severe toxemia, tight nuchal cord, cord prolapse, and intrapartum maternal death. In 1985, using data from the Collaborative Perinatal Project, Nelson and Ellenberg[42] reported an increased incidence of CP in low birthweight infants. In term neonates, prolonged depression of Apgar scores was also significantly associated with CP.

Data in animals further implicated fetal asphyxia as a cause of neurologic injury. In 1943, Windle and Becker[43] demonstrated clinical and histopathologic evidence of neural damage in experimentally asphyxiated fetal guinea pigs. Windle later studied fetal rhesus monkeys[44,45] and reported clinical and histopathologic changes associated with prolonged total anoxia, hypercapnia, severe acidosis, and hypotension. Total anoxia for less than 8 min did not consistently result in neuronal injury, whereas anoxia for more than 10 min invariably produced neuropathology. There were no survivors beyond 20–25 min of anoxia. In his animal models, total anoxia produced a pattern of neural necrosis in the brainstem, thalamus, and basal ganglia, with relative sparing of the cerebral cortex. These injuries manifested clinically as seizures, ataxia, and athetosis. They could not account for the more common subtypes of CP, which involve cerebral injury, mental retardation, and spasticity. In the late 1960s and early 1970s, Myers[46,47] demonstrated that, unlike total anoxia, prolonged partial asphyxia in monkeys produced acidosis, late FHR decelerations, and neuropathologic defects consistent with the more common clinical and histopathologic findings in CP. In addition to lesions in the thalamus and basal ganglia, prolonged partial asphyxia resulted in generalized cerebral necrosis or focal necrosis in the parasagittal regions and the border zones between the parietal and occipital lobes.

Although the conclusions of some of the early epidemiologic studies have been called into question,[48–51] they created and fostered the long-held assumption that birth-related asphyxial insults were the primary cause of CP. In reality, these studies demonstrated that perinatal asphyxia is *one* cause of CP. More recently, attention has been focused on the relative contributions of possible prenatal factors, including congenital central nervous system (CNS) abnormalities, infections, mercury toxicity, in-utero strokes, maternal hyperthyroidism, and maternal proteinuria.[51,52] Nelson and Ellenberg[51] performed a multivariate analysis of risk in 189 cases of cerebral palsy from the Collaborative Perinatal Project. After correcting for major non-CNS congenital malformations, birthweight of 2000 g or less, microcephaly, and alternative explanations for CP, they reported that only 9% of all cases were associated with birth asphyxia (defined as one or more of the following: lowest FHR $\leq$ 60 bpm, 5-min Apgar score $\leq$ 3, time to first cry $\geq$ 5 min).

In 1988, Blair and Stanley[50] reported their findings in 183 CP cases and 549 matched controls, reaching very similar conclusions. The diagnosis of birth asphyxia was assigned to all infants with "fetal distress" and a 1-min Apgar less than 7 or a spontaneous respiration time of more than 2 min. Fetal distress was defined as any of the following: (1) meconium, (2) FHR greater than 160 bpm or less than 120 bpm, (3) "abnormal" FHR tracing, or (4) documentation of "fetal distress" not otherwise specified. Using these criteria, they demonstrated that birth asphyxia nearly tripled the odds of developing CP (odds ratio, 2.84; 95% confidence interval, 1.85–4.37). However, most infants with birth asphyxia did *not* develop CP. Furthermore, of 183 cases of CP, birth asphyxia was considered to be present in only 13. They concluded that only 8.2% of CP was potentially attributable to birth asphyxia.

The prevalence of CP in school-age children is approximately 2–2.5 per 1000, and has not decreased appreciably over several decades.[53] Recently, in fact, the prevalence of CP has increased in Japan,[54] Australia,[55] Finland,[56] and the United Kingdom,[57] primarily as a result of improved survival of low birthweight infants at increased risk for CP.[58] In the face of increasing perinatal survival and improvements in prenatal and intrapartum care, the lack of a significant decline in the incidence of CP provides additional evidence that the association with adverse intrapartum events is weaker than originally believed. Although the cause in most cases of CP is unknown, adverse *pre*natal events appear to play a greater role than previously recognized.

Imaging modalities such as ultrasound, computed tomography, magnetic resonance imaging, and technetium scanning provide new insights into the prenatal origins of neurologic injury. Following an asphyxial insult, neuronal necrosis produces characteristic changes that evolve over the course of days to weeks. These changes are frequently detectable with imaging techniques, and can aid in establishing the timing of the injury. The location of the abnormality may also play a role in the timing of the injury. Periventricular leukomalacia is usually seen with injuries between 28 and 34 weeks of gestation. On the other hand, parasagittal neuronal damage is generally a phenomenon of term infants. Greater understanding of the etiologic factors involved in the development of CP should help to rectify the prevailing misconception of birth asphyxia as the sole cause of the disorder. This, in turn, should lead to more realistic expectations of the possible benefits of intrapartum fetal monitoring.

Asphyxia and FHR Abnormalities

As early as the nineteenth century, researchers using auscultation recognized that certain FHR patterns were associated with poor perinatal outcome. Kennedy,[10] in 1833, related Bodson's description of fetal distress in association with a FHR that exhibited "slowness of its return when a contraction is passing on." In 1838, Schwartz[59] recommended frequent counting of the fetal heart tones in labor, and implicated "asphyxic intoxication" as a cause of alterations in their "individual normal frequency." Referring to Schwartz's description of the relationship between FHR decelerations and uterine contractions, Gültekin-Zootzmann[60] noted that "in those cases in which the heart sounds returned slowly to their earlier rhythm, or when the attenuations persisted or deteriorated during the pauses, the result would be a weak, moribund or dead fetus."

Seitz[14] in 1903 described three progressively ominous stages of FHR deceleration. He attributed the first two stages to irritation and paralysis of the vagal centers, and the third to paralysis of all extracardiac nerve centers, concluding that it was possible to detect early signs of compromise before the fetus was actually in danger.

The introduction of EFM and fetal scalp blood sampling in the 1960s provided additional tools for evaluating the fetus. In 1967, Hon and Quilligan[61] proposed a system for classification of FHR decelerations, and in 1969 Kubli et al.[62] demonstrated the relationship between the fetal scalp pH and the type and severity of FHR deceleration. They reported that fetuses with no decelerations, early decelerations, or mild variable decelerations had average scalp pH values of 7.29 or higher, while those with severe variable or late decelerations had pH values lower than 7.16. In 1975, Paul et al.[63] found that, despite the presence of late decelerations, neonates with average FHR variability had significantly higher scalp pH values and Apgar scores than those with decreased variability.

In 1982, Clark et al.[64] reported that FHR accelerations of 15 bpm for 15 s in response to fetal scalp stimulation predicted a scalp pH of 7.19 at least. Smith et al.[65] in 1986 reported a similar relationship between fetal scalp pH and the FHR response to vibroacoustic stimulation with an artificial larynx applied to the maternal abdomen over the fetal head for 1 to 3 s. Of 30 fetuses with FHR accelerations in response to this stimulus, all had scalp pH values of 7.25 or higher. This and other work has helped to shed some light on the complex relationship between fetal biochemistry and the neurologic regulation of FHR.

Electronic Fetal Monitoring Versus Traditional Auscultation

With EFM rapidly replacing the traditional practice of intermittent intrapartum FHR auscultation, a series of nonrandomized studies[66–76] in the mid-1970s reported significantly decreased perinatal mortality in electronically monitored patients. These studies are often criticized on the basis of nonrandomized design and nonconcurrent controls, citing rapidly improving neonatal care and falling perinatal mortality rates as possible sources of bias. MacDonald and Grant[77] pointed out that, over the time period of these studies, hospitals not using EFM experienced rates of improvement in perinatal outcome similar to those seen in hospitals that were using EFM. Nevertheless, these studies had the effect of validating the use of EFM.

In 1976, the first of a series of randomized controlled trials was published, comparing EFM to intermittent auscultation of the FHR during labor. To date, there have been nine such studies; five in high-risk populations,[24,78–81] two in low-risk populations,[82,83] and two in combined low- and high-risk populations.[84,85] These trials are summarized in Table 7.1.

Randomized Controlled Trials of EFM Versus Auscultation

In 1976, Haverkamp and associates,[78] in Denver, reported the first prospective, randomized study of 483 high-risk obstetric patients, comparing EFM with intermittent FHR auscultation in labor. A point-rating system[86] was used to assess risk status, with a score of 6 or less indicating high risk. Patients in labor who had meconium-stained amniotic fluid, required oxytocin, or had FHR abnormalities on admission were also included. In the EFM group, a scalp electrode was placed as soon as possible. Auscultation in the control group was performed every 15 min in the first stage of labor, and every 5 min in the second stage, for 30 s after uterine contractions. Electronic monitoring was employed in both groups, but was blinded in the control group.

In the EFM group, FHR patterns were evaluated using the criteria of Kubli and Hon.[62] In patients with late decelerations or severe variable decelerations that persisted after 15 min of corrective measures (oxygen, positional changes, correction of hypotension), delivery was effected. Fetal distress in the control group was diagnosed by the presence of bradycardia to 100 bpm after three or more consecutive contractions. Delivery was effected if fetal distress was not relieved within 15 min. There were no significant differences in outcome as measured by perinatal mortality, Apgar scores, cord blood pH values, neurologic signs in the neonate, or neonatal nursery morbidity between the EFM and control groups. The monitored group, however, had significantly higher rates of cesarean delivery overall (16.5% versus 6.8%) and of cesarean for fetal distress (7.4% versus 1.2%).

Questions have been raised concerning the comparability of the two groups. For instance, review of the monitor tracings revealed a higher incidence of abnormal FHR patterns early in labor in the study group. Furthermore, the study group had a higher incidence of maternal postpartum infectious morbidity (13.2% versus 4.6%), which was not explained by the increased rate of cesarean birth. These findings suggest that the study group may have represented a higher risk population than did the control group and that effective randomization was not achieved.

The second study, by Renou et al.[79] in Melbourne, Australia, in 1976, randomized 350 high-risk patients into EFM and auscultation groups. High-risk patients were

TABLE 7.1. Prospective randomized clinical trials of EFM versus intermittent FHR auscultation

Authors	Year	N (total)	Risk status	Perinatal mortality	Neonatal neurologic signs	Cesarean section rate
Haverkamp et al.[78]	1976	483	High	0[a]	0	e
Renou et al.[79]	1976	350	High	0	f	e
Kelso et al.[82]	1978	504	Low	0	0	e
Haverkamp et al.[24]	1979	690	High	0	0	e
Wood et al.[83]	1981	989	Low	0	0	0
MacDonald et al.[84]	1985	12,964	Combined	0	f	0
Neldam et al.[85]	1986	969	Combined	0	0	0
Luthy et al.[80]	1987	246	High (PTL)	0	0	0
Vintzileos et al.[81]	1993	1,428	High	f	0	0

[a] Symbols:
0 = No difference.
f = Lower in EFM group.
e = Higher in EFM group.

defined as those with a poor obstetric history, a medical or obstetric complication, an abnormal FHR detected by auscultation, or meconium in the amniotic fluid. Continuous EFM was performed in the study group, and scalp pH was measured if the FHR tracing was judged to be abnormal. Abnormalities were defined as a slowing of the FHR in relation to the contraction cycle, a baseline FHR less than 100 bpm, or loss of normal beat-to-beat variability.[87] The protocol for auscultation in the control group was not reported. Criteria for obstetric intervention were not specified in either group. In this study, there were no significant differences between the groups with respect to perinatal mortality, Apgar scores, or maternal or neonatal infection. Patients in the monitored group, however, had significantly higher cord blood pH values and significantly fewer neonatal intensive care unit (NICU) admissions and abnormal neonatal neurologic findings. The cesarean rate was significantly higher in the monitored group than in the control group (22.3% versus 13.7%); however, the indications for intervention were not specified, making this difference difficult to interpret. The authors commented that the difference in cesarean rates was not statistically significant after removal of 6 patients in the monitored group who had had a previous cesarean birth. The rationale for removing these patients on the basis of their previous operations is unclear. The rates of cesarean for fetal distress were not reported.

In 1978, Kelso et al.[82] in Sheffield, England, published the first randomized controlled trial comparing EFM and intermittent auscultation in 504 low-risk patients. Women with risk factors such as multiple gestation, breech presentation, hypertension, diabetes, and other medical and obstetric complications were excluded from the study. Continuous EFM was employed in study patients; a fetal scalp electrode was placed as early as possible. Auscultation in the control group was performed at least every 15 min for 1 min during and immediately following a contraction. Crossover was not permitted, and scalp pH determination was not utilized. The dip area[88] was used as a measure of fetal distress in the EFM group; however, criteria for intervention were not specified. In the control group, a FHR higher than 160 bpm or lower than 120 bpm was considered indicative of fetal distress. There were no significant differences between the groups with respect to perinatal mortality, low Apgar scores, cord blood pH values, NICU admissions or length of stay, neonatal or maternal infections, or abnormal neonatal neurologic findings. There was a significantly higher cesarean rate in the monitored group (9.5% versus 4.4%); however there was no difference in the incidence of cesarean for fetal distress (EFM, 1.6%; control, 1.2%).

In 1979, Haverkamp and associates[24] published a second randomized controlled trial in high-risk patients that was similar in design to the first. It included additional measures of infant status as well as the option to perform fetal scalp pH determination during labor. Blinded EFM in the control group was not performed in this trial. A total of 690 high-risk patients were randomized into three groups. In the first group, fetal assessment during labor was accomplished by intermittent auscultation; the second group received continuous EFM alone, and the third group received continuous EFM with the option to measure scalp blood pH as needed. Risk assessment guidelines, auscultation protocols, and criteria for the diagnosis of fetal distress were the same as in their previous study. Among the three groups, there were no significant differences in perinatal mortality, Apgar scores, cord blood pH values, maternal or neonatal infectious morbidity, NICU admissions, or neonatal neurologic abnormalities. A significant increase in the incidence of cesarean birth was demonstrated in the group with EFM alone (EFM alone, 18%; EFM with the option to scalp sample, 11%; auscultation, 6%). The option to perform scalp sampling resulted in an intermediate cesarean rate that was not significantly different from either of the other groups. When analyzed together, electronically monitored patients had a significantly higher rate of cesarean for fetal distress than did controls (5.2% versus 0.43%).

The fifth trial was published in 1981 by Wood et al.[83] in Melbourne, Australia. A total of 989 low-risk patients (890 at one hospital and 99 at another) were randomized to receive EFM or intermittent auscultation. Women with previous preterm births, meconium-stained amniotic fluid, fetal tachycardia or bradycardia, renal disease, hypertension, diabetes, or other medical or obstetric complications were excluded from participation. Monitored patients had placement of a fetal scalp electrode as early as possible. The protocol for auscultation in the control group was not described. Scalp pH measurements were obtained as needed. Fetal distress was diagnosed as in the previous study by Renou et al.[79,87] The criteria for operative intervention were not specified. There were no significant differences between the groups with respect to perinatal mortality, Apgar scores, cord blood pH values, NICU admissions, or neonatal neurologic abnormalities. In this study, cesarean rates were not significantly different between the groups (4% in the monitored group and 2% in the auscultated group), although the overall rate of operative intervention (including forceps) was significantly higher in the monitored group. Rates of cesarean delivery for fetal distress were not reported. It should be noted that the randomization process was compromised at the larger study hospital, requiring subsequent data manipulation.

In 1985, MacDonald et al.,[84] in Dublin and Oxford, published a randomized controlled trial comparing EFM with intermittent FHR auscultation in 12,964 pregnancies. It was the first study to prospectively calculate the sample size needed to demonstrate statistically significant differences between the groups. Before initiation of the study, estimates were made of the anticipated frequencies of intrapartum stillbirths, neonatal deaths, neonatal seizures in survivors, and other severe abnormal neurologic characteristics. They calculated that 13,000 patients would be needed to demonstrate a 50% reduction in the combined incidence of intrapartum stillbirths, neonatal deaths, and neonatal seizures in survivors (power, 75%; $p \leqslant 0.05$). A trial of that size would have a 50% chance of detecting a 50% reduction in the rate of neonatal seizures, alone. Of the study participants, 22.5% were identified as high risk.

Amniotomy was performed within 1 h of admission in all patients, and those with no amniotic fluid or with moderate to dense meconium were excluded from the study. It is unclear whether later passage of meconium resulted in removal from the study of a patient once she had been included.

In the EFM group, a fetal scalp electrode was applied as early as possible, and scalp pH measurements were used as needed. Criteria for evaluation of the FHR tracings were similar to those of Kubli and Hon.[62] Suspicious or ominous tracings were those with marked tachycardia or bradycardia, moderate tachycardia or bradycardia with decreased variability, absent or minimal variability, late decelerations, moderate to severe variable decelerations, and FHR patterns that were difficult to interpret. In the first stage of labor, a scalp pH measurement was performed if a suspicious or ominous FHR pattern persisted for at least 10 min. A scalp pH less than 7.20 was an indication for delivery, regardless of the FHR pattern. If the pH was between 7.20 and 7.25 with a persistently suspicious or ominous FHR pattern, delivery was effected. If the scalp pH was higher than 7.25, but the tracing remained suspicious or ominous, the pH was repeated within 30–60 min. In the second stage of loabor, delivery was effected if FHR abnormalities persisted for at least 10 min.

In the control group, FHR auscultation was performed every 15 min for 60 s in the first stage of labor, and between each contraction during the second stage. If the FHR was less than 100 bpm or more than 160 bpm during three contractions and could not be corrected with conservative measures, the scalp pH was measured and the patient was managed as described previously. Blood sampling was also performed at unspecified intervals in the control group when labor exceeded 8 h. There were no significant differences between the groups with respect to perinatal mortality, low Apgar scores, neonatal trauma, resuscitation requirement, NICU admissions, or infectious morbidity.

Among the 28 perinatal deaths, asphyxia was considered to be the primary cause in 7 cases in each group. There were significantly more cases of neonatal seizures and persistent neurologic abnormalities (>1 week) in the control group; however, no differences with respect to neurologic abnormality remained at 1-year and 4-year follow-up (3 cases in each group). Labor was significantly shorter in the EFM group, and analgesia (meperidine) was required less often. Scalp sampling was used more frequently in the EFM group (4.4% versus 3.5%), and twice as many fetuses with low scalp pH (<7.20) were identified.

The cesarean rate in the EFM group (2.4%) was not significantly different from that in the auscultated group (2.2%). Overall rates of operative delivery were higher in the EFM group (10.6% versus 8.5%), as a result of a higher incidence of forceps delivery (8.2% versus 6.3%). Rates of cesarean delivery for fetal distress were not significantly different (EFM, 0.4%; control, 0.2%).

The overall frequency of seizures in the patients who were classified as high risk was 4.3 per 1000. This was significantly higher than in the low-risk group (2.6 per 1000). The incidence of seizures in surviving neonates was the same in both groups (2.3/1000). Electronic monitoring did not reduce the seizure incidence to a greater extent in high-risk patients than in low-risk patients. In this study, the largest to date, EFM was associated with no increase in maternal morbidity.

In 1986, Neldam and associates[85] in Copenhagen, Denmark, reported a randomized controlled trial of EFM versus intermittent auscultation in 969 combined low- and high-risk patients. The study excluded those with non-gestational diabetes mellitus. In the EFM group, monitoring was initiated when the patients no longer desired to ambulate. A scalp electrode was placed as soon as possible thereafter.

In the control group, fetal heart tones were auscultated twice an hour for at least 15 s at a cervical dilatation of 5 cm or less, every 15 min from 5 cm until the second stage of labor, and for 30 s after each contraction or at least every 5 min during the second stage. Scalp pH sampling was optional, and was performed only five times (EFM, three; control, two). In the EFM group, intervention was considered if FHR abnormalities remained unresolved after 15 min of corrective measures. Abnormalities included bradycardia (<120 bpm), tachycardia (>160 bpm), late decelerations, variable decelerations (not further specified), silent FHR pattern (beat-to-beat variability ≤5 bpm), and saltatory pattern (variability ≥25 bpm). Intervention was considered in the control group if the FHR was less than 100 bpm following three or more consecutive contractions. No statistical differences were detected between the groups with respect to perinatal mortality, low Apgar scores, seizures, NICU admissions or length of stay. Significantly more pathologic FHR patterns were detected in the EFM group; however, there was no difference in the incidence of cesarean delivery between the groups.

The eighth study, by Luthy et al.[80] in Seattle (Washington) and Vancouver (British Columbia) in 1987, compared EFM and auscultation in 246 high-risk patients with preterm labor. Inclusion criteria were preterm labor, singleton gestation, cephalic presentation, estimated gestational age of 26–32 weeks, and estimated fetal weight of 700–1750 g. Patients with preterm premature rupture of the membranes were not excluded. In the EFM group, external monitoring was used until advanced cervical dilatation (7 cm), at which time amniotomy was performed and a scalp electrode was placed. In those with ruptured membranes, a scalp electrode was placed once delivery was inevitable. Ominous FHR patterns were those with persistent late decelerations with at least three successive contractions in the absence of correctable cause, FHR greater than 180 bpm with total loss of variability persisting more than 15 min, FHR less than 100 bpm for more than 3 min, or severe variable decelerations persisting for more than 30 min. An ominous FHR pattern lasting more than 30 min or a scalp pH less than 7.20 was an indication for delivery.

In the control group, auscultation was performed for at least 30 s, at least every 15 min in the first stage of labor, and at least every 5 min in the second stage. Ominous patterns were those with a FHR less than 100 bpm for more than 30 s after three or more consecutive contractions, or baseline FHR greater than 180 bpm for more than 15 min or less than 100 bpm for more than 60 s. Scalp pH was used as clinically indicated in both groups. Fetal scalp pH values below 7.20 or ominous FHR patterns in the absence of a correctable cause were considered indications for delivery. The groups did not differ with respect to the use of tocolytics, corticosteroids, oxytocin, or regional anesthesia. There were no differences in perinatal mortality, low Apgar scores, cord pH values, neonatal seizures, respiratory distress syndrome, or intracranial hemorrhage. Cesarean rates were similar (EFM, 15.6%; controls, 15.2%). There was no difference in the incidence of cesarean for fetal distress (EFM, 8.2%; controls, 5.6%).

The most recent randomized trial, published in 1993 by Vintzileos et al.,[81] was conducted in Athens, Greece and compared EFM and intermittent auscultation in 1428 patients in a population with a high baseline perinatal mortality rate (20.4–22.6 per 1000). The relatively high incidence of the outcome measure to be studied (perinatal death) markedly improved the likelihood of detecting a statistically significant effect of EFM. Using an average incidence of 21 perinatal deaths per 1000, they prospectively calculated that a sample of 2210 patients would have an 80% chance of detecting a 67% reduction in perinatal mortality at the 5% level of significance. Reviews were conducted every 3 months, and the study was ended after the third review in light of a statistically significant fivefold decrease in perinatal mortality in the EFM group.

All subjects had a singleton living fetus with an estimated gestational age of 26 weeks at least. Fetuses with known congenital or chromosomal anomalies were excluded.

In the EFM group, external monitoring was used as long as satisfactory tracings were obtained. Scalp electrodes were placed as needed. In the control group, FHR auscultation was performed every 15 min during the first stage of labor and every 5 min during the second stage. The FHR was counted during contractions and for at least 30 s immediately afterward. Nonreassuring patterns in the EFM group included late decelerations, prolonged decelerations to 80 bpm or lower for more than 2 min, severe variables to 70 bpm or lower for 60 s or more, variable decelerations with a rising baseline and loss of variability, tachycardia with decreased variability (<5 bpm), persistent decreased variability, or a sinusoidal pattern. In the auscultated group, nonreassuring patterns included a FHR less than 100 bpm during and immediately after a contraction, and a persistent FHR less than 100 bpm or more than 160 bpm. Scalp sampling was not used in either group, and crossover was not permitted.

In both groups, delivery was effected if nonreassuring FHR patterns failed to resolve after 20 min of conservative measures. There were significantly fewer perinatal deaths in the EFM group (2.6/1000 versus 13/1000). Furthermore, no cases of hypoxia-related perinatal death occurred in the EFM group, whereas six cases occurred in the auscultated group (0.9%). This difference was also statistically significant. The groups did not differ significantly with respect to low Apgar scores, NICU admissions or length of stay, ventilator requirements, neonatal hypoxic-ischemic encephalopathy, intraventricular hemorrhage, seizures, hypotonia, necrotizing enterocolitis, or respiratory distress syndrome. Although the incidence of cesarean for fetal distress was significantly higher in the EFM group (5.3% versus 2.3%), the overall incidence of cesarean birth was not significantly different between the EFM and control groups (9.5% versus 8.6%).

Potential Benefits of Electronic Fetal Monitoring

When EFM was introduced in the 1960s, proponents anticipated marked reductions in perinatal mortality and neonatal neurologic injury. Regarding the former, eight of nine randomized clinical trials conducted during the past 20 years have failed to detect a statistically significant reduction in perinatal mortality with EFM, compared to traditional intermittent FHR auscultation (see Table 7.1). However, it is crucial to point out that only one randomized trial[81] had sufficient statistical power to demonstrate such a difference. In that trial, electronically monitored patients had a statistically significant fivefold improvement in perinatal mortality compared to those monitored with intermittent auscultation. In comparison to the high perinatal mortality rates (20.4–22.6/1000) in the study by Vintzileos,[81] MacDonald et al.[84] calculated the combined anticipated frequencies of intrapartum stillbirths and neonatal deaths in their population to be 3 per 1000. In such a population, a study with an 80% likelihood of detecting a 50% reduction in perinatal mortality ($p < 0.05$), would require more than 33,000 patients.[89] The total number of patients in all nine studies, combined, was 18,623.

With regard to neonatal neurologic injury, the problem of study size was the same. Seven of the nine trials[24,78,80–83,85] showed no beneficial influence of EFM (see Table 7.1). Two trials[79,84] reported fewer neonatal seizures in the EFM groups. The only study to examine long-term neurodevelopment,[84,90] however, found no difference between the groups in the incidence of neurologic abnormality at 1 or 4 years of age. Assuming that the school-age incidence of CP is approximately 2 per 1000, and that approximately 10% of CP is attributable to birth asphyxia, the anticipated incidence of asphyxia-related CP is roughly 0.2 per 1000. A study large enough to detect a 50% reduction in the incidence of asphyxia-related CP (power, 80%; $p \leqslant 0.05$) would require more than 500,000 patients.[89] Thus, it is not surprising that the randomized trials to date have failed to detect a statistically significant reduction in CP with the use of EFM.

Potential Risks of Electronic Fetal Monitoring

Early concerns regarding the potential for maternal or neonatal infections in electronically monitored patients have proven to be unfounded. Only one study[78] demonstrated an increased risk of maternal infectious morbidity in patients randomized to EFM. These results are very difficult to interpret in light of the fact that fetal scalp electrodes were used in both the EFM and control groups (FHR tracings were recorded in the control group, but clinicians did not have access to them). The largest randomized trial to date[84] revealed no increased infectious morbidity in electronically monitored patients. Current evidence does not support an association between EFM and increased infectious morbidity.

Current data suggest that the effect of EFM on cesarean delivery rates is minimal. While four early randomized trials[24,78,79,82] reported significantly more cesarean deliveries in electronically monitored patients, the five most recent studies[80–85] have shown no such difference. Furthermore, as the use of EFM in the United States has expanded to include nearly three-fourths of all births, the incidence of cesarean delivery for fetal distress has increased by only 1.8%, from 0.36% in 1974[2] to 2.16% in 1991.[1] This increase reflects less than 15% of the total rise in the rate of cesarean birth. Since the introduction of EFM, 20 years of research and clinical experience have refined the interpretation of FHR patterns. Additional tools such as fetal scalp pH determination, fetal scalp stimulation, and vibroacoustic stimulation help to confirm fetal well-being in the presence of nonreassuring tracings that previously may have prompted operative intervention. Logic dictates that better understanding of the capabilities and limitations of EFM should lead to better perinatal outcome and fewer unnecessary cesareans. Although this contention appears to be borne out in the literature, the debate will no doubt continue.

Despite the lack of consensus on many points, EFM has been clearly shown to be at least as effective in identifying fetal compromise as is the practice of frequent FHR auscultation with intensive, one-on-one nursing. While this level of individualized nursing care may be available in some settings, most delivery units will find the personnel requirements to be impractical and cost prohibitive.

Management of Intrapartum Fetal Distress

Timely diagnosis and appropriate management of intrapartum fetal distress remain among the most challenging tasks facing the obstetrician. During the intrapartum period, uterine contractions subject the fetus and the uteroplacental exchange unit to intermittent episodes of diminished maternal blood flow. These brief interruptions in oxygen delivery are usually well tolerated by the healthy fetus; however, repetitive or prolonged hypoxic stress may overwhelm the fetal compensatory mechanisms and lead to acidosis and asphyxia. In such cases, the fetus will usually exhibit FHR patterns indicative of stress or distress. Clear understanding of the pathophysiology of these patterns is essential to optimize fetal outcome and minimize unnecessary operative deliveries.

Interpretation of the FHR tracing is based upon (1) the baseline FHR, (2) variability in the FHR, and (3) periodic patterns that are classified according to their temporal relationship to uterine contractions.

Baseline Fetal Heart Rate

The normal FHR baseline ranges from 120 to 160 bpm. It is commonly closer to 160 bpm in early pregnancy, declining as gestational age advances. Likewise, the FHR may gradually decrease toward 120 bpm during the course of labor. In general, a FHR baseline below 120 bpm is termed bradycardia, and a rate in excess of 160 bpm is termed tachycardia. Abnormalities in the FHR baseline may have very different causes and consequences. It is important, therefore, to characterize the underlying etiology as accurately as possible and to institute appropriate therapy at the earliest possible time.

Bradycardia

Bradycardia is defined as an abnormally low baseline FHR (<120 bpm), and must be differentiated from the abrupt FHR changes characteristic of decelerations. Although FHR decelerations are very common, true fetal bradycardia is not. A bradycardic FHR baseline between 100 bpm and 120 bpm in association with otherwise reassuring FHR patterns most likely represents a normal variant. Rarely, fetal bradycardia may be seen in association with maternal beta-blocker therapy, hypothermia, hypoglycemia, hypothyroidism, or fetal cardiac conduction defects (i.e., congenital atrioventricular block).

Documentation of fetal heart block should prompt a search for structural fetal cardiac abnormalities, which are present in 20% of cases. Other possible causes of heart block include viral infections (i.e., cytomegalovirus) and maternal systemic lupus erythematosis with anti-Ro (SSA) antibodies. Most congenital causes of fetal bradycardia do not present as acute changes in the FHR and rarely require emergency intervention. Any abrupt decline in the FHR to less than 120 bpm more likely represents a deceleration than a change in the baseline, and should be considered pathologic until proven otherwise.

Tachycardia

Fetal tachycardia has many possible etiologies. Most commonly, it is the result of decreased vagal or increased sympathetic outflow, associated with fever, infection, fetal anemia, or fetal hypoxia. Other possible causes include maternal hyperthyroidism, sympathomimetic medications (ritodrine, terbutaline), parasympatholytic medications (atropine, phenothiazines), and fetal cardiac arrhythmias. The underlying etiology should be identified and treated, when possible.

The source of any maternal fever must be aggressively sought. Specifically, intra-amniotic infection must be ruled out. The diagnosis of chorioamnionitis requires intrapartum antibiotic therapy. Possible causative medications should be discontinued, and maternal hyperthyroidism should be excluded. Fetal cardiac arrhythmias may require ultrasonographic evaluation to rule out structural lesions and cardiac failure. Antiarrhythmic therapy may be instituted if deemed necessary. Although tachycardia alone does not necessarily indicate fetal distress, it is commonly seen in association with other FHR patterns suggestive of hypoxia, including loss of variability and repetitive late decelerations. In such cases, consideration should be given to fetal scalp pH determination or delivery.

Sinusoidal Pattern

The sinusoidal FHR pattern is an uncommon FHR baseline abnormality. As the name implies, it is sinusoidal in appearance, with an amplitude of 5–15 bpm and a frequency of 2–5 cycles per minute. Variability is usually decreased, and accelerations are absent. Although the pathophysiologic mechanism is not known, this pattern is classically associated with fetal hypoxia secondary to severe fetal anemia. It may also occur in association with amnionitis,[91] fetal sepsis, or administration of narcotic analgesics (Stadol, Demerol, Nisentil). A persistent sinusoidal pattern that is not attributable to medications is a concerning finding. Labor should be allowed to continue only in the presence of a normal fetal scalp pH or other FHR evidence of fetal well-being.

Fetal Heart Rate Variability

Variability in the FHR results from constant interplay between the sympathetic and parasympathetic arms of the fetal autonomic nervous system. Modulation of vagal tone occurs in response to blood pressure changes detected by aortic arch baroreceptors. Oxygen and carbon dioxide fluctuations, detected by chemoreceptors, similarly influence vagal outflow. Continual adjustments in vagal tone are manifested in the FHR tracing as "short-term" ("beat-to-beat") variability superimposed upon broader, cyclical (3–5 cycles/min) "long-term" variability. In clinical use, the term "FHR variability" refers to a composite of the two.

Average FHR variability (6–25 bpm) represents a normally oxygenated vagal connection between the fetal central nervous system (CNS) and the cardiac conduction system. Increased variability (>25 bpm), or "saltatory" FHR pattern, is uncommon and most often represents an exuberant autonomic response of a normal fetus. On occasion, it may reflect increased catecholamine release in the early stages of fetal hypoxia. Careful evaluation of the associated FHR findings should help to clarify such cases. Decreased variability (0–5 bpm) most often reflects decreased fetal CNS activity associated with fetal sleep states, fetal anomalies, medications (analgesics, magnesium sulfate, benzodiazepines, phenothiazines), or fetal acidosis. Parasympatholytic medications (atropine) may decrease FHR variability by blocking vagal influence on the sinoatrial node. Persistent decreased variability in a nonanomalous fetus (not attributable to medications or fetal sleep state) is an ominous sign, particularly when associated with other FHR patterns suggestive of fetal hypoxia. In such cases, demonstration of fetal well-being is desirable if labor is to be permitted to continue.

Reassurance may be obtained by several different means, including fetal scalp pH determination. A fetal scalp pH of 7.25 or higher provides evidence that the fetus is not acidotic. In the presence of a persistently nonreassuring FHR tracing, however, the pH should be repeated every 30–60 min. A scalp pH of 7.20–7.25 is suspicious, and should be repeated within 30–60 min, regardless of other FHR findings. A scalp pH less than 7.20 more, when confirmed, is an indication for delivery. A more practical and less invasive approach employs fetal scalp stimulation or vibroacoustic stimulation. Frequently, these stimuli will provoke FHR accelerations and improve FHR variability, thereby providing reassurance that the fetus is not acidotic. If reassurance cannot be obtained, delivery should be considered.

Periodic Patterns

The FHR baseline is commonly interrupted by accelerations or decelerations in rate. These periodic patterns have important clinical implications regarding the well-being of the fetus.

Accelerations

Accelerations in FHR occur in association with fetal movement, probably as a result of increased catecholamine release and decreased vagal stimulation of the heart. Starting at approximately 30–32 weeks gestation, they normally occur during the fetal wake state at a rate of 15–20 per hour. The occurrence of at least two qualifying FHR accelerations (15 bpm for 15 s) in a period of 10 min constitutes a "reactive" FHR pattern. A reactive FHR pattern is a reassuring finding, reflecting a normally oxygenated, nonacidotic CNS–cardiac axis. The persistent absence of spontaneous accelerations, on the other hand, is abnormal and may reflect fetal compromise. A nonreactive FHR tracing must be carefully interpreted in the context of the clinical presentation and other FHR characteristics. During labor, the frequency and amplitude of FHR accelerations may be diminished by fetal sleep states, commonly used medications (narcotics, $MgSO_4$, atropine), or fetal acidosis. In the absence of spontaneous accelerations, fetal scalp stimulation or vibroacoustic stimulation often provokes fetal movement and FHR accelerations. If these measure fail to induce FHR accelerations, and if other FHR characteristics are concerning, consideration should be given to fetal scalp pH determination or delivery.

Decelerations

Decelerations in the FHR are most commonly encountered during the intrapartum period. They are generally divided into three categories ("early, variable, and late") on the basis of their temporal relationship to the onset of a uterine contraction.

Early Decelerations

Early decelerations are typically uniform, shallow dips in the FHR (rarely below 100 bpm) that start when a contraction starts and end when the contraction ends. They probably result from fetal head compression and a reflex augmentation of vagal tone. Perinatal outcome is not adversely affected by these decelerations, and they are considered to be clinically benign.

Variable Decelerations

Variable decelerations result from umbilical cord compression, and have a variable temporal relationship to uterine contractions. Initially, umbilical vein compression decreases fetal venous return and causes reflex FHR elevation ("shoulder"). Subsequent umbilical arterial compression dramatically increases fetal peripheral resistance and produces a rapid-onset baroreceptor-mediated slowing of the heart rate. Maximum vagal tone may result in a junctional or idioventricular escape rhythm which appears as a relatively stable rate of 60–70 bpm. As the cord is decompressed, this sequence of events occurs in reverse. Isolated variable decelerations are usually not clinically significant. Repetitive severe variables (<70 bpm for >60 seconds), however, may not allow sufficient fetal recovery between decelerations, resulting in persistent hypoxemia, hypercapnia and respiratory acidosis. Prolonged tissue hypoperfusion may lead to metabolic acidosis and, ultimately, fetal death. In animal models, Clapp et al.[92] reported that frequent episodes of hypoxemic stress, produced by intermittent umbilical cord occlusion over a period of hours, produced fetal injury even in the absence of acidosis.

When repetitive, severe variable decelerations are present, prolapse of the umbilical cord must be ruled out by digital examination. Maternal positional changes may relieve cord compression. Uterine hypertonus or tachysystole may be relieved by discontinuing oxytocin, administering uterine relaxants ($MgSO_4$, terbutaline), or both. Restoration of normal amniotic fluid volume in patients with oligohydramnios can decrease the frequency and severity of variable decelerations as well as the incidence of fetal distress as diagnosed by the FHR tracing.[93–95] Transcervical amnioinfusion with 300–500 ml of warmed saline should be instituted as needed, and ideally an amniotic fluid index (AFI) of at least 10 cm should be maintained. Repetitive severe variable decelerations that persist despite these maneuvers must be evaluated in the context of the associated FHR patterns. In the absence of reassuring FHR findings, delivery should be expedited.

Occasionally, variable decelerations fail to promptly return to baseline, and may be more accurately termed "prolonged decelerations." Prolonged decelerations are usually the result of cord compression (cord prolapse, tight nuchal cord) or other acute interruption of uteroplacental transfer of oxygen (tetanic contraction, uterine rupture, maternal hypotension, maternal apnea, or placental abruption). Umbilical cord prolapse must be excluded. Maternal positional changes or manual elevation the fetal head may relieve cord compression, if present. Documentation of separate maternal and fetal heart rates is necessary. Heart block may be ruled out by ultrasound confirmation of identical fetal atrial and ventricular rates. Acute maternal hypotension (i.e., epidural anesthesia) may respond to positional changes, fluids, and ephedrine, if necessary. Tetanic uterine contractions are often relieved by discontinuing oxytocin; however, uterine relaxants ($MgSO_4$, terbutaline) are occasionally required. Oxygen is administered to the mother by face mask. If these measures fail to result in resolution of the prolonged deceleration, rapid delivery is indicated.

Late Decelerations

Late decelerations reflect inadequate uteroplacental transfer of oxygen during uterine contractions. Typically, these are uniform decelerations that start after the onset of a contraction and end after the contraction ends. During uterine contractions, disruption of maternal uteroplacental perfusion causes a decline in fetal Po_2. When the fetal Po_2 falls below a critical threshold (17–18 mm Hg), a complex chemoreceptor- and baroreceptor-mediated reflex is initiated. Centralization of blood volume (favoring perfusion of the brain, heart, and adrenals) occurs via vasoconstriction in the vascular beds of the limbs and gut. The resulting increase in peripheral resistance provokes a reflex deceleration in the FHR. Isolated late decelerations within an otherwise normal tracing are usually of minimal clinical significance. However, continuing hypoxic stress, as evidenced by repetitive late decelerations, may lead to metabolic acidosis, asphyxia, and eventually fetal death.

The goal in treating late decelerations is to improve uteroplacental perfusion and oxygen delivery to the fetus. Usual measures include (1) the left lateral decubitus position, to improve maternal venous return and cardiac output, (2) oxygen by face mask, (3) an intravenous fluid bolus of 250–500 ml of crystalloid to restore the maternal intravascular volume and improve cardiac output, and (4) discontinuation of oxytocin. In the presence of late decelerations with decreased FHR variability, reassurance of fetal well-being is imperative to justify continuation of labor. A normal fetal scalp pH (>7.25), or an acceleratory FHR response to fetal scalp or vibroacoustic stimulation provides reassurance that acidosis is not present. A pattern of repetitive late decelerations without such reassurance strongly suggests fetal distress and is an indication for delivery.

Conclusion

Dramatic changes in intrapartum management have taken place during the past two decades, highlighted by the tremendous proliferation of electronic fetal monitoring, the decline in maternal and perinatal mortality, and the soaring utilization of cesarean birth. Many widely held beliefs have been challenged. Large case-control studies have demonstrated the limited contribution of "birth asphyxia" to the overall incidence of CP. Randomized trials have dampened some of the early enthusiasm regarding the potential benefits of EFM (although it may be rekindled by the most recent study). Nevertheless, the objective of intrapartum management remains the same: to optimize the outcome for both mother and fetus by (1) preventing intrapartum fetal asphyxia and possible long-term sequelae, and (2) avoiding unnecessary operative deliveries. To that end, the most effective resource available to the obstetrician is a thorough understanding of the FHR patterns associated with normal and abnormal fetal physiologic states.

References

1. National Center for Health Statistics. Rates of Cesarean delivery—United States, 1991. Morb Mortal Wkly Rep 1993;42:285–289.
2. Rosen MG, Chairman. Consensus Task Force on Cesarean Childbirth. National Institutes of Health Publication No. 82-2067. Washington, DC: US Dept of Health and Human Services, 1981:49, 125, 275.
3. Speert H. A pictorial history of gynecology and obstetrics. Philadelphia: FA Davis, 1973:297.
4. Morris RC, ed. Obstetrics. Philadelphia: WB Saunders, 1895:917.
5. Sachs BP, Yeh J, Acker D, Driscoll S, Brown DAJ, Jewett JF. Cesarean section-related maternal mortality in Massachusetts, 1954–1985. Obstet Gynecol 1988;71:385–388.
6. Douglas RG, Landesman R. Recent trends in cesarean section. Am J obstet Gynecol 1950; 59:96–107.
7. Phelippeaux. Notice biographique et bibliographique sur Philippe Le Goust. Archives de Tocologie des Maladies des Femmes, Paris 1879;6:304, quoted by Gültekin-Zootzmann.[60]
8. Mayor H. Biblioth Univ. De Geneve, November 9, 1818. Quoted by Thomas H. Classical Contributions to obstetrics and gynecology. Springfield, Illinois: CC Thomos, 1935.
9. Le Jumeau JA (de Kergaradec). Mémoire sur l'Auscultation appliquée à l'Etude de la Grossesse ou Recherches sur deux nouveaux signes propres à faire reconnaître plusieurs circonstances de l'Etat de Gestation; lu à l'Académie royale de médecine, dans sa séance générale du 26 décembre 1821. Paris, 1822, quoted by Gültekin-Zootzmann.[60]
10. Kennedy E. Observations of obstetrical auscultation. Dublin: Hodges and Smith, 1833.
11. Schwartz H. Arch Gynaekol 1870;1:361.
12. Winckel F. Lehrbuch der Geburtshilfe. Leipzig: 1893, Veit.
13. Kilian. Quoted by Jaggard WW, in Hirst BC, ed. A system of obstetrics. Philadelphia: Lea Broth, 1888.
14. Seitz L. Die fetalen Herztone unter der Geburt. Munchen: Habil-Schrift, 1903.
15. Hillis DS. Attachment for the stethoscope. JAMA 1917;68:910.
16. DeLee JB. Ein nues stethoskopf fur die Geburtshilfe besonders geeignet. Zentralbl Gynaekol 1922;46:1688.
17. Cremer MV. Ueber die direckte ableitung der aktionsstrome des menchlichen herzens vom oesophagus und ueber das elektrokardiogramm des fetus. Münch Med Wochenschr 1906;53:811.

18. Smyth CN. Lancet 1953;2:1124.
19. Sureau CL. Recherches electrocardiographiques foetal au cours de la gestation et du travail: premiers resultal d'une nouvelle technique d'enregistrement par electrodes endouterines. Gynecol Obstet 1956;55:21.
20. Kaplan S, Toyama S. Fetal electrocardiography: utilizing abdominal and intrauterine leads. Obstet Gynecol 1958;11:391.
21. Williams RL, Hawes WE. Cesarean section, fetal monitoring and perinatal mortality in California. Am J Public Health 1979;69:864.
22. National Center for Health Statistics. Monthly Vital Statisties Report: Advance Report of Maternal and Infant Health Data from the Birth Certificate, 1990. Vol. 42, No. 2 (suppl), July 8, 1993.
23. Kirschbaum TH. Editorial: diagnosis of fetal distress. Obstet Gynecol 1969;34:721–728.
24. Haverkamp AD, Orleans M, Langendoerfer S, McFee J, Murphy J, Thompson HE. A controlled trial of the differential effects of intrapartum fetal monitoring. Am J Obstet Gynecol 1979;134:399–408.
25. Haesslein HC, Niswander KR. Fetal distress in term pregnancies. Am J Obstet Gynecol 1980;137:245–253.
26. The American Heritage Dictionionary of the English Language, Third Edition Boston, Massachusetts, Houghton Mifflin, 1992.
27. Dorland's illustrated medical dictionary, 26th Ed. Philadelphia: WB Saunders 1985; 129:990.
28. Parer JT, Livingston EG. What is fetal distress? Am J Obstet Gynecol 1990;162:1421–1427.
29. Gilstrap LC, Leveno KJ, Burris J, Williams ML, Bertis BL. Diagnosis of birth asphyxia on the basis of fetal pH, Apgar score, and newborn cerebral dysfunction. Am J obstet Gynecol 1989;161:825–830.
30. Committee on Obstetrics, Maternal and Fetal Medicine of the American College of Obstetrics. Utility of umbilical cord blood acid-base assessment (ACOG Committee Opinion #91, February 1991). Washington, DC: American College of Obsktricians and Gynecologists, 1991.
31. Johnson MV. Cellular alterations associated with perinatal asphyxia. Clin Invest Med 1993;16(2):122–132.
32. Nelson KB, Ellenberg JH. Epidemiology of cerebral palsy. In: Schoenberg BS, ed. Advances in neurology, Vol. 19. New York: Raven, 1978:421–435.
33. Alberman E. Main causes of major mental handicap: prevalence and epidemiology. In: DIBA Foundation symposium 59, major mental handicap: methods and costs of prevention. Amsterdam, Elsevier-Excerpta Medica, 1978.
34. National Institutes of Health. On causes of mental retardation and cerebral palsy: task force on joint assessment of prenatal and perinatal factors associated with brain disorders. Pediatrics 1985;76:457.
35. Little WJ. On the influence of abnormal parturition, difficult labours, premature birth and asphyxia neonatorum, on the mental and physical condition of the child, especially in relation to deformities. Trans Obstet Soc Lond 1862;3:293.
36. Schreiber F. Apnea of the newborn and associated cerebral injury: a clinical and statistical study. JAMA 1938;111:1263.
37. Lilienfeld AM, Parkhurst E. Study of association of factors of pregnancy and parturition with development of cerebral palsy: preliminary report. Am J Hyg 1951;53:262–270.
38. Lilienfeld AM, Pasamanick BP. The association of maternal and fetal factors with the development of cerebral palsy and epilepsy. Am J Obstet Gynecol 1955;70:93–101.
39. Eastman NJ, DeLeon M. The etiology of cerebral palsy. Am J Obstet Gynecol 1955; 69:950–961.
40. Eastman NJ, Kohl SG, Maisel JE, Kavaler F. The obstetrical background of 753 cases of cerebral palsy. Obstet Gynecol Surv 1962; 17:459–500.
41. Steer CW, Bonney W. Obstetric factors in cerebral palsy. Am J Obstet Gynecol 1962; 83:526–531.
42. Nelson KB, Ellenberg JH. Antecedents of cerebral palsy. I. Univariate analysis of risks. Am J Dis Child 1985;139:1031–1038.
43. Windle WF, Becker RF. Asphyxia neonatorum: an experimental study in the guinea pig. Am J Obstet Gynecol 1943;45:183–199.
44. Ranck JB, Windle WF. Brain damage in the monkey, *Maccaca mulatta*, by asphyxia neonatorum. Exp Neurol 1959;1:130–154.
45. Faro MD, Windle WF. Transneuronal degeneration in brains of monkeys asphyxiated at birth. Exp Neurol 1969;24:38–53.
46. Myers RE. Fetal asphyxia and perinatal brain damage affecting human development. Publication No. 185. Washington, DC: Pan American Health Organization, 1969:205–214.

47. Myers RE. Two patterns of perinatal brain damage and their conditions of occurrence. Am J Obstet Gynecol 1972;122:246–276.
48. Freud S. Die Infantile Cerebrallähmung. Nothnagels Spez Path u Therapie (Wien) 1957;IX:2.
49. Haddow KM, Gage RP. Neurologic lesions in relation to asphyxia of the newborn and focators of pregnancy: long-term follow-up. Pediatrics 1960;26:616–622.
50. Blair E, Stanley FJ. Intrapartum asphyxia: a rare cause of cerebral palsy. J Pediatr 1988; 112:515–519.
51. Nelson KB, Ellenberg JH. Antecedents of cerebral palsy: multivariate analysis of risk. N Engl J Med 1986;315:81–86.
52. Paneth N. The causes of cerebral palsy. Recent evidence. Clin Invest Med 1993;16: 95–102.
53. Stanley FJ, Blair E. Why have we failed to reduce the frequency of cerebral palsy. Med J Aust 1991;154:623.
54. Takesita K, Ando Y, Ohtani K, et al. Cerebral palsy in Tottori, Japan. Neuroepidemiology 1989;8:184–192.
55. Stanley F. Survival and cerebral palsy in low birthweight infants: implications for perinatal care. Paediatr Perinatol Epidemiol 1992;6:298–310.
56. Riikonen R, Raumavirta S, Sinivuori E, et al. Changing pattern of cerebral palsy in the south-west region of Finland. Acta Paediatr Scand 1989;78:581–587.
57. Pharoah POD, Cooke T, Cooke RWI, et al. Birthweight specific trends in cerebral palsy. Arch Dis Child 1990;65:602–606.
58. Ellenberg JH, Nelson KB. Birth weight and gestational age in children with cerebral palsy or seizure disorders. Am J Dis Child 1979; 133:1044.
59. Schwartz H. Die vorzeitgen Athembewegungen. Leipzig, 1838, quoted by Gültekin-Zootzmann.[60]
60. Gültekin-Zootzmann B. The history of monitoring the human fetus. J Perinat Med 1975;3:135–144.
61. Hon EH, Quilligan EJ. The classification of fetal heart rate. Conn Med 1967;31:779.
62. Kubli FW, Hon EH, Khazin AF, Takemura H. Observations on heart rate and pH in the human fetus during labor. Am J Obstet Gynecol 1969;104:1190–1206.
63. Paul RH, Suidan AK, Yeh S, Hon EH. The evaluation and significance of intrapartum baseline FHR variability. Am J Obstet Gynecol 1975;123:206–210.
64. Clark SL, Gimovsky ML, Miller FC. The scalp stimulation test: a clinical alternative to fetal scalp blood sampling. Am J Obstet Gynecol 1984;148:274–277.
65. Smith CV, Nquyen HN, Phelan JP, Paul RH. Intrapartum assessment of fetal well-being: a comparison of fetal acoustic stimulation with acid-base determinations. Am J Obstet Gynecol 1986;155:726–728.
66. Chan WH, Paul RH, Toews J. Intrapartum fetal monitoring: maternal and fetal morbidity and perinatal mortality. Obstet Gynecol 1973;41:7–13.
67. Kelly VC, Kulkarni D. Experiences with fetal monitoring in a community hospital. Obstet Gynecol 1973;41:818–824.
68. Tutera G, Newman RL. Fetal monitoring: its effect on the perinatal mortality and caesarean section rates and its complications. Am J Obstet Gynecol 1975;122:750–754.
69. Edington PT, Sibanda J, Beard RW. Influence on clinical practice of routine intra-partum fetal monitoring. Br Med J 1975;3:341–343.
70. Shenker L, Post RC, Seiler JS. Routine electronic monitoring of fetal heart rate and uterine activity during labor. Obstet Gynecol 1975;46:185–189.
71. Koh KS, Greves D, Yung S, et al. Experience with fetal monitoring in a university teaching hospital. Can Med Assoc J 1975;112:455–462.
72. Lee WK, Baggish MS. The effect of unselected intrapartum fetal monitoring. Obstet Gynecol 1976;47:516–520.
73. Paul RH, Huey JR, Yaeger CF. Clinical fetal monitoring—its effect on cesarean section rate and perinatal mortality: five-year trends. Postgrad Med 1977;61:160–164.
74. Amato JL. Fetal monitoring in a community hospital: a statistical analysis. Obstet Gynecol 1977;50:269–274.
75. Johnstone FD, Campbell DM, Hughes GJ. Antenatal care: has continuous intrapartum monitoring made any impact on fetal outcome? Lancet 1978;1:1298–1300.
76. Hamilton LA, Gottschalk W, Vidyasagar D, et al. Effects of monitoring on perinates. Int J Gynaecol Obstet 1978;15:483–490.
77. MacDonald D, Grant A. Fetal surveillance in labour—the present position. In: Bonnar J, ed. Recent advances in obstetrics and gynaecology, Vol. 15. London: Churchill Livingstone, 1987:83–100.

78. Haverkamp AD, Thompson HE, McFee JG, Cetrullo C. The evaluation of continuous fetal heart rate monitoring in high-risk pregnancy. Am J Obstet Gynecol 1976;125: 310–320.
79. Renou P, Chang A, Anderson I, Wood C. Controlled trial of fetal intensive care. Am J Obstet Gynecol 1976;126:470–476.
80. Luthy DA, Kirkwood KS, van Belle G, et al. A randomized trial of electronic fetal monitoring in preterm labor. Obstet Gynecol 1987;69:687–695.
81. Vintzileos AM, Antsaklis A, Varvarigos I, Papas C, Sofatzis I, Montgomery JT. A randomized trial of intrapartum electronic fetal heart rate montoring versus intermittent auscultation. Obstet Gynecol 1993;81: 899–907.
82. Kelso IM, Parsons RJ, Lawrence GF, Arora SS, Edmonds DK, Cooke ID. An assessment of continuous fetal heart rate monitoring in labor: a randomized trial. Am J Obstet Gynecol 1978;131:526–532.
83. Wood C, Renou P, Oats J, Farrell E, Bleischer N, Anderson I. A controlled trial of fetal heart rate monitoring in a low-risk obstetric population. Am J Obstet Gynecol 1981;141: 527–534.
84. MacDonald D, Grant A, Sheridan-Pereira M, Boylan P, Chalmers I. The Dublin randomized controlled trial of intrapartum fetal heart rate monitoring. Am J Obstet Gynecol 1985; 152:524–539.
85. Neldam S, Osler M, Hansen PK, Nim J, Smith SF, Hertel J. Intrapartum fetal heart rate monitoring in a combined low- and high-risk population: a controlled clinical trial. Eur J Obstet Gynecol Reprod Biol 1986;23: 1–11.
86. Goodwin JW, Dunne JT, Thomas RW. Antepartum identification of the fetus at risk. Can Med Assoc J 1969;101:458.
87. Renou P, Wood C. Interpretation of the continuous fetal heart rate record. Clin Obstet Gynaecol 1974;1:191–215.
88. Shelley T, Tipton RH. Dip area: A quantitative measure of fetal heart rate patterns. J Obstet Gynecol Br Commonw 1971;78: 694–701.
89. Fleiss. Statistical methods for rates and proportions, 2nd Ed. New York: Wiley, 1981: 38–45.
90. Grant A, O'Brien N, Joy M, Hennessy E, MacDonald D. Cerebral palsy among children born during the Dublin randomized trial of intrapartum monitoring. Lancet 1989;2: 1233–1236.
91. Gleicher H, Runowicz C, Brown B. Sinusoidal fetal heart rate patterns in association with amnionitis. Obstet Gynecol 1980;56:109.
92. Clapp JF, Peress NS, Wesley M, Mann LI. Brain damage after intermittent partial cord occlusion in the chronically instrumented fetal lamb. Am J Obstet Gynecol 1988;159: 504–509.
93. Miyazaki FS, Taylor NA. Saline amnioinfusion for relief of variable prolonged decelerations. Am J Obstet Gynecol 1983;146: 670–674.
94. Miyazaki FS, Nevarez F. Saline amnioinfusion for relief of repetitive variable decelerations: a prospective randomized study. Am J Obstet Gynecol 1985;153:301–306.
95. Strong TH, Hetzler G, Sarno AP, Paul RH. Prophylactic intrapartum amnioinfusion: a randomized clinical trial. Am J Obstet Gynecol 1990;162:1370–1375.

8
Fetal Macrosomia

HUNG N. WINN AND JOHN C. HOBBINS

Macrosomia, arbitrarily defined as a birthweight of more than 4000 g at term, complicates about 10% and 25% of term and postterm pregnancies, respectively.[1–4] It remains an important cause of perinatal morbidity and mortality and maternal morbidity, which arise mainly from birth injury and asphyxia and increased rate of cesarean section, respectively.[5] In addition to the significant immediate complications, the long-term impact of less severe perinatal asphyxia on the mental development of the affected children is unknown. The major obstacle in reducing the perinatal morbidity and mortality associated with fetal macrosomia has been the inability to predict with certainly which fetuses will sustain birth injury before delivery. To minimize the adverse perinatal and maternal outcomes associated with this problem, it is important to have a well-planned management scheme for patients with fetal macrosomia, recognizing that there is currently no perfect solution. In the following discussion, the current understanding of fetal macrosomia with regard to risk factors, diagnosis, and intrapartum management is presented.

Risk Factors

The first step in the management of fetal macrosomia is the identification of patients at risk for this condition. It has been shown that patients with such characteristics as diabetes, obesity, advanced age, multiparity, postterm pregnancy, and prior maternal and sibling birth weights of more than 4000 g are at risk of having fetal macrosomia.[2,5–8]

In general, maternal diabetes mellitus doubles the risk of having neonatal macrosomia compared to that of the nondiabetic population. Maternal hyperglycemia, especially during the period of 29 to 32 weeks of gestation, has been associated with fetal macrosomia.[9–14] Interestingly, it has been demonstrated that the risk of fetal macrosomia also rises in untreated patients with one abnormal value on a glucose tolerance test (GTT),[15–17] even though these patients may be classified as normal if the criterion of having two abnormal values is used for diagnosing gestational diabetes mellitus.

The mechanism of fetal macrosomia in the case of maternal diabetes mellitus as originally proposed by Pederson et al.[18] is the acceleration of fetal growth from fetal hyperinsulinemia in response to fetal hyperglycemia, which is in turn affected by maternal hyperglycemia. This hypothesis is supported by a high degree of correlation between cord serum levels of C-peptide and fetal macrosomia in diabetic patients.[19] Fetal macrosomia and organomegaly of the placenta, liver, and heart have been observed in the chronically maintained euglycemic and hyperinsulinemic rhesus monkey.[20] The direct impact of maternal and fetal hyperglycemia per se on fetal macrosomia remains unclear.[19,21,22] It is possible that macrosomic fetuses maintain euglycemia by increased production of insulin. This could explain the similar levels of maternal glycosylated hemoglobin but higher levels of C-peptide in cord serum of macrosomic fetuses compared with non-macrosomic fetuses in diabetic patients. It has also been demonstrated that the exogenous animal insulin (bovine or porcine) in combination with the anti-insulin antibody can cross the placenta as an insulin–antibody complex and that there is a significant correlation between fetal macrosomia and elevated cord serum levels of animal insulin.[23] Thus, the exogenous insulin administered to pregnant patients may independently accelerate fetal growth. Metabolic factors other than maternal hyperglycemia may also play a role in causing fetal macrosomia.

Because the fetus may continue to grow beyond 37 weeks of gestation, the incidence of fetal macrosomia increases with advancing gestational age with an overall rate of about 25%–30% at 41 weeks gestation or greater.[1,4] Maternal obesity (prepregnancy weight of at least 90 kg) is a strong indicator of fetal macrosomia and is associated with an increased risk of at least 1.5 fold.[24–27] The increased incidence of fetal macrosomia in obese patients could be caused by multiple factors, such as altered metabolic homeostasis with an inherent risk of abnormal glucose metabolism and heredity. Advanced maternal age (more than 35 years old) is also a risk factor for fetal macrosomia even after correction for parity and gestational diabetes.[28] Genetic influence may also account for the high correlation between maternal and sibling birth weights and the current neonate's birthweight.

Diagnosing Fetal Macrosomia

The next step in the management of patients at risk for fetal macrosomia at term is estimating fetal weight. This can be done by clinical assessment or ultrasound examination. Estimation of fetal weight by clinical examination of fundal height is usually inaccurate, especially at the extremes of fetal sizes. A fundal height that is larger than expected may point to the potential for fetal macrosomia. Ultrasonography provides a more accurate means of predicting fetal macrosomia. One approach utilizes the macrosomic index, which is the difference between the fetal chest circumference and the biparietal diameter (BPD), to identify macrosomic fetuses in pregnancies complicated by diabetes mellitus. In one study, 87% of macrosomic infants had an average chest diameter that exceeded the BPD by 1.4 cm or greater.[29]

If the gestational age is known, ultrasound examination during the third trimester can be used to identify large-for-gestational-age (LGA) fetuses who are at risk for being macrosomic at delivery. If the estimated fetal weight (EFW) or the fetal abdominal circumference (AC), the latter reflecting insulin-sensitive tissue, are above the 90th percentile, macrosomia can be correctly diagnosed in 74% and 88.8% of infants of diabetic mothers, respectively.[30] In pregnancies complicated by diabetes mellitus, accelerated growth of abdominal circumference (above the 90th percentile), which may become apparent as early as 24 weeks of gestation, is associated with increased risk of fetal macrosomia at term.[31]

A growth of abdominal circumference of 1.2 cm or more per week during the gestational interval of 32–39 weeks is noted in about 80% of macrosomic infants of diabetic pregnancies.[32] Optimal clinical management of fetal macrosomia requires more than just identification of fetuses at risk for macrosomia; it requires an accurate estimation of fetal weight. In general, by using the fetal AC and the BPD, one can estimate fetal weights on the basis of the published formulas to within 10% and 5% of the actual weight in 80% and 50% of normal fetuses, respectively.[33] It appears that the formula of Hadlock et al.,[34] which uses the AC and femur length (FL) [$\log_{10}(\text{weight}) = 1.304 + 0.05281\ \text{AC} + 0.1938\ \text{FL} - 0.004\ \text{AC} \times \text{FL}$]

provides the best estimation of birth weight of macrosomic fetuses.[35] At present, accuracy in predicting birthweight in macrosomic fetuses remains limited. In fact, one study demonstrated a positive predictive value for fetal macrosomia of only 67% when the birthweight was greater than 4100 grams.[36]

Failure to take into account the increased fetal body fat in macrosomic fetuses, at least among diabetic pregnancies, may contribute to the overestimation of fetal weight by ultrasound utilizing available formulas.[37] An innovative model taking into account many variables such as maternal height, gravity, parity, gestational age, fundal height, amniotic fluid index, and fetal subcutaneous tissue in addition to the commonly used parameters such as BPD, AC, head circumference, and FL, has reduced the error of predicting EFW in macrosomic fetuses to only 4.7%.[38] In pregnancies at risk for fetal macrosomia, serial ultrasound examinations every 2–3 weeks for EFW and AC starting at about 36.0 weeks of gestation may be helpful in detecting fetal macrosomia. It is important to recognize that the error of estimating fetal weight by ultrasound utilizing the available formulas could be as great as 10%.

Intrapartum Management

Once fetal macrosomia is suspected, delivery should be considered after fetal lung maturity is documented. In most situations, a trial of labor can be attempted unless there are contraindications such as absolute cephalopelvic disproportion (CPD), placenta previa, vasa previa, malpresentation of the fetus, previous classical cesarean section, an EFW greater than 5000 g, or documentation of substantial body-to-head disproportion in a fetus of more than 4500 g. Fetal lung maturity should be documented before induction of labor. Amniocentesis for the lecithin/sphingomyelin (L/S) ratio and phosphatidylglycerol (PG) is generally utilized to document fetal lung maturity in the following situations: (1) uncertain gestational age, (2) gestational age less than 38 weeks, (3) and gestational age of less than 40.0 weeks in diabetic pregnancies.

Induction of Labor

One of the major concerns about induction of labor in the presence of an unfavorable cervix, that is, a Bishop score of 5 or less, is failed induction. To facilitate the induction, cervical ripening can be accomplised by the use of intracervical artificial laminaria (Lumicil) and low-dose continuous oxytocin infusion overnight. The oxytocin infusion is started at 0.5 mu/min and doubled hourly to a maximum dose of 4 mu/min. Mild uterine contractions usually are obtained at this low dose. This combination of laminaria and low-dose oxytocin infusion is usually sufficient to make the cervix favorable for a full induction of labor the next morning. Some advocate that amniotomy by preformed as soon as possible and that this represents the critical point of induction by allowing the release of arachidonic acid from the membranes for conversion to prostaglandins. All our patients reach the active phase of labor, and the majority do so within a normal length of time, when this protocol is followed.

Accurate assessment of uterine contractions and optimal fetal heart rate monitoring are important whenever induction of labor is attempted. This becomes even more important when induction of labor is carried out for impending fetal macrosomia in view of the increased risk of CPD and birth injury. Labor course should be closely monitored for arrest or protraction disorders.[22,39,40] Thus, intrauterine pressure and direct fetal electronic monitoring should be initiated as soon as it is feasible and safe to do so.

In the United States, intravenous oxytocin is the drug of choice for induction of labor. Recently there has been a shift to a lower total dose by starting at a lower concentration and increasing the concentration at longer intervals and in lesser increments. This change in the usage of oxytocin for induction of labor has been made because (1) the steady-state oxytocin concentration is not reached until approximately 40 min from the last dose,[41,42] (2) effective uterine contractions can occur with an oxytocin infusion rate between 4.0 and 8.0 mu/min in the majority of patients (in fact, approximately 2 SD of patients will develop progressive cervical dilatation with an oxytocin infusion rate of no more than 8.0 mu/min),[42–44] and (3) the incidence of uterine hypercontractility and the resulting abnormal heartrate are reduced.

The following is a suggested protocol for the continuous infusion of oxytocin for induction of labor: The oxytocin infusion is started at 0.5 mu/min and then doubled every hour up to the rate of 8 mu/min. Thereafter, the dose is increased 4 mu/min every hour until the maximum dose of 20 mu/min is reached. The dose of oxytocin should not be raised further when adequate uterine contractions, that is, a frequency of 2–3 min, durations of 40–90 s, and intensities of 40–90 mmHg, or progressive cervical dilatation and descent, are obtained regardless of how low the oxytocin dose may be. A lower starting dose of oxytocin reduces the risk of uterine hyperstimulation, which can occur at very low doses.[45,46]

Birth Injury and Shoulder Dystocia

The major complication associated with delivery of a macrosomic fetus is birth trauma, such as clavicle fracture and subsequent brachial plexus injury from a difficult vaginal delivery.[39,47–50] The overall risk of shoulder dystocia rises sharply from 3% for birthweights less than 4000 g to 10.3% and 23.9% for birthweights ranging between 4000 and 4500 g and those greater than 4500 g, respectively. When birthweights are greater than 3500 g, the incidence of shoulder dystocia generally doubles in diabetic patients compared to nondiabetic mothers for similar birthweights. Thus, the combination of macrosomia and diabetes places the patient at a very high risk for neonatal shoulder dystocia.[47] Although there is a strong correlation between birthweight and brachial plexus injury, the latter occurs as a consequence of shoulder dystocia in most patients.[51] Asphyxia and brachial palsy occur in as many as 42% of infants with true shoulder dystocia.[52]

Shoulder dystocia is a dreadful experience for many obstetricians and their patients because it is associated with significant perinatal morbidity and mortality and usually occurs unexpectedly. It would be ideal if shoulder dystocia could be anticipated and thus could be avoided in pregnancies at risk for this condition. In general, fetal weight or pelvimetry is a poor indicator of CPD or birth trauma. The fetal-pelvic index (FPI) was recently introduced to address this issue.[7] First, the anteroposterior and transverse diameters of the fetal head and abdomen and the maternal inlet and midpelvis are determined. Fetal and maternal measurements are obtained by ultrasonography and x-ray pelvimetry, respectively. The corresponding circumferences are calculated from the two perpendicular diameters using the formula $C = (TD + APD) \times 3.14/2$, where C is the circumference, APD the anteroposterior diameter, and TD the transverse diameter. The FPI is the sum of the two most positive circumference differences (fetopelvic differences: HC − IC and AC − MC). A positive or negative FPI indicates the presence or absence of CPD/shoulder dystocia, respectively. The sensitivity and specificity of the FPI in the diagnosis of CPD/shoulder dystocia are 94%. Further study is needed to determine the predictive value of the FPI in detecting CPD/shoulder dystocia in pregnancies complicated by fetal macrosomia.

Another approach to the detection of shoulder dystocia has utilized fetal biometric evaluation. Studies of neonates have demonstrated significant differences in many neonatal anthropometric measurements between newborn infants with and without dystocia. The mean shoulder circumference was significantly larger when shoulder dystocia had complicated delivery.[53] A recent study demonstrated that the neonatal bisacromial diameter is well correlated with the ultrasonographically measured circumferences of the fetal chest and arm.[54] Thus, the relationship between fetal chest circumference and other fetal parameters such as head circumference or abdominal circumference may be potentially useful in predicting shoulder dystocia in pregnancies at risk for this condition.

Once a trial of labor is under way, the course of labor should be closely followed by a cervicographic analysis. Protraction and arrest disorders of labor may be associated with an increased risk for shoulder dystocia. Operative vaginal delivery by itself or in combination with an abnormal course of labor may predispose to shoulder dystocia in the case of fetal macrosomia.[45,52,55,56] Thus, one should be very cautious when selecting a midforceps delivery to correct disorders of descent in the presence of fetal macrosomia, especially in a diabetic mother.

To minimize the perinatal morbidity and mortality associated with shoulder dystocia once it occurs, one should be ready to initiate the appropriate steps and maneuvers to effect the delivery in a timely fashion. Time is a critical factor with regard to perinatal mortality and morbidity. The cardinal diagnostic sign of shoulder dystocia is retraction of the delivered head against the maternal perineum caused by the obstruction to the passage of both shoulders at the pelvic inlet. This situation more likely arises when both shoulders present to the pelvic inlet from the anterior-posterior position instead of an oblique one and when the shoulder circumferences are large.[53,56,57] Direct traction on the fetal head may be ineffective unless therapeutic fracture of the clavicles is associated. Because the fetus naturally passes through the birth canal with a screwlike motion, rotation of the shoulder to reduce the shoulder circumference, or rotation to an oblique position, are the underlying mechanisms of most effective maneuvers. These are described next.[56,58]

McRobert's Maneuver

This easily performed maneuver involves a sharp flexion of the patient's thighs onto the abdomen. It accomplishes the following: (1) straightening of the maternal sacrum relative to the lumbar spine, thus eliminating the sacral promontory as an obstruction; and (2) anterior rotation of the pubic symphysis, thus assisting in bringing the fetal posterior shoulder through the pelvic inlet by simultaneously displacing the fetal anterior shoulder cephalad.[59]

Woods' Maneuver

Woods was the first author to describe the principle of rotation of the shoulders rather than direct traction on the fetal head in resolving shoulder dystocia. In this classical approach, delivery of the posterior shoulder is accomplished by pressing on the anterior aspect of the posterior shoulder toward the fetal back while a gentle pressure is placed on the uterus fundus. The posterior shoulder can usually be delivered under the pubic symphysis after a rotation of 180°. The remaining shoulder is similarly delivered by rotating an arc of 180°, but in the opposite direction with pressure on its anterior aspect.[58]

Rubin's Maneuver

Rubin's maneuver applies the same principle of rotation but differs from the Woods' maneuver in that adduction instead of abduction of the shoulders is carried out. Adduction is accomplished by pushing the shoulder toward the fetal chest. It should be reemphasized that the adduction is performed with direct pressure on the fetal shoulder, not by rotating the fetal head. This maneuver results in reduction of the shoulder transverse diameter and circumference as well as rotation of the shoulder to the more oblique position.[50]

Delivery of the Posterior Arm

Performance of this maneuver requires locating the fetal elbow with the operator's hand in the vagina. The fetal forearm is then flexed, drawn across the fetal chest, and delivered through the vaginal opening. It would be easier for the operator to use his or her left or right hand, depending on whether the fetal back is facing the maternal left or right side, respectively.

Zavanelli Maneuver

The Zavanelli maneuver should be used as the last resort when all other maneuvers to effect a vaginal delivery have failed. The maneuver involves returning the fetal head to the vagina by reversing the sequence of the normal birthing process. The fetus is subsequently delivered by cesarean section from the cephalic presentation. Excellent neonatal outcomes have been reported with this maneuver.[60]

Elective Cesarean Section

Although macrosomia is associated with an increased incidence of cesarean section delivery as a result of abnormal labor such as arrest of descent or cervical dilatation, an estimated fetal weight in the range of 4000–4900 g may not be an absolute indication for elective cesarean section. In fact, a trial of labor in patients with ultrasonographic diagnosis of fetal macrosomia (estimated fetal weight above 90th percentile) may result in vaginal delivery in about 60% of cases.[61] Furthermore, vaginal birth after previous cesarean section occurs in 58% and 43% of patients with infant birthweight ranges of 4000–4499 g and 4500 g or greater, respectively, without a higher risk of uterine rupture.[62] However, elective cesarean delivery may have a role in the management of pregnancies with fetal macrosomia in certain situations. For example, when the EFW is more than 5000 g, the birthweight is likely to be at least 4500 g, assuming 10% error is the estimation. On the other hand, it could be more than 5500 g. In this range of birthweights (>4500 g), the risks for shoulder dystocia are 22.6% and 50% for nondiabetic and diabetic patients, respectively.[47] One study showed that 76% of shoulder dystocia occurs in infants of diabetic mothers with birthweights of 4250 g or greater.[63] Thus, it would not be unreasonable to perform an elective cesarean section for delivery of the macrosomic fetus in this setting. Of course, vaginal delivery should be attempted even in this weight range if the patient has a history of previous vaginal delivery of other infants of similar size or larger.

Another situation involves breech presentation of a macrosomic fetus, for which elective cesarean section delivery after failed external version may be indicated. The American College of Obstetricians and Gynecologists places the upper limit for fetal size of 4000 g when vaginal delivery of a breech fetus is attempted.[64] The patient's prior obstetric history and the obstetrician's experience with breech delivery certainly play a significant role in the decision.

It should be emphasized that individualization is key to the management of macrosomic fetuses in these clinical settings, because good data are lacking. Regardless of the method of delivery, patients should be well informed of the risks for shoulder dystocia as well as the available options when macrosomia is suspected.

Suggested Management Protocol

The best approach to the management of fetal macrosomia remains to be determined because the available data on this subject are retrospective and the error of estimating fetal weight by ultrasonography and predicting shoulder dystocia is still large. These limitations should be considered in the clinical management of fetal macrosomia. The patient should also be so informed. An ultrasound determination of fetal weight and fetal presentation should be made as early as 37 weeks of gestation in patients at risk for fetal macrosomia. If there has been a question of gestational age, an ultrasound examination should be made prior to 20 weeks of gestation to confirm the date of confinement.

Once fetal macrosomia is suspected, fetal lung maturity should be documented before delivery is attempted. Elective cesarean section may be considered in cases in which the estimated fetal weight is greater than 5000 g is normal pregnancies, greater than 4500 g in diabetic patients, or greater than 4000 g in breech presentation with the exact cutoff level being influenced by the perceived accuracy of sonographic estimates of fetal weight. External version of the breech fetus may be attempted unless there are contraindications such as bleeding, placenta previa or abruption, certain fetal anomalies, oligohydramnios, or premature rupture of membranes.[65,66] The alternative approach would be to obtain the FPI and select the mode of delivery depending on whether the FPI is positive or negative. It should be noted that the FPI has not been tested in a large population of patients to determine its predictive value.

If the fetus has an estimated weight between 4000 and 4900 g in nondiabetic patients and presents as vertex, a trial of labor may be conducted provided there are no contraindications. Internal monitoring of fetal heartrate and uterine pressure should be initiated as soon as it is feasible. The course of labor should be closely followed, keeping in mind that an abnormal course of labor may herald a potential CPD or shoulder dystocia. Operative vaginal delivery is not recommended because it may be associated with an unusually high risk of shoulder dystocia and subsequent birth injury.

If shoulder dystocia occurs in spite of the foregoing precautions, timely application of a preplanned sequence of steps and effective maneuvers should be carried out. Time is critical because the risk of fetal brain injury and death increases as the duration of shoulder dystocia advances. Arbitrarily, we recommend trying the McRobert's maneuver first because of its simplicity and effectiveness, followed by Rubin's maneuver, Woods' maneuver, suprapubic pressure, and delivery of the posterior arm in decreasing order of preference. The Zavanelli maneuver should be attempted if these procedures to effect vaginal delivery have failed and sufficient time has elapsed since the occurrence of shoulder dystocia to raise concern about neonatal asphyxia or death.

Conclusion

The management of fetal macrosomia with an increased risk of shoulder dystocia and its attendant morbidity and mortality remains a challenge for the modern clinician. In spite of many attempts to identify the patient at risk for fetal macrosomia, the perfect predictor has not yet emerged. The physician must therefore rely on clinical skills in raising his or her index of suspicion for fetal macrosomia; once it is considered, ancillary diagnostic testing should help to confirm the diagnosis. The summation of this information should then be weighed carefully as the plan for delivery is constructed. A well-orchestrated attempt at vaginal delivery or a timely operative delivery can be carried out with a high probability of excellent maternal and fetal outcomes. In cases in which macrosomia with shoulder dystocia is encountered without antecedent warning, foreknowledge of the serial maneuvers that will enable the delivery of the impacted shoulders can minimize intrapartum trauma.

References

1. Arias F. Predictability of complications associated with prolongation of pregnancy. Obstet Gynecol 1987;70:101.
2. Boyd ME, Usher RH, McLean FH. Fetal macrosomia: prediction, risks, proposed management. Obstet Gynecol 1983;61:715.
3. Golditch IM, Kirkman K. The large fetus. Management and outcome. Obstet Gynecol 1978;52:26.
4. Chervenak JL, Divon MY, Hirsch J, et al. Macrosomia in the postdate pregnancy: Is

routine ultrasonographic screening indicated? Am J Obstet Gynecol 1989;161:753.
5. Spellacy WN, Miller S, Winegar A, et al. Macrosomia—maternal characteristics and infant complications. Obstet Gynecol 1985;66:158.
6. Klebanoff MA, Mills JL, Berendes HW. Mother's birth weight as a predictor of macrosomia. Am J Obstet Gynecol 1985;153:253.
7. Morgan MA, Thurnau GR. Efficacy of the fetal-pelvic index for delivery of neonates weighing 4000 grams or greater: a preliminary report. Am J Obstet Gynecol 1988;158:1133.
8. Larsen CE, Serdula MK, Sullivan KM. Macrosomia: Influence of maternal overweight among a low-income population. Am J Obstet Gynecol 1990;162:490.
9. Coustan DR, Imarah J. Prophylactic insulin treatment of gestational diabetes reduces the incidence of macrosomia, operative delivery, and birth trauma. Am J Obstet Gynecol 1984;150:836.
10. Goldberg JD, Franklin B, Lasser D, et al. Gestational diabetes: impact of home glucose monitoring on neonatal birth weight. Am J Obstet Gynecol 1986;154:546.
11. Lin C-C, River J, River P, et al. Good diabetic control early in pregnancy and favorable fetal outcome. Obstet Gynecol 1986;67:51.
12. Morris MA, Grandis AS, Litton JC. Glycosylated hemoglobin concentration in early gestation associated with neonatal outcome. Am J Obstet Gynecol 1985;153:651.
13. Willman SP, Leveno KJ, Guzick DS, et al. Glucose threshold for macrosomia in pregnancy complicated by diabetes. Am J Obstet Gynecol 1986;154:470.
14. Combs CA, Gunderson E, Kitzmiller JL, et al. Relationship of fetal macrosomia to maternal postprandial glucose control during pregnancy. Diabetes Care 1992;15:1251.
15. Langer O, Brustman L, Anyaegbunam A, et al. The significance of one abnormal glucose tolerance test value on adverse outcome in pregnancy. Am J Obstet Gynecol 1987;157:758.
16. Lindsay MK, Graves W, Klein L. The relationship of one abnormal glucose tolerance test value and pregnancy complications. Obstet Gynecol 1989;73:103.
17. Tallarigo L, Giampietro O, Penno G, et al. Relation of glucose tolerance to complications of pregnancy in nondiabetic women. N Engl J Med 1986;315:989.
18. Pederson J, Bojsen-Moller B, Paulson H. Blood sugar in newborn infants of diabetic mothers. Acta Endocrinol 1954;15:33.
19. Sosenko IR, Kitzmiller JL, Loo SW, et al. The infant of the diabetic mother. N Engl J Med 1979;301:859.
20. Susa JB, McCormick KL, Widness JA, et al. Chronic hyperinsulinemia in the fetal rhesus monkey. Diabetes 1979;28:1058.
21. Widness JA, Schwartz HC, Thompson D, et al. Glycohemoglobin (HbA_{1c}): a predictor of birth weight in infants of diabetic mothers. J Pediatr 1978;92:8.
22. Yatscoff RW, Mehta A, Dean H. Cord blood glycosylated (glycated) hemoglobin: correlation with maternal glycosylated (glycated) hemoglobin and birth weight. Am J Obstet Gynecol 1985;152:861.
23. Menon RK, Cohen RM, Sperling MA, et al. Transplacental passage of insulin in pregnant women with insulin-dependent diabetes mellitus. N Engl J Med 1990;323:309.
24. Edwards LE, Dickes WF, Alton IR, et al. Pregnancy in the massively obese. Course, outcome, and obesity prognosis of the infant. Am J Obstet Gynecol 1978;131:479.
25. Gross T, Sokol RJ, King K. Obesity in pregnancy. Risks and outcome. Obstet Gynecol 1980;56:4.
26. Philipson EH, Kalhan SC, Edelbert SC, et al. Maternal obesity as a risk factor in gestational diabetes. Am J Perinatol 1985;2:4.
27. Wolfe HM, Zador IE, Gross TL, et al. The clinical utility of maternal body mass index in pregnancy. Am J Obstet Gynecol 1991;164:1306.
28. Kirz DS, Dorchester W, Freeman RK. Advanced maternal age. The mature gravida. Am J Obstet Gynecol 1985;152:7.
29. Elliott JP, Garite TJ, Freeman RK, et al. Ultrasonic prediction of fetal macrosomia in diabetic patients. Obstet Gynecol 1982;60:159.
30. Tamura RK, Sabbagha RE, Depp R, et al. Diabetic macrosomia—accuracy of third trimester ultrasound. Obstet Gynecol 1986;67:828.
31. Keller JD, Metzger BE, Dooley SL, et al. Infants of diabetic mothers with accelerated fetal growth by ultrasonography: are they all alike? Am J Obstet Gynecol 1990;163:893.
32. Landon MB, Mintz MC, Gabbe SG. Sonographic evaluation of fetal abdominal growth: Predictor of large-for-gestational-age infant in pregnancies complicated by diabetes

mellitus. Am J Obstet Gynecol 1989;160: 115.

33. Shepard MJ, Richards VA, Berkowitz RL, et al. An evaluation of two equations for predicting fetal weight by ultrasound. Am J Obstet Gynecol 1982;142:47.
34. Hadlock FB, Harrist RB, Sharman RS, et al. Estimation of fetal weight with the use of head, body, and femur measurements—a prospective study. Am J Obstet Gynecol 1985;151:333.
35. Hirata GI, Medearis AL, Horenstein J, et al. Ultrasonographic estimation of fetal weight in the clinically macrosomic fetus. Am J Obstet Gynecol 1990;162:238.
36. Miller JM, Haywood LB, Oscar FK, et al. Ultrasonographic identification of the macrosomic fetus. Am J Obstet Gynecol 1988;158:1110.
37. Bernstein IM, Catalano PM. Influence of fetal fat on the ultrasound estimation of fetal weight in diabetic mothers. Obstet Gynecol 1992;79:561.
38. Farmer RM, Medearis AL, Hirata GI, et al. The use of a neural network for the ultrasonographic estimation of fetal weight in the macrosomic fetus. Am J Obstet Gynecol 1992;166:1467.
39. Oppenheim WL, Davis A, Growdon WA, et al. Clavicle fractures in the newborn. Clin Orthop Relat Res 1990;250:176.
40. Turner MJ, Rasmussen MJ, Turner JE, et al. The influence of birth weight on labor in nulliparas. Obstet Gynecol 1990;76:159.
41. Amico JA, Seitch J, Robinson AG. Studies of oxytocin in plasma of women during hypocontractile labor. J Clin Endocrinol & Metab 1984;58:274.
42. Seitchik J, Castillo M. Oxytocin augmentation of dysfunctional labor. I. Clinical data. Am J Obstet Gynecol 1982;144:899.
43. Baxi LV, Petrie RH, Caritis SN. Induction of labor with low-dose prostaglandin F_{2a} and oxytocin. Am J Obstet Gynecol 1980;136:28.
44. Seitchik J, Castillo M. Oxytocin augmentation of dysfunctional labor. III. Multiparous patients. Am J Obstet Gynecol 1983;145:777.
45. Blakemore KJ, Qin NG, Petrie RH, et al. A prospective comparison of hourly and quarter-hourly oxytocin dose increase intervals for the induction of labor at term. Obstet Gynecol 1990;75:757–761.
46. Mercer B, Pilgrim P, Sibai B. Labor induction with continuous low-dose oxytocin infusion: a randomized trial. Obstet Gynecol 1991;77:659–663.
47. Acker DB, Benjamin PS, Friedman EA. Risk factors for shoulder dystocia. Obstet Gynecol 1985;66:762.
48. Acker DB, Gregory KD, Sachs BP, et al. Risk factors for Erb-Duchenne palsy. Obstet Gynecol 1988;71:389.
49. Gordon M, Rich H, Deutschberger J, et al. The immediate and long-term outcome of obstetric birth trauma. I. Brachial plexus paralysis. Am J Obstet Gynecol 1973;117:1.
50. Rubin A. Management of shoulder dystocia. JAMA 1964;189:11.
51. McFarland LV, Raskin M, Daling JR, et al. Erb/Duchenne's palsy: a consequence of fetal macrosomia and method of delivery. Obstet Gynecol 1986;68:784.
52. Gross SJ, Shime J, Farine D. Shoulder dystocia: predictors and outcome. A five-year review. Am J Obstet Gynecol 1987;156:334.
53. Modanlou HD, Komatsu G, Dorchester W, et al. Large-for-gestational-age neonates: anthropometric reasons for shoulder dystocia. Obstet Gynecol 1982;60:417.
54. Winn HN, Grasso J, Holcomb W, et al. The potential of fetal chest circumference in the prenatal diagnosis of shoulder dystocia. Proceedings of the 9th Annual Society of Perinatal Obstetricians Meeting, New Orleans, 1989.
55. Benedetti TJ, Gabbe SG. Shoulder dystocia: a complication of fetal macrosomia and prolonged second stage of labor with midpelvic delivery. Obstet Gynecol 1978;52:526.
56. Hopwood HG. Shoulder dystocia: fifteen years' experience in a community hospital. Am J Obstet Gynecol 1982;144:162.
57. Smeltzer JS. Prevention and management of shoulder dystocia. Clin Obstet Gynecol 1986;29:2.
58. Woods CE. A principle of physics as applicable to shoulder delivery. Am J Obstet Gynecol 1943;45:796.
59. Gonik B, Stringer CA, Held B. An alternate maneuver for management of shoulder dystocia. Am J Obstet Gynecol 1983;145:7.
60. Sandberg ED. The Zavanelli maneuver: a potentially revolutionary method for the resolution of shoulder dystocia. Am J Obstet Gynecol 1985;152:479.
61. Delpapa EH, Mueller-Heubach E. Pregnancy outcome following ultrasound diagnosis of macrosomia. Obstet Gynecol 1991; 78:340.

62. Flamm BL, Goings JR. Vaginal birth after cesarean section: is suspected fetal macrosomia a contraindication? Obstet Gynecol 1989;74:694.
63. Langer O, Berkus MD, Huff RW, et al. Shoulder dystocia: Should the fetus weighing >4000 grams be delivered by cesarean section? Am J Obstet Gynecol 1991;165:831.
64. American College of Obstetricians and Gynecologists. Management of the breech presentation. ACOG Technical Bulletin No. 95. Washington, DC: American College of Obstetricians and Gynecologists, 1986.
65. Marchick R. Antepartum external cephalic version with tocolysis: a study of term singleton breech presentations. Am J Obstet Gynecol 1988;158:1339.
66. Morrison JC, Myatt RE, Martin JN, et al. External cephalic version of the breech presentation under tocolysis. Am J Obstet Gynecol 1986;154:900.

9 Twin Gestation and Multiple Births

Cornelia R. Graves and Frank H. Boehm

The incidence of twins, triplets, and high-order multifetal pregnancies has increased significantly since the introduction of artificial stimulation of ovulation and in vitro fertilization.[1] These pregnancies are at a higher risk than singletons for perinatal morbidity and mortality. Although much has been written about the mode of delivery for these gestations, there is little consensus over whether cesarean section improves the outcome. Despite little evidence to support cesarean section for multiple gestations, the cesarean section rate has increased from 3% in 1972 to 51% in 1985.[2] The cesarean rate for triplets increased from 16% in 1977 to almost 94% in 1988.[3] The purpose of this chapter is to discuss the management of multiple gestations and to make recommendations concerning possible modes of delivery.

Management of Delivery of Vertex/Vertex Twins

Twin gestations make up approximately 1% of all pregnancies.[4] Although dizygotic twinning varies with race, age, parity, and other factors, the monozygotic twinning rate is constant, occurring at a rate between 3 and 4 per 1000 births.[5] Morbidity and mortality are increased remote from delivery. Complications include spontaneous abortion, intrauterine growth retardation, preeclampsia, placental abnormalities, and preterm labor.[6]

There is consensus regarding the safety of vaginal delivery of twins when both twins are vertex, regardless of gestational age.[7] Cesarean section should be reserved for usual indications, that is, fetal stress or distress and cephalopelvic disproportion.

Management of the Delivery of Vertex/Nonvertex Twins

Cesarean section has been recommended by some investigators as the mode of delivery in vertex/nonvertex presentations to avoid hazardous breech extraction of the second twin.[8] During the past several years numerous studies have suggested, however, that the second, nonvertex twin weighing more than 1500 g can be delivered vaginally with as much safety as by cesarean section.[9] In 1986, Rabinovici et al.[10] managed 60 twin deliveries after the 35th gestational week with vertex-breech and vertex-transverse presentations according to a randomized protocol; 33 patients were allowed to undergo vaginal delivery, and 27 were delivered by cesarean section. There were no significant differences between 1- and 5-min Apgars or the incidence of neonatal morbidity between the second twin in both study groups. As expected, maternal febrile morbidity was significantly higher in the cesarean section group.

Because vaginal breech delivery is not stressed in many training programs, external cephalic version can be attempted after delivery of the first twin to convert the second twin to a vertex presentation so as to allow for vaginal delivery. Tchabo and Tomai[11] reported 11 of 12 conversions from transverse to vertex and 16 of 18 successful breech conversions. No complications were noted in this study. Other authors, however, have found increased maternal and neonatal morbidity using external version.[12] Wells et al.[13] noted that patients undergoing external version were more likely to undergo abdominal delivery for the second twin than those in which a breech extraction was performed. The external version candidates also had a higher incidence of emergent anesthesia for cord prolapse and persistent fetal bradycardia.

Recent studies have also supported that vaginal delivery of the nonvertex second twin is safe. Fisherman et al.[14] in 1993 published a retrospective study that reviewed 781 twins at 20 weeks or more. Of the 309 liveborn vaginal second twins, 207 were vertex and 183 were breech. There were no significant differences in any of the neonatal outcome measures, including Apgar scores at 5 min and neonatal intensive care unit (NICU) admissions when stratified by birthweight.

If vaginal delivery of the nonvertex second twin is to be undertaken, there must be a skilled obstertrician available who has been trained in the techniques of internal and external version, an anesthesiologist knowledgeable about uterine relaxation techniques, and the ability to convert to an operative delivery mode within minutes. There should also be a second assistant skilled in ultrasonography and able to assist with delivery if needed.

The Vanderbilt University Hospital Labor and Delivery guidelines for vertex/nonertex twins weighing more than 1500 g are as follow:

1. Amniotomy is performed when it can be accomplished and internal electrode devices are placed.

2. If the patient fails to progress in labor, an intrauterine pressure catheter is inserted and oxytocin augmentation is begun.

3. Epidural anesthesia is advised. This allows for a comfortable labor and delivery as well as being an effective method of pain control should fetal manipulation or cesarean section be needed.

4. Ultrasound is used to note the position of the second twin, as well as for guidance and monitoring of the second twin.

5. After the first twin delivers, an attempt is made to convert a transverse lie or breech presentation into a vertex presentation.

6. If this cannot be accomplished, the second twin is then delivered by complete breech extraction.

In a review of 50 vertex/nonvertex deiveries of twins more than 1500 g at Vanderbilt University, 55% were delivered vaginally versus 45% by cesarean. Outcomes of both first and second twins, as defined by Apgar scores at 1 and 5 min and NICU admissions, were not statistically different.

A dilemma in obstetrics is how to manage the second twin when the birthweight is less than 1500 g. In the Fisherman study,[14] no conclusion could be reached for twins under 1500 g. Adam et al.[15] concluded that there was increased perinatal morbidity for the second twin with a birthweight of less than 1500 g delivered vaginally, with morbidity being measured in terms of intraventricular hemorrhage (IVH), seizures, respiratory distress syndrome (RDS), and Apgar scores less than 7. In addition, Barrett et al.[16] retrospectively evaluated the relationship between type of delivery and perinatal morbidity and mortality in twins of birthweight less than 2000 g, and found that vaginally delivered second twins weighing between 601 and 999 g had an increased risk of neonatal mortality when compared to their siblings. The study also found that, among twins who weighed 1000–1499 g, vaginally delivered second twins had significantly lower Apgar scores and increased risks of neonatal morbidity in comparison to their siblings, whereas second twins who were delivered by cesarean section had no difference in Apgar scores or morbidity from those of their siblings. The authors concluded that cesarean section was the optimal route of delivery of all twins expected to have a birth weight of less than 1500 g.[16]

Rydstrom and Ingemarrson,[17] however, in 1991 failed to reveal any significant impact of abdominal birth on fetal outcome for low birthweight twins, even when fetal presentation was taken into consideration. Greig et al.[4] in 1992 also failed to find support for routine cesarean section of the nonvertex twin, regardless of birthweight.

In 1985, Chervenak et al.[9] outlined a protocol for the management of twin deliveries. During the intrapartum period, fetal weight is assessed by ultrasound. If the estimated fetal weight is less than 2000 g, external version after delivery of the first twin is attempted if the second twin is nonvertex; if this is unsuccessful, cesarean section is performed. If the second twin weighs more than 2000 g, external version is attempted; however, if unsuccessful, a vaginal breech delivery is attempted.

While there are few data to support cesarean section for the nonvertex second twin weighing less than 1500 g, it is believed by many perinatologists that because of possible entrapment of the aftercoming head, and a possible increased incidence of IVH and traumatic delivery, cesarean section should be performed.[18,19] It should be emphasized that if cesarean section is performed, an adequate uterine incision, preferably a low vertical incision that can be extended into a classical incision, is advised to avoid birth trauma.[9]

Management of the Nonvertex-Presenting Twin

The incidence of interlocking twins is approximately 1 in every 1000 twin deliveries, with a fetal mortality rate of 31%.[20] This condition occurs exclusively in breech-vertex presentations when the chins directly overlie each other. Some believe that in carefully selected cases and in circumstances in which the first twin is nonvertex, vaginal delivery may be safely undertaken. Many authors agree, however, that cesarean section is recommended if the first twin is nonvertex, because there are no studies documenting the safety of vaginal delivery in this group.[21]

Management of Monoamniotic Twin Gestations

It has been assumed in the foregoing discussion that the pregnancies being managed were diamniotic. Monoamniotic twinning is an uncommon event widely recognized to have increased antenatal and perinatal morbidity and mortality. Double survival is rare.[22]

While most authors are of the opinion that these pregnancies should undergo cesarean delivery to prevent cord entanglement and the risk of fetal interlocking,[22,23] Tessen and Zlatnick[24] reported no fetal deaths and only 1 case of fetal distress requiring emergent cesarean section in 20 cases of labor and delivery of monoamniotic twins. Prophylactic preterm delivery has been advocated by 32 weeks to prevent cord-related deaths;[22] however, recent data reveal little support for early intervention.[24]

Triplets and Other Multiple Gestations

Naturally occurring multiple births of more than two infants are rare;[25] however, the discovery of ovulation induction agents has significantly increased their incidence. Virtually no guidelines exist in the United States for optimal intrapartum care. Because preterm labor in these patients is a frequent complication, cesarean section is undertaken based on the gestational age of the infants.[26] Loucopoulos et al.[27] in 1982 reviewed their experience with multifetal pregnancies. Cesarean section was performed in 42% while the vaginal route was the mode of delivery in 58%. Most infants were delivered by breech extraction (32%) or spontaneously (27%). Cesarean section improved only the 1-min Apgar for the third neonate. Ron-El et al.[28] noted that fetal outcome was similar in the vaginal and cesarean section deliveries among each gestational age group in their review of triplet and quadruplet pregnancies.

More recently, Creinin et al.,[3] after reviewing the outcome of 13 sets of triplets in a retrospective analysis, recommend cesarean section as the optimal mode of delivery. Because of potential problems with malpresentation, changes in presentation that can occur in active labor and delivery, the increased risk of cord prolapse, and complications with fetal monitoring, most authors recommend cesarean section. Indeed, since 1981 the trend has been toward cesarean section. However, some authors still advocate vaginal delivery as an option. Itzkowic[29] and Deale and Cronje[30] observed that a gestational age of more than 34 weeks and an estimated fetal weight of 2000 g were favorable for vaginal delivery.

As there are no randomized trials for cesarean section versus vaginal delivery in triplet and high-order multiple deliveries, it is the author's opinion that these pregnancies should be delivered by cesarean section, except for special circumstances. If vaginal delivery is undertaken, the obstetrician should be skilled in version maneuvers, and the capacity to proceed with immediate cesarean section should be available.

References

1. Holberg G, Biale Y, Lewenthal H, Insler V. Outcome of pregnancy in 31 triplet gestations. Obstet Gynecol 1982;59:472.
2. Bell D, Johansson D, McLean FH, Usher RH. Birth asphyxia, trauma and mortality in twins: has cesarean improved outcome? Am J Obstet Gynecol 1986;167:901.
3. Creinin M, Katz M, Laros R. Triplet pregnancy: changes in morbidity and mortality. J Perinatol 1991;11:207.
4. Greig P, Veille J-C, Morgan T, Henderson L. The effect of presentation and mode of delivery on neonatal outcome in the second twin. Am J Obstet 1992;167:901.
5. Marivate M, Norman RJ. Twins. Clin Obstet Gynaecol 1982;9:783.
6. Kohl SG, Casey G. Twin gestation. Mt Sinai J Med 1975;42:523.
7. Chervenak FA. The controversy of mode of delivery in twins: the intrapartum management of twin gestation (Part II). Semin Perinatol 1986;10:44.
8. Cetrulo CL. The controversy of mode of delivery in twins: The intrapartum management of twin gestation (Part I). Semin Perinatol 1986;10:39.
9. Chervenak F, Johnson R, Voucha S, Hobbins J, Berkowitz R. Intrapartum management of twin gestation. Obstet Gynecol 1985;65:119.
10. Rabinovici J, Barkai G, Reichman B, Serr DM, Mashiach S. Randomized management of the second nonvertex twin: vaginal delivery or cesarean section. Am J Obstet Gynecol 1987;156:52.
11. Tchabo JG, Tomai T. Selected intrapartum external cephalic version of the second twin. Obstet Gynecol 1992;79:421.
12. Trofalter KF. Management of delivery. Clin Perinatol 1988;15:93.
13. Wells S, Thorp J, Bowes W. Management of the second nonvertex twin. Surgery 1991:172.
14. Fisherman A, Grubb D, Kovacs B. Vaginal delivery of the nonvertex second twin. Am J Obstet Gynecol 1993;168:861.
15. Adam C, Allen AC, Baskett TF. Twin delivery: influence of presentation and method of delivery on the second twin. Am J Obstet Gynecol 1991;165:23.
16. Barrett J, Staggs SM, Van Houydonk JE, Growden JH, Killam AP, Boehm F. The effect of type of delivery upon neonatal outcome in premature twins. Am J Obstet Gynecol 1982;143:360.
17. Rydstrom H, Ingemarrson I. A case control study of the effects of birth by cesarean section on intrapartum and neonatal mortality among twins weighing 1500–2400 gm. Br J Obstet Gynaecol 1991;98:249.
18. Cetrulo CL, Ingardia CL, Sbarra AJ. Management of multiple gestation. Clin Obstet Gynecol 1990;23:533.

19. Kelsick F, Minkoff H. Management of the breech second twin. Am J Obstet Gynecol 1982;144:873.
20. Khanda S. Locked twins. Obstet Gynecol 1972;39:453.
21. Gabbe S, Neibyl J, Simpson J. Obstetrics, normal and abnormal pregnancies, 2nd Ed. New York: Churchill Livington, 1991.
22. Rodis JF, Vintzileos AM, Campbell DJ. Antenatal diagnosis and management of monoamniotic twins. Am J Obstet Gynecol 1987;157:1255–1257.
23. Carr SR, Aronson MP, Coustan DR. Survival rates of monoamniotic twins do not increase after 30 weeks' gestation. Am J Obstet Gynecol 1990;163:719–722.
24. Tessen J, Zlatnick F. Monoamniotic twins: a retrospective controlled study. Obstet Gynecol 1991;77:832.
25. James JN, Berbos BF, King A. Quintuplet pregnancy. JAMA 1964;188:113.
26. Caspi E, Ronen J, Schreyer P. The outcome of pregnancy after gonadotropin therapy. Br J Obstet Gynaecol 1976;83:967.
27. Loucopoulos A, Jewelewic Z. Management of multifetal pregnancies: sixteen years' experience at the Sloan Hospital for Women. Am J Obstet Gynecol 1982;143:902.
28. Ron-El R, Eliahu C, Schreyer P, Weinraub Z, Arleli S, Goldberg M. Triplet and quadruplet pregnancies and management. Obstet Gynecol 1981;57:458.
29. Itzkowic D. A survey of 59 triplet pregnancies. Br J Obstet Gynaecol 1979;86: 23.
30. Deale CJC, Cronje HS. A review of 367 triplet pregnancies. S Afr Med J 1984;66: 92.

10
Genital Herpes: Contemporary Management

ZANE A. BROWN

To discuss the consequences of genital herpes during pregnancy, particularly as it relates to the route of delivery, it is necessary to review the classification of this infection.

Infectious Stages

Primary First Episode

A primary first episode is the initial genital herpes virus infection in an individual who lacks antibodies to both HSV-1 and HSV-2.

Nonprimary First Episode

A nonprimary first episode is the initial genital herpes virus infection in an individual who has antibodies to the heterologous strain of the virus. The most common example is an individual who has antibodies to HSV-1 as a result of orolabial herpes earlier in life and is infected with HSV-2 for the first time. Less commonly, an individual may have preexisting HSV-2 antibodies and is infected with HSV-1 (either orally or genitally). The designation nonprimary first episode is justified by data suggesting that heterologous antibody not only reduces the risk of acquiring a herpes infection, but modifies the severity of the initial episode.[1–3]

Recurrent or Reactivation Infection

Patients in whom the HSV viral type that is recovered from the lesion or genital tract is the same as the serotype of the HSV antibodies present in their sera are said to have recurrent or reactivation infection.

Asymptomatic Shedding

Asymptomatic shedding is defined as the presence of infectious virus in the genital tract in the absence of either symptoms perceived by the patient or lesions observed by a trained observer.

It is important to point out that recent studies have shown that as many as 85% of first episodes, whether primary or nonprimary, are asymptomatic or at least have symptoms that are unrecognized and therefore unreported by the patient.[4–6] The remaining 15% show bilateral genital lesions, moderate to severe local pain, dysuria, sacral paresthesia, tender regional lymph node enlargement, central nervous system (CNS) symptoms, and the systemic signs and symptoms of a viremia such as fever and malaise. Similarly, approximately 75% of HSV-2 antibody-positive individuals do not report genital lesions or a history compatible with recurrent genital herpes.[7] However, with education, more than 50% of these "asymptomatic" seropositive individuals are able to identify and report genital lesions.[8] The frequency, duration, and severity of symptomatic recurrences increases as pregnancy progresses.[6] Recurrences when symptomatic are usually much milder than either primary or nonprimary first episodes. The lesions last several days instead of weeks, are usually unilateral, and are infrequently associated with systemic signs and symptoms.

A first-ever recurrence occurring during pregnancy (in an individual with longstanding HSV-2 seropositivity), however, can have a spectrum of involvement ranging from very mild to a severe illness suggesting first-episode disease. Because of its broad spectrum of presentation, it is unreliable, particularly in pregnancy, to classify genital herpes infection using clinical presentation alone. In a recent study of 29 women presenting in pregnancy with a first clinical episode of genital herpes, 50% had bilateral lesions, 40% had bilateral inguinal lymphadenopathy and reported malaise, 15% reported a headache, and 10% reported myalgias.[9] However, after serotyping them with a type-specific HSV serologic assay, only 4 of the 29 (14%) were true first episodes (2 primary and 2 nonprimary). The remainder were first-ever recurrences among seropositive individuals. In addition, a detailed survey of recurrent genital symptoms is as inadequate and does not improve upon the simple question "Have you ever had genital herpes?" in eliciting a history of genital herpes in HSV-2 seropositive patients.[10] Therefore, the appropriate classification of genital herpes depends on the accurate serotyping of the patient using type-specific serologic assays.

Detection of Antibodies

Type-specific HSV serologic assays have been developed that detect antibodies directed against viral protein epitopes which are specific for each viral type.[11] This avoids the extensive cross-reactivity between HSV serotypes that renders the commonly used enzyme immunoassay and indirect immunofluorescence assay unreliable and therefore unusable in distinguishing HSV-1 from HSV-2 antibodies, even though most proprietary laboratories report the results as "titers" to HSV-1 and HSV-2. Assays such as the Western blot, which use HSV type-specific antigens, are 98% sensitive and 100% specific for detecting the presence of HSV-2 antibodies even in the presence of HSV-1 antibodies, and would enable the practitioner to detect HSV-2 seroconversion during pregnancy and identify the HSV-2 seropositive but asymptomatic or unaware patient. Unfortunately, these type-specific assays are currently available only through regional research facilities. With increased demand, it is hoped that these assays will be offered by commercial laboratories.

In a recent study at the University of Washington in which cultures of the labia and cervix were taken from almost 24,000 women at the onset of labor, the rate of asymptomatic shedding was 0.4% or about 1.2% of the HSV-2-seropositive population.[12] When the genital secretions of a subset of these HSV culture-negative, asymptomatic HSV-2-seropositive women were examined by a polymerase chain reaction (PCR), 20% showed evidence of HSV-2 DNA.[13] None of the newborns were infected with the HSV. The significance of this finding is at present unclear. It may represent noninfectious viral fragments or it may represent infectious virus in a titer that is below the level of sensitivity of viral isolation techniques. Whatever the explanation for this observation, it is apparent that asymptomatic shedding of the HSV at the onset of labor probably occurs much more frequently than has been appreciated in the past using viral isolation techniques. It also implies that neonatal transmission of HSV-2 from a seropositive mother during labor and delivery is probably an exceedingly rare event.

Therefore, HSV is more likely to be present in the genital tract at the onset of labor as a result of asymptomatic rather than symptomatic disease (whether first episode or recurrent). It is a result of this asymptomatic shedding of virus that almost all neonatal infections occur.[12,14–17] It is difficult to understand the rationale for the cesarean delivery of women with symptomatic recurrent genital herpes at the onset of labor when most neonatal infections occur as a result of maternal asymptomatic shedding.

Neonatal Herpes

Genital herpes is the most common ulcerative sexually transmitted disease in North America.[18,19] The major concerns related to genital herpes infection during pregnancy are HSV transmission to the infant and the cesarean section rate among women who are beginning pregnancy with symptomatic recurrent genital herpes. The incidence of neonatal herpes appears to increase at least in some populations probably as a result of parallel increases of the infection in the adult population.[20,21] In Seattle/King County, Washington between 1966 and 1985, the incidence of neonatal herpes rose from 3.2 to approximately 15 cases per 100,000 live births.[14] This occurred in spite of a cesarean section rate over the same period of time approaching 70% for women entering pregnancy with symptomatic recurrent genital HSV-2.[22] Since 1985, the rates have decreased to about 11 cases per 100,000 live births and have remained at that level to the present. Rates of neonatal herpes, however, vary widely throughout the United Sates and Europe. For example, the current rate of neonatal HSV at the University of Washington is approximately 1 in 1800 live births, a rate that is higher than that found in most other areas of the country. Nationally, the reported rates of neonatal herpes range from 1 case per 2000–10,000 live births.[12,21,23,24] Although the mortality and morbidity of neonatal herpes has decreased with the advent of antiviral chemotherapy, more than 40% of neonates with the infection still die or are impaired.[25]

While neonatal HSV is a relatively infrequent occurrence, its impact on the delivery of obstetric care is extensive. Concern for exposing the newborn to infected genital secretions during labor and vaginal delivery has led to very high rates of cesarean section with its attendant increase in costs, morbidity, and mortality. In a recent study in the three most populous counties in Washington State, the obstetric records of all pregnant women identified from birth certificates as having genital herpes were reviewed and matched to pregnant women without genital herpes delivering at the same time and in the same hospital as the index case.[22] Women with recurrent symptomatic genital herpes (n = 909) had five times as many cesarean sections as women without genital herpes (75% versus 15%) (Table 10.1). Of the women with a history of genital herpes, 80% were without lesions at delivery and more than 50% were without lesions within 7 days of delivery; only 1.3% were culture positive at the time of delivery. Thus, most did not have active genital herpes at the onset of labor but were still delivered by cesarean section. Associated with the increased cesarean section rate was a 2.7-fold increase in postpartum endometritis. We can make a rough and conservative estimate of the impact of this excess cesarean

TABLE 10.1. Route of delivery among women with and without genital herpes[a]

	First episode (n = 71)	History of recurrence (n = 650)	Lesion at less than 1 week from delivery (n = 290)	No Hx HSV (n = 917)
Primary cesarean section				
HSV indication	46.5	45.4	78.6	0
Other indication	14.1	13.1	9.7	10.3
Repeat cesarian section	1.4	4.0	4.8	7.6
Vaginal delivery	38.0	37.5	6.9	82.1

[a] From Wolf.[22] Data are percentages.

section rate nationally by making several assumptions. If we assume a delivery rate in the United States of approximately 3 million births per year and an incidence of symptomatic genital herpes of 5% in the reproductive-aged population, then 150,000 women will deliver annually with a history of symptomatic recurrent genital herpes. If we further assume a baseline cesarean section rate of 20%, then the excess cesarean section rate of 50% attributable to genital herpes will result in about 75,000 cesarean sections for the indication of genital herpes. If we assume that a cesarean section costs approximately $3000 more than a vaginal delivery, then the national costs for these excess cesarean sections resulting from recurrent genital herpes may approximate $225 million annually. To this must be added the indirect costs, such as the increased recovery time and increased time lost from work, the cost of repeat cesarean sections, and the myriad of social consequences to the immediate family and the community. Thus, current obstetric practice has not only failed to reduce the incidence of neonatal herpes, but has significantly increased the rate of cesarean section and complications among women entering pregnancy with a history of recurrent genital herpes.[26]

Until recently, strategies to prevent maternal fetal transmission centered about pregnant women with recurrent HSV-2 with the assumption that it was this population of women who by reactivating their genital herpes near parturition were at greatest risk of transmitting the infection to their infants. It was thought that women whose disease was reactivated in the last several weeks before the onset of labor were at risk of having infectious virus persist in their genital tracts at the onset of labor and were therefore delivered by cesarean section. This thinking led to the practice of weekly HSV antepartum cultures from 34 weeks to term, and was in large part responsible for the high rates of cesarean section in women with symptomatic recurrent genital herpes. A recent study has demonstrated that antepartum asymptomatic shedding of the HSV as determined by weekly HSV cultures does not predict whether the HSV will be present in the genital tract at the onset of labor.[27] In addition, the duration of asymptomatic shedding is brief, seldom more than 3 days.[6]

Recent studies have demonstrated that most of the pregnancy morbidity and the majority of cases of the neonatal herpes are not the result of reactivation of genital herpes in women at the onset of labor, but result from women acquiring genital herpes, frequently asymptomatically, in late pregnancy.[12,28,29] This is consistent with the observations that women with first-episode disease are devoid of type-specific homologous antibody with which to transplacentally protect their offspring and that they excrete HSV from the cervix in higher titers and for longer periods of time than do women with recurrent genital herpes. In a recent study performed at the University of Washington,[12] herpes cultures were obtained from the cervix and labia of 24,887 asymptomatic women within 48 h of delivery (Figure 10.1). Of these, 97 patients (0.4%) demonstrated asymptomatic shedding as a result of either first-episode ($n = 24$) or recurrent disease ($n = 73$). Of the 24 patients with first-episode disease, 6 had primary while 18 had nonprimary first-episode infections. Three of the 6 mothers (50%) with asymptomatic shedding as a result of primary disease and 5 of the 18 mothers (24%) with nonprimary first-episode disease transmitted the infection to their newborns. Therefore, 8 of 24 mothers (33%) with first-episode disease transmitted the infection to their infants, with two neonatal deaths. In contrast, of the 73 women with asymptomatic shedding secondary to recurrent disease, only 2 (3%) infected their infants. Of interest, of the 73 women with asymptomatic shedding secondary to recurrent genital herpes, 67 were HSV-2 and 6 were HSV-1. None of the 67 infants exposed to HSV-2 were infected whereas 2 of the 6 infants (30%) exposed to HSV-1 became infected. Therefore, it would appear that HSV-1 transmits significantly more readily to the newborn than HSV-2. However, recent studies have demonstrated

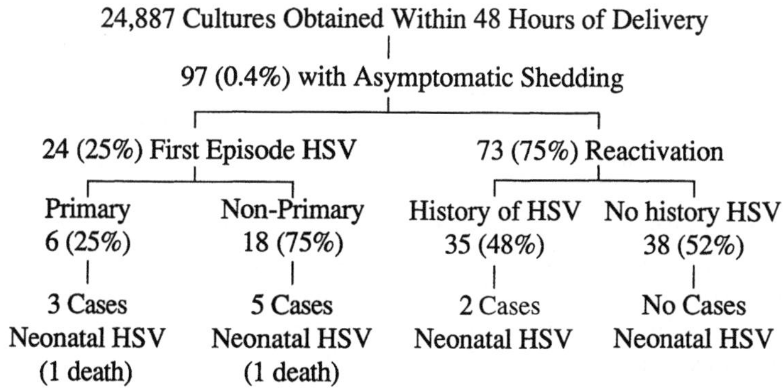

FIGURE 10.1. Asymptomatic shedding of genital HSV during labor. (From Brown et al.,[12] by permission of the *New England Journal of Medicine*.)

that when infection with HSV-1 occurs, the neurodevelopmental consequences are significantly less severe than those from infections caused by HSV-2.[25] One study[12] also demonstrated that the use of fetal scalp electrodes in laboring mothers with a history of recurrent genital herpes was a risk factor for neonatal herpes.

In a companion study of 5742 patients, sera for HSV antibodies were obtained at the first prenatal visit and again at the onset of labor.[29] Among women who were HSV seronegative at the first prenatal visit, the rate of HSV-2 seroconversion was 2.5% during the mean observation period of 28 weeks. When the mean interval of observation was adjusted for the entire pregnancy, the seroconversion rate was 3.7%. In a small subset of these patients in which partner sera were also obtained as part of a pilot study, the rate of seroconversion for HSV seronegative women with HSV-2 seropositive partners was 33%.[30] In addition to the significantly increased risk of neonatal herpes, as has been described, women who HSV-2 seroconvert during pregnancy (whether symptomatically or asmptomatically) have an increased risk of late fetal loss, premature labor, and intrauterine growth retardation.[28]

It is important to point out that many cases of neonatal herpes cannot be prevented, even with the best care.[31] From the foregoing, it is evident that infants acquire their infection by contacting infectious virus in the genital secretions of asymptomatic mothers at the onset of labor; most commonly as a result of maternal first-episode disease acquired in the latter half of pregnancy. Even if every patient entering pregnancy with symptomatic recurrent genital herpes was delivered by cesarean section, only a very small number of the cases of neonatal herpes would be prevented. It would not prevent those cases caused by asymptomatic reactivation at the onset of labor among HSV-2 seropositive women without knowledge of their disease or maternal first-episode disease in late pregnancy with asymptomatic shedding at the onset of labor. Cesarean section may even decrease the transplacental transmission of protective anti-HSV antibodies.[32] In addition, approximately 20%–30% of newborns who develop neonatal HSV are delivered by cesarean section; of these, about one-fourth had intact fetal membranes.[21] Furthermore, about 5% of infants with neonatal HSV are born with the disease, suggesting that the infection was acquired in utero.[33]

Possible Strategies to Reduce the Rate of Excess Cesarean Sections for Genital Herpes

First, the most important strategy of reducing cesarean delivery would be to restrict the use of cesarean section to women with lesions or symptoms of a recurrence during labor. Currently, most of the cesarean sections performed for genital herpes are for recurrences "close to" but not at the onset of labor. Cesarean delivery should be discouraged as a prophylactic measure in the absence of lesions or symptoms. HSV cultures obtained during the antepartum period have not been shown to be predictive of the status of the cervix at the onset of labor, and HSV cultures obtained from the maternal genitalia on admission in early labor have a poor sensitivity for predicting infants who will subsequently develop neonatal herpes.[12] Therefore, HSV cultures either during pregnancy or at the onset of labor are of little clinical use and are labor intensive to obtain as well as expensive and uncomfortable for the patient. Patients with a history of recurrent genital herpes admitted in early labor should have their external genitalia thoroughly examined with a bright light on an examination table. All too often, a pelvic exam, if it takes place, occurs on a labor bed in the muted light of a modern labor room. In addition, patients and partners should be questioned about the presence of symptoms suggesting recurrent genital herpes even if there is no history of genital herpes.

Second, as a result of studies in progress, acyclovir may be shown to suppress symptomatic reactivation of genital herpes at term and therefore obviate the need for cesarean section.

Acyclovir is a nucleoside analog whose selective activity inhibits the replication of HSV at concentrations as much as 3000 fold less than those that inhibit mammalian cellular functions. Because of its remarkable specificity for cells already infected with HSV, it has proven remarkably safe in mammalian fetal test systems.[34] Several large retrospective series have been unable to demonstrate any evidence of fetal teratogenicity or toxicity.[35,36] Even though acyclovir is cleared by the kidney, recent pharmacokinetic studies demonstrated that the steady-state acyclovir levels of pregnant women for doses of 200 mg or 400 mg every 8 h are comparable to those expected for nonpregnant woment.[37] Surprisingly, the increased volume of distribution of pregnant women and the increased renal blood flow in late pregnancy do not alter the steady-state pharmacokinetics as would have been expected. Acyclovir is highly concentrated in the amniotic fluid and milk[38] but does not accumulate in the fetus.[38] The mean maternal: infant plasma ratio at delivery is 1.3.

A number of studies have demonstrated the efficacy of acyclovir in suppressing symptomatic recurrences in nonimmunosuppressed, nonpregnant adults.[39] However, asymptomatic shedding of the virus may continue in spite of clinically effective suppression of symptomatic recurrences.[40] In pregnancy, preliminary data from several prospective trials currently in progress suggest that acyclovir is effective in suppressing symptomatic recurrences.[41–42] As in the nonpregnant woman, asymptomatic shedding may continue with a risk of neonatal infection.[40] Until the prospective studies of prophylactic acyclovir in late pregnancy that are now in progress have been completed and evaluated for safety and efficacy, routine prophylactic use of acyclovir in late pregnancy cannot be recommended.

Third, a rapid test for HSV antigen could be used at the onset of labor to identify those patients with HSV in their genital tracts. Theoretically, this would permit a more selective and specific use of cesarean section to prevent contact between genital secretions containing infectious virus and the fetus during parturition.

Such a rapid test is not currently available and will not become available in the foreseeable future for several reasons. It would have to be available within several hours and have a specificity of greater than 99.5% for the presence of infectious virus to avoid a substantial increase in unnecessary cesarean sections. This type of specificity may be impossible to achieve, because antigen detection systems will detect some noninfectious cases in which viral fragments or noninfectious virus are present in the specimen.[31] The "rapid" tests currently available use monoclonal antibodies that bind to HSV proteins. They were designed to be used with material from lesions, and lack the sensitivity to separate low levels of virus from background activity.

Summary

The major consequence of entering pregnancy with a diagnosis of recurrent genital herpes is the increased risk of cesarean section. Most cases of neonatal herpes are caused by first episodes of genital herpes acquired late in pregnancy. The most promising means of reducing the extraordinary rates of cesarean section for genital herpes remains a careful history and examination on admission to the labor unit, with cesarean section being reserved for only those individuals who have lesions or symptoms suggestive of a recurrence. For women with frequent recurrences, the use of prophylactic acyclovir in late pregnancy is promising. However, safety and efficacy have not as yet been proven. Most importantly, given the technology and resources available in clinical practice today, most cases of neonatal herpes cannot be prevented.

References

1. Mertz GJ, Benedetti J, Ashley R, Selke S, Corey L. Risk factors for the sexual transmission of genital herpes. Ann Intern Med 1992;116:197–202.
2. Bryson Y, Dillon M, Bernstein DI, Randolf J, Zakowski R, Garratty E. Risk of acquisition of genital herpes simplex virus type 2 in sex partners of persons with genital herpes: a prospective couple study. J Infect Dis 1993; 167:942–946.
3. Corey L, Holmes KK. Genital herpes simplex virus infections: current concepts in diagnosis, therapy and prevention. Ann Intern Med 1983;98:973–983.
4. Mertz GJ, Schmidt O, Jourden JL, et al. Frequency of acquisition of first-episode genital infection with herpes simplex virus from symptomatic and asymptomatic source contacts. Sex Transm Dis 1985;12:33–39.
5. Boucher FD, Yasukawa LL, Bronzan RN, Hensleigh PA, Arvin AM, Prober CG. A prospective evaluation of primary genital herpes simplex virus type 2 infections acquired during pregnancy. Pediatr Infect Dis J 1990; 9:499–504.
6. Brown ZA, Vontver LA, Benedetti J, et al. Genital herpes in pregnancy: risk factors associated with recurrences and asymptomatic viral shedding. Am J Obstet Gynecol 1985;153:24–30.
7. Koutsky LA, Stevens CE, Holmes KK, et al. Underdiagnosis of genital herpes by current clinical and viral-isolation procedures. N Engl J Med 1992;326:1533–1539.
8. Langenberg A, Benedetti J, Jenkins J, Ashley R, Winter C, Corey L. Development of clinically recognizable genital lesions among women previously identified as having "asymptomatic" herpes simplex virus type 2 infection. Ann Intern Med 1989;110:882–887.
9. Hensleigh P, Andrews W, Brown Z, et al. Medical history and clinical presentation does not identify primary genital herpes infections in pregnant women. Presented at the 15th Annual Meeting, Society of Perinatal Obstetricians. January 23–28, 1995. Atlanta Hilton, Atlanta, Georgia.
10. Brown ZA, Benedetti J, Watts DH, et al. A comparison between detailed and simple histories in the diagnosis of genital herpes complicating pregnancy. Am J Obstet Gynecol 1994 (in press).
11. Ashley RA, Militonia J, Lee F, Nahmias A, Corey L. Comparison of Western blot (immunoblot) and glycoprotein G-specific immunodot enzyme assay for detecting antibodies to herpes simplex virus types 1 and 2 in human sera. J Clin Microbiol 1988;26: 662–667.
12. Brown ZA, Benedetti J, Ashley R, Burchett S, et al. Neonatal herpes simplex virus infection in relation to asymptomatic maternal infection at the time of labor. N Engl J Med 1991;324:1247–1252.
13. Cone RW, Hobson AC, Brown ZA, et al. Frequent detection of genital herpes simplex virus DNA by polymerase chain reaction among pregnant women. JAMA 1994;272: 797–796.
14. Sullivan-Bolyai J, Hull HF, Wilson C, Corey L. Neonatal herpes simplex virus infection

in King County, Washington: increasing incidence and epidemiologic correlates. JAMA 1983;250:3059–3062.
15. Prober CG, Hensleigh PA, Boucher FD, Yasukawa LL, Au DS, Arvin AM. Use of routine viral cultures at delivery to identify neonates exposed to herpes simplex virus. N Engl J Med 1988;318:887–891.
16. Whitley RJ, Corey L, Arvin A, et al. Changing presentation of herpes simplex virus infection in neonates. J Infect Dis 1988;158:109–116.
17. Arvin AM, Yeager AS, Bruhn FW, Grossman M. Neonatal herpes simplex infection in the absence of mucocutaneous lesions. J Pediatr 1982;100:715–721.
18. Genital herpes infection—United States, 1966–1984. Morbid Mortal Wkly Rep 1986; 35:402–403.
19. Quinn TC, Cates W. Epidemiology of sexually transmitted diseases in the 1990s. In: Quinn TC, ed. Sexually transmitted diseases. Raven, New York, 1992:1–37.
20. Nahmias AJ, Lee FK, Beckman-Nahmias S. Sero-epidemiological and -sociological patterns of herpes simplex virus infection in the world. Scand J Infect Dis [suppl] 1990;69: 19–36.
21. Stone KM, Brooks CA, Guinan ME, Alexander ER. National surveillance for neonatal herpes simplex virus infections. Sex Transm Dis 1989;16:152–156.
22. Wolf M. Unpublished thesis. Obstetrical course and complications of women with symptomatic genital herpes: a population-based analysis, 1990.
23. Garland SM. Neonatal herpes simplex: Royal Women's Hospital 10 year experience with management guidelines for herpes in pregnancy. Aust NZ J Obstet Gynecol 1992;32: 331–334.
24. Chuang T. Neonatal herpes incidence, prevention and consequences. Am J Prev Med 1988;4:47–53.
25. Whitley R, Arvin A, Prober C, et al. Predictors of morbidity and mortality in neonates with herpes simplex virus infections. N Engl J Med 1991;324:450–454.
26. Randolph AG, Washington AE, Prober CG. Cesarean delivery for women presenting with genital herpes lesions. JAMA 1993;270: 77–82.
27. Arvin AM, Hensleigh PA, Prober CG, et al. Failure of antepartum maternal cultures to predict the infant's risk of exposure to herpes simplex virus at delivery. N Engl J Med 1991;324:1247–1252.
28. Brown ZA, Vontver LA, Benedetti J, et al. Effects on infants of a first episode of genital herpes during pregnancy. N Engl J Med 1987;317:1246–1251.
29. Brown Z, Selke S, Zeh J, et al. HSV seroconversion during pregnancy and its impact on pregnancy outcome and infant development. In: Program and abstracts of the 32nd Interscience Conference on Antimicrobial Agents and Chemotherapy, October 11–14, 1992, Anaheim, California.
30. Brown ZA, Annholm A, Ashley R, et al. HSV serological discordancy among sexual partners and rates of seroconversion during pregnancy. In: Program and abstracts of the Infectious Disease Society for Obstetrics and Gynecology, August 1993, Stowe, Vermont.
31. Prober CG, Corey L, Brown ZA, et al. The management of pregnancies complicated by genital infections with herpes simplex virus. Clin Infect Dis 1992;15:1031–1038.
32. Bujko M, Sulovic V, Sbutega-Milosevic G, Zwanovic V. Mode of delivery and level of passive immunity against herpes simplex virus. Clin Exp Obstet Gynecol 1989;16: 6–8.
33. Nahmias AJ, Keyserling HH, Kerrick GM. Herpes simplex. In: Remington JS, Klein JO, eds. Infectious diseases of the fetus and newborn infant, 2nd Ed. Philadelphia: WB Saunders, 1983:636–678.
34. Gnann JW, Barton NH, Whitley RJ. Acyclovir: mechanism of action, pharmacokinetics, safety and clinical applications. Pharmacotherapy 1983;3:275–283.
35. Kingsley S. Fetal and neonatal exposure to acyclovir. Paris, 1986: unpublished data.
36. Andrews EB, Yankaskas BC, Cordero JF, Schoeffler K, Hampp S, and the Acyclovir in Pregnancy Registry Advisory Committee. Acyclovir in pregnancy registry: six years experience. Obstet Gynecol 1992;79:7–13.
37. Frenkel LM, Brown ZA, Bryson YJ, et al. Pharmacokinetics of acyclovir in the term human pregnancy and neonate. Am J Obstet Gynecol 1991;164:569–576.
38. Lau RJ, Emery MG, Galinsky RE. Unexpected accumulation of acyclovir in breast milk. Obstet Gynecol 1987;69:468–471.
39. Kaplowitz LG, Baker D, Gelb L, et al. Prolonged continuous acyclovir treatment of normal adults with frequently recurring

genital herpes simplex virus infection. JAMA 1991;265:747–751.
40. Strauss SE, Seidlin M, Takiff HE, et al. Effect of oral acyclovir treatment on symptomatic and asymptomatic virus shedding in recurrent genital herpes. Sex Transm Dis 1989; 16:107–113.
41. Stray-Pedersen B. Acyclovir in late pregnancy to prevent neonatal herpes simplex. Lancet 1990;336:756.
42. Scott L, Jackson G, Sanchez P, Castaneda Y, Hall M, Wendel G. Prevention of cesarean section for recurrent genital herpes simplex virus (HSV) using acyclovir suppressive therapy. Presented at the Society of Gynecologic Investigation, Abstract #S223, 1993, Orlando, Florida.
43. Haddad J, Langer B, Astruc D, Messer J, Lokiec F. Oral acyclovir and recurrent genital herpes in late pregnancy. Obstet Gynecol 1993;82:102–104.

11
Methods for Safe Reduction of Cesarean Section Rates

NORBERT GLEICHER, RICHARD H. DEMIR, JEANNE B. NOVAS, and STEPHEN A. MYERS

This chapter is presented under the assumption that cesarean section rates in the United States should be reduced. An understanding of the reasons that lead obstetricians to perform cesarean sections should provide the necessary information needed to effect such a change in practice patterns. Such a recommendation was already made by a National Consensus Development Conference in 1980, when the national rate was still below 17%.[1] Since then, it has further increased at an annual rate of approximately 1% until a plateau of approximately 24% was reached in the early 1990s (Figure 11.1).[2,3] Some physicians have nevertheless challenged the effort toward a reduction in cesarean section rates,[4,5] and a level of general hesitance within the profession seems to remain. Efforts by a number of authoritative bodies[6–8] seem to have had no significant effect on reducing the rate. This chapter thus needs to be seen against a background of at least subliminal resistance toward efforts directed at the reduction in cesarean section rates. Any such reduction must therefore first and foremost consider the causes of this resistance.

Cesarean sections, like most other invasive procedures, are principally intended to improve medical outcome. Because pregnancy outcome relates to both mother and fetus, cesarean sections can improve either maternal or fetal outcome or both. It was the belief that cesarean section could beneficially affect perinatal outcome that led to an initial increase in popularity of this surgery in the 1970s. A very dramatic improvement in the safety of the procedure, as a consequence of better anesthetic management, blood banking, and antimicrobial therapies of pregnant women, reduced many of the most significant risks, lowering the risk-benefit ratio in favor of surgical deliveries. When cesarean deliveries were reported to result in improvement in perinatal outcome, the national cesarean section rate was, however, still in the low single digits. In 1970 only 5.5% of deliveries were by cesarean section. By 1987, this rate had increased to 24.4% (see Figure 11.1).[2] At an utilization level of 5.5%, cesarean delivery, in selected high-risk situations, could be expected to improve perinatal outcome. The error consequently perpetrated was the assumption that a continuous benefit on perinatal outcome could be achieved if more and more indications for cesarean section deliveries were added. In the later 1970s it almost became a dictum in U.S. teaching departments that cesarean section rates had to approach 20%. If that was not achieved, departmental quality assurance processes kicked in and often led to a further expansion in cesarean section indications.

It was during that decade of practice that a mind-set was put into place which principally suggested that one can never be wrong in performing a cesarean section. In contrast, the nonperformance of cesareans was constantly subject to quality assurance

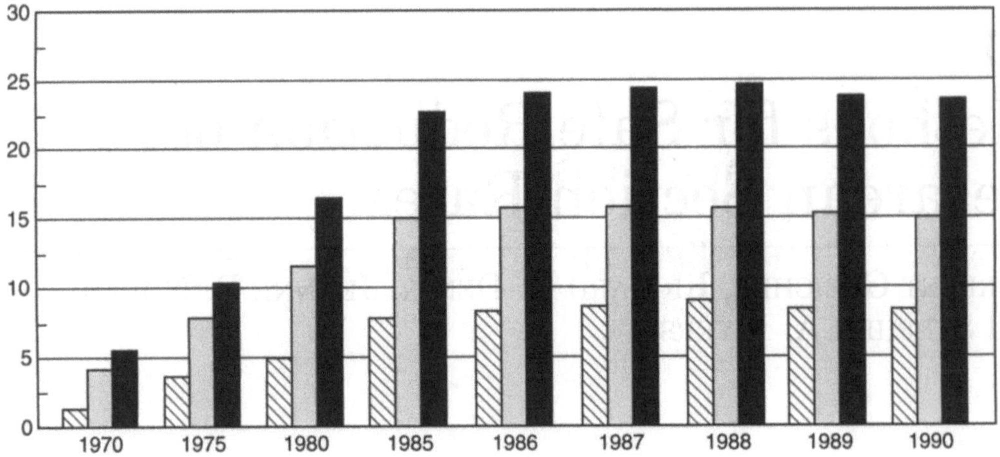

FIGURE 11.1. Cesarean section rates, 1970–1990, in the United States. *Hatched bars*, repeat; *shaded bars*, primary; *solid bars*, total.

reviews. The trend toward increased cesarean utilization was enhanced when a presumption was made by the legal community that the liberal performance of cesarean sections had become a new standard of care. The average practitioner was now virtually forced to abandon the traditional practice of obstetrics in favor of a cesarean section-driven approach. After all, "one never got sued for performing a cesarean." However, even this dictum has since been refuted because even a cesarean section cannot protect from legal exposure. An attitude of practicing defensive medicine has in increasing fashion perpetrated the practice of obstetrics during the past two decades. Surveys of obstetricians have persistently indicated that the profession considers the current medical-legal situation the principal contributing factor to the national cesarean section crisis.[1,9,10]

Attempts to document such an impact through scientific investigation have largely failed, however,[11–13] although one recent study supported a possible positive association between malpractice claims risk and the rate of cesarean delivery.[14] It is therefore questionable whether cesarean section rates would, in fact, significantly decrease if the profession's liability situation changed. One can nevertheless not overlook the psychologic impact of such a widely held belief on attempts to reduce the national cesarean section rate. It is simply not enough to tell the profession that all reasonable scientific evidence suggests that medical-legal considerations do not significantly impact on physician behavior in regards to cesarean sections. Instead, one has to provide the practicing physician with the necessary reassurances that the nonperformance of a nonindicated cesarean section will not increase their liability exposure.

This reassurance can be provided at the local level by establishing departmental practice guidelines, which, if properly followed, will represent an excellent defensive tool should the need for one arise. On the national level one would welcome a similar approach. Unfortunately, national organizations have been slow and rather timid in responding to the increase in national cesarean section rates. While the previously mentioned Consensus Development Conference of 1980 already considered the national cesarean section rates at that time to be excessive,[1] the American College did not publish a first policy change in regards to cesarean sections until 1984,[6] and more up-to-date vaginal birth after cesarean (VBAC) guidelines till 1988.[7] What organized medicine apparently fails to understand is the psychologic need of the profession to be actively supported in the attempt to reduce the national cesarean section rate. This can best be done by giving the practicing obstetrician clearly outlined recommendations for the performance of cesarean sections.

The cesarean section is the most frequently performed surgical procedure in the United States.[15] Like any other surgical intervention, it should be only performed to improve (fetal or maternal) outcome. In the age of meta-analyses, it appears relatively simple to define indications for cesarean sections.[16] Almost 15 years after the Consensus Development Conference,[1] the only existing cesarean section guidelines refer to vaginal delivery after previous cesarean (VBAC)[7] and even those are timid and do not reflect, for example, published experience with more than one previous cesarean section.[17–20] There are still no clear guidelines for breech delivery, cesarean section for dystocia/cephalopelvic disproportion, or even fetal distress. How can one then be surprised if many obstetricians choose the path of least resistance—cesarean delivery.

The shift toward increased cesarean utilization also resulted in yet another (probably unexpected) consequence of considerable significance. A whole generation of physicians have departed their training programs not only indoctrinated with the wrong dogmas but also untrained in alternative delivery methods. If the chairman of a major teaching department in 1992 still can pride himself of "never having allowed the vaginal delivery of a breech in his department," then one can easily calculate the number of graduating residents who have entered practice unable to deliver a vaginal breech, even if they wished to do so. The same is true for forceps deliveries and, probably, vacuum extractions.

A philosophy that sees a cesarean section delivery as the solution to basically every obstetric problem has thus also given us a generation of physicians untrained in alternative delivery methods. Cesarean sections then rather automatically became a self-fulfilling prophecy. Either the practitioner has no choice but to perform a cesarean (for lack of training in alternative approaches) or untrained individuals perform alternative procedures with (expected) bad outcome. They then only falsely reemphasize the alleged benefit of cesarean sections through their inexperience.

One cannot expect individuals who are untrained or inexperienced with operative vaginal or vaginal breech deliveries to embrace such delivery modalities without further training. However, not even one postgraduate course has addressed this lack of training over the last few years. One can contrast this to the plethora of postgraduate activities surrounding the concept of operative laparoscopy. A rather disturbing conclusion is then reached as to where the emphasis is placed within the profession in regards to the training of manual skills.

As a consequence, patients undergo cesarean sections because physicians lack the theoretical and practical skills to perform alternative procedures. Paradoxically, lack of such skills is generally accepted as an indication for a cesarean section delivery. This stands in contrast to other obstetric situations, where an apparent lack of skills usually results in a mandated referral to a perinatologist. In most institutions a vaginal breech at term or the need for a possible forceps delivery are, however, *not* considered indications for perinatal consultation. This should not surprise because there is considerable evidence that the establishment of perinatology as a subspecialty by itself contributed significantly to the mind set that has given us the current cesarean section rates.[21]

The national cesarean section crisis can then be seen as the consequence of a highly successful education system which, during a span of approximately 20 years, has indoctrinated a whole generation of obstetricians in the use (and misuse) of cesarean sections and, concomitantly, has created a generation of subspecialists who often reemphasize this practice pattern through their own behavior. Because even many subspecialists are not trained in alternative delivery modalities, a vicious cycle has been created that tends to inhibit attempts to affect the national cesarean section rate.

For example, newly published guidelines for the education of residents mandate exposure and participation in alternative delivery methods to cesarean sections.[22] This effort should be applauded. Mandated educational efforts, however, do not work without proper faculty support. This has been demonstrated when a mandate to teach breast disease was introduced into ob/gyn residencies.[23] The profession faces a similar failure in regards to vaginal breech and forceps deliveries unless a concentrated effort is made to enlist faculty support for such approaches. Lomas et al.[24] demonstrated convincingly the importance of opinion leaders in implementing cesarean section guidelines.

Insecurities about cesarean section utilization are not restricted to the United States. A recent British study, the consequence of rising cesarean section rates in an English teaching hospital, revealed a not entirely unexpected pattern. On retrospective analysis, 30% of all cesarean sections performed were found to be unnecessary. More interesting, however, was the considerable disagreement among auditors on when to perform a cesarean. The auditors disagreed with themselves in 25% of cases when shown identical information a second time.[25] One might expect similar findings in U.S. teaching departments.

Well-trained opinion leaders are therefore of crucial importance at the departmental and the national level. Only if a committed leadership actively supports the effort to reduce cesarean section rates will an impact be observed. What still remains to be determined is whether such an effort will be directed by physicians or by bureaucrats from government and insurance agencies. The upcoming revolution in U.S. health care cannot overlook the most frequently performed surgical procedure. Government organizations are already touting the cost savings from just stabilizing the national cesarean section rate. Because a reduction in rate can impact on national health care costs quite significantly,[3] considerable efforts can be expected from both government and the insurance industry in this area. We are well advised to prepare as a profession and try to be leaders rather than followers.

In the remainder of this chapter we attempt to demonstrate how cesarean section rates can be safely reduced at the local, departmental level.

Background

The assumption that the national cesarean section rate can be safely reduced stems from a number of observations. Perhaps the most important one is that despite a constant increase in the national cesarean section rate since the 1960s, maternal and child health status has *not* improved in parallel.[3] Cesarean section rates in other developed nations are dramatically lower than in the United States, with perinatal outcome statistics that frequently are superior to those in this country.[26] The U.S. rate is the third highest among 21 reporting countries, exceeded only by Brazil (where cesareans are often performed for cosmetic reasons) and Puerto Rico.[3] Within the United States, significant regional differences can be found in the incidence in cesarean section utilization, with the South being the clear leader (27.6% cesarean section rate), followed by the Northeast (22.6%), the Midwest (21.8%), and the West (19.8%).[3] Cesarean section rates are uniformly higher in older women, in proprietary hospitals (than nonprofit or government hospitals), in smaller hospitals (fewer than 300 beds), for patients with private insurance (than charity patients and women with public health support), and for private services (in comparison to teaching services, even within the same institutions).[3,27]

Because neither geographic variation nor most of the patient characteristics just noted also define an increased risk for adverse outcome, these variations in cesarean section rates cannot be explained on the basis of medical necessity. This is further supported by the observation that institutions with a disproportionate number of high-risk patients, so-called perinatal centers (level III institutions), usually demonstrate lower cesarean section rates than providers of lower levels of care.[28] All these observations suggest that at least some cesarean section rates throughout the country can be reduced without adverse impact on either mother or child.

Further evidence in support of this contention comes from a careful review of published obstetric experience. Table 11.1 demonstrates that approximately two-thirds of cesarean sections are presently performed as either repeat cesareans or because of a diagnosis of dystocia/cephalopelvic disproportion. Any attempt to reduce cesarean section rates has therefore to contend first with those two indications because they represent almost two-thirds of all indications for cesarean section deliveries.[3] Evidence in the literature suggests that at most only one-third of previous cesarean sections require a repeat cesarean if patients are allowed a proper trial of labor.[29–31] In 1991, the most recently available data from the Office of Vital and Health Statistics Systems, National Center for Health Statistics, Center for Disease Control, suggest that only 24.2% of all previous cesarean section patients had a vaginal delivery.[3] While these numbers indicate an increase in VBAC utilization (from 8.5% in 1986), they still clearly lag behind a reasonable level of expected national utilization. One can therefore conclude that a further increase in VBAC deliveries is possible, is safe, and should be the goal of each department.[32] On a national level this means that of approximately 338,100 repeat cesarean sections performed annually (35% of 966,000 total cesareans), only 112,700 are really needed.

A number of studies have demonstrated that the cesarean section indication of dystocia can be limited to no more than 2% of all deliveries.[33–35] At present, slightly fewer than 9% of pregnant women undergo cesarean sections with a primary diagnosis of dystocia/cephalopelvic disproportion, more than four times the recommended rate.[3]

Similar calculations can also be made for less frequent indications for cesarean delivery. For example, convincing data in the literature suggest that in carefully selected breech presentations delivery by cesarean does not improve outcome over vaginal delivery.[36,37] In fact, these data refer not only to the classical singleton frank breech[38,39] at term but also to premature breeches,[40] nonfrank breeches,[41,42] and breeches as part of a multiple birth.[43] If one summarizes these data, the conclusion is that a safe vaginal delivery of breech presentations can be accomplished in as many as two-thirds of cases. Considering the recently reported success with external ver-

TABLE 11.1. Current national and recommended cesarean section rates, according to indication (%)

Indication	Percentage of all cesarean sections[1]	Current national rates as a percentage of all deliveries[a]	Deliveries by cesarean section in Porreco's study[36]	Quilligan's low recommendation[45]	Mt. Sinai target rate[46]
Previous cesarean section	30	6.8	1.5	2.0	2.2
Dystocia	30	6.8	1.5	2.0	2.2
Breech presentation of fetus	15	3.4	1.1	1.3	1.1
Fetal distress	5	1.1	0.6	1.5	1.1
Other	20	4.5	1.0	1.0	4.5
Total	100	22.7	5.7	7.8	11.1

[a] Calculated at a national cesarean section rate of 22.7% in 1985 (with a rounding error of 0.1 in the total).

sions of breeches, one can further conclude that a cesarean section because of breech delivery should be a rare event.

In 1991, 11.7% of all cesarean sections were performed because of breech presentation.[3] This represents 113,022 women or approximately 2.8% of all deliveries. Because breech presentations represent approximately 3.5% of term pregnancies,[3] one can calculate an 80% national cesarean section rate for breeches. This is clearly still reflective of an almost universally held opinion that breeches should be delivered by cesarean section, even though valid studies statistically supporting this practice are rare.[37]

Cesarean sections for a fetal distress indication reflect only a small part of the total spectrum of cesarean section indications. The most recent data (1991), in fact, suggest that at 9.2% of all cesarean section indications, fetal distress is a precipitating diagnosis even less frequent than breech presentation.[3] Because most medical-legal liability situations involve the allegation of fetal brain damage in utero, the proper response to a presumption of fetal distress (possibly by cesarean section) has obviously great potential legal ramifications. The fact that cesarean sections for fetal distress represent only such a small fraction of all cesarean section indications further supports the contention that medical-legal considerations affect national cesarean section rates only marginally.

Even if cesarean section rates for fetal distress and all other (less frequent) indications remained at their current level of utilization, one could still calculate a safe target rate for a national cesarean section rate of approximately 11%.[44] We did this calculation in consideration of published experience by Porreco[34] and based on a detailed evaluation of published literature by Quilligan (see Table 11.1),[45] both of whom actually suggested an even lower goal (6% and 8%, respectively).

Based on a recent survey by the New York-based Metropolitan Life Insurance Co., the average cost differential between a cesarean section and vaginal delivery is \$6193.[46] A *safe* 11% national cesarean section rate would reduce the number of these surgical procedures (based on 1991 data) from 966,000 to approximately 440,000, with a cost savings of \$3.26 billion. The political significance of this fact is obvious.

A review of published experience thus confirms that a safe reduction in cesarean section rates can be achieved. This led us[44] and others[47,48] to implement specific cesarean section reduction programs at a departmental level.

The Mount Sinai–Chicago Experience

Even though the departmental cesarean section rate had been considerably below the national average, the leadership of the department in 1985 reached the conclusion that only a well-defined program would affect the departmental practice pattern as a whole. Our national research at that point had confirmed the departmental impression that the well-thought-out recommendations of the National Consensus Development Conference were not effective in changing practice patterns.[32,49] We therefore decided to design a cesarean section reduction initiative that appropriated the universally accepted recommendations of the Consensus Development Conference but, in addition, took the step from "recommendation" to peer-reviewed "implementation."

Peer review was critically important. Not only did the Health Care Quality Improvement Act of 1986 attempt to bolster the concept of peer review,[50] but we believed that a well-designed peer review process could be the key factor in controlling cesarean section rates. When extramural funding for a computer system was secured, we were able to put into place a computer-assisted data-collection and analysis system that allowed us to implement a comprehensive and fair-minded peer review system which, in fact, became the cornerstone of our cesarean section reduction initiative. Data collection was not based on maximal completeness but on the concept of minimal essential data. This turned out to be a wise decision because maintenance of the system required online data input. A requirement for too many data would have reduced compliance and therefore reduced the reliability of our data. Appendices I and II demonstrate neonatologic and obstetric outcome data-collection summaries.

We then analyzed the literature (as previously described) to determine a target cesarean section rate that we as the department wanted to achieve (see Table 11.1). Our departmental rate in 1985 was approximately 17%. We felt confident that a rate of 11%–12% could be safely achieved.[44] The Mount Sinai cesarean section reduction initiative was started on a departmental level in January of 1986 and involved on a voluntary basis all members of the department, whether geographic fulltime faculty or voluntary attending staff. Participation of the voluntary staff was crucial because approximately 50% of all departmental deliveries were under their control.

Cooperation from the neonatal service also was crucial. Any attempt to change a cesarean section practice pattern has to carefully control for neonatal outcome. Because the failure to do so had in our opinion led to an unwarranted increase in cesarean section utilization, it seemed of critical importance *not* to make the same mistake in attempting to influence cesarean section rates downward. We therefore chose to tabulate 35 different neonatal morbidity factors in addition to standard perinatal mortality tabulations (see Appendix I). Because birthweight and gestational age are the most important determinants of neonatal outcome, we also ensured that our neonatal outcome system evaluation permitted us to differentiate the effects of route of delivery from, for example, prematurity. Such a capability is crucial because disproportionate weight distribution in allegedly comparable patient populations, especially in regards to breech deliveries, have repeatedly resulted in blatantly false statistical outcome evaluations in the literature.[37]

As noted before, the program was voluntary. It was principally based on the establishment of clinical practice guidelines (CPGs) for the most frequent cesarean section indications. While CPGs have since become part of the medical vernacular, in 1986 the concept of CPGs required careful explanation to all departmental physicians. Appendix III summarizes the departmental CPGs for repeat cesarean section, dystocia/cephalopelvic disproportion, breech presentation, and fetal distress. The CPGs basically followed recommendations already made by the National Consensus Development Conference 6 years earlier,[1] and were therefore neither experimental nor department specific.

Departmental CPGs were reviewed on an ongoing basis as part of the quality assurance process. Obstetric management patterns that did not follow departmental CPGs were discussed in weekly conferences, preferably in the presence of the responsible attending physician. In addition, the departmental data bank allowed the establishment of physician-specific practice patterns. Each attending and resident physician thus received at regular intervals a personal practice profile (Figure 11.2). These practice profiles showed an individual's cesarean section rate in relationship to the rest of the department. If an individual fell outside 2 SD from the departmental mean, their practice pattern was reviewed in a private conference with either the department chairman or chief of maternal-fetal medicine. Not a single disciplinary action was taken in the first 2 years of the program.[44,51]

The department's cesarean section reduction initiative also mandated a second opinion from a nonaffiliated board-certified member of the department in all nonemergency cesarean sections. This mandate was initially actively reviewed. It became, however, increasingly apparent that this second opinion requirement was unnecessary in view of other program components.

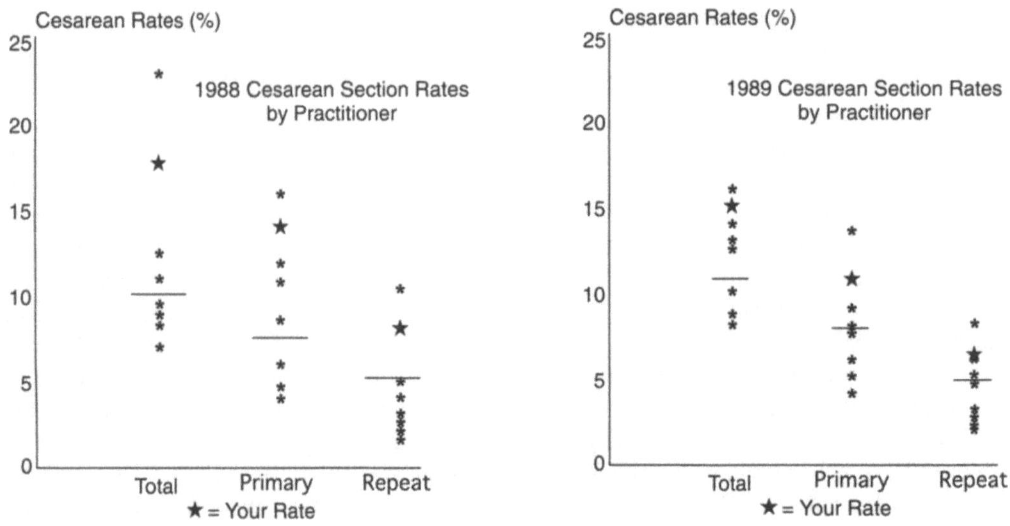

FIGURE 11.2. Scattergram of cesarean section rates for selected Mount Sinai—Chicago physicians in comparison to departmental mean rate (horizontal line). The 1988 data reflect more outlier physicians than the 1989 statistics. Each report is individualized (by star) to a physician. (From Myers and Gleicher,[51] by permission of *The Quality Letter for Healthcare Leaders.*)

Enforcement of the requirement was therefore abandoned after the first 2 years.

The first 2 years of experience with the program were reported in 1988.[44] The departmental cesarean section rate fell in that time period from 17.5% to 11.5% and neither fetal nor neonatal mortality or morbidity were affected. This report received considerable attention, not the least because of an accompanying editorial that implied that a reduction of cesarean section rates may after all affect neonatal outcome adversely.[52] Specifically, the editorial noted the occurrence of adverse outcomes with some vaginal breech deliveries. The editorial writers apparently did not know, however, that the *New England Journal of Medicine* had required much more detailed outcome data submission during the review process than they were willing to publish in the final manuscript. Moreover, a reduction in cesarean section utilization is not necessarily in itself a tool to improve neonatal outcome. Its purpose is primarily to maintain neonatal outcome while achieving a less invasive and less costly method of delivery for the mother. A level of neonatal mortality and morbidity can therefore be expected. Our charge was to determine whether that (prior) level was adversely affected by the cesarean section reduction initiative. We conclusively demonstrated that this was not the case, although adverse outcomes obviously continued to occur.

The Mount Sinai–Chicago experience was thus the *first* departmental cesarean section reduction initiative that conclusively demonstrated the feasibility of such programs. Up to that point, efforts to reduce cesarean section rates were either within the realm of theoretical considerations,[1,45,49] involved only single indications,[29,38,40] or represented the effort of single practitioners.[34] Moreover, the program also demonstrated that a significant impact on cesarean section rates could be made by establishing rather simple program components as CPGs, with an accompanying quality assurance program.

It was probably exactly that simplicity that caused a considerable degree of skepticism in the profession after the initial Mount Sinai experience was published. Arguments were made that the experience was not duplicable in the "real" world because such a program could only be implemented in a teaching hospital with residents. Others argued that an inner-city population was not representative, and, finally, the argument was repeatedly made that private practitioners in a private hospital setting just did not have an adequate emergency backup to instigate a successful cesarean section reduction initiative. As is demonstrated in the following section, none of these arguments holds up to scrutiny. In fact, every serious attempt ever made to affect cesarean section rates within an institution has been successful, whether in inner-city teaching hospitals or private or public institutions, whether driven by faculty physicians or private practitioners.

Cesarean Section Reduction Programs

While the Mount Sinai experience demonstrated the feasibility of a successful cesarean section program within a teaching institution, the argument was made that experiences gained from a disproportionally large service population were not applicable to more private settings. This argument was further supported by the repeatedly made observation that even within the same teaching institutions, the teaching service routinely achieved lower cesarean section rates than a private service.[12,48,53]

Porreco's pioneering efforts demonstrated this fact preceding the Mount Sinai experience,[34] although the Mount Sinai Program achieved a greater reduction in the private cesarean section rate (from 20% to 12.4%) than the faculty rate (from 15% to 11.7%) and thus statistically equalized the cesarean section utilization between the two services.[44]

In the same city, Northwestern Memorial Hospital, a primarily private tertiary university facility with a much smaller service population than Mount Sinai, initiated a cesarean section reduction program in 1986 when the departmental rate had reached 27.3%. Their experience has been reported in two publications.[48,54] Total, primary, and repeat cesarean section rates, respectively, were reduced from 27.3%, 18.2%, and 9.1% in 1986 to 16.9%, 10.6%, and 6.4% in 1991 by "encouraging" VBAC deliveries, distributing physicians' cesarean section rate information annually to all staff (as part of a peer review process) and by implementing an active management program for labor. While thus successful in their overall goal to diminish the departmental cesarean section rate, their impact on private versus teaching services varied greatly. As Sandmire[55] in a discussion of these data correctly asked, can a private service cesarean section rate of almost twice the clinic rate be really regarded as success, or does it have to be seen as an acknowledgment of failure to influence one's peers?

The feasibility to affect cesarean section patterns in a strictly private practice setting was first shown by Gellman and associates[31] who demonstrated quite early during the national debate on the topic that VBAC deliveries could be safely performed in a private setting. Even though the number of studies from the private sector has (not surprisingly) remained small, enough evidence has, in fact, been presented to support the contention that the private practice of obstetrics is just as receptive to properly guided change as the academic world. Rayner[56] demonstrated this by obtaining a very respectable VBAC rate in a small rural community (Gellman and associates[31] reported their effort from New York City). In a Canadian rural community hospital, Iglesias et al.[57] were able to reduce their overall cesarean section rate from 23% to 12% over 4 years. This was accomplished by implementing VBAC and breech delivery guidelines and criteria for the diagnosis of dystocia, following a report from the National Consensus Conference on Aspects of Cesarean Birth (NCCACB), released in Canada in 1986.[8] Those were very similar to recommendations by the U.S. National Consensus Development Conference in 1980.[1]

Peck[58] recently reported the experience of Lutheran Hospital in La Crosse, Wisconsin, a community-based tertiary-care center with a wide perinatal referral area. Their cesarean section rate during 4 years had remained in the 10%–11% range with a perinatal mortality of 11 in 1000. Peck attributed their success to 12 specific factors: an active midwifery program delivered 30% of their newborns with a cesarean section rate of less than 5 percent for this very low risk population. (Butler et al.[59] recently once again quite convincingly demonstrated that nurse-widwife care is associated with a reduced incidence of cesarean sections.) Eighty percent of women with previous cesarean section underwent a trial of labor, and among those 80 percent delivered vaginally (almost the same experience as reported by Mount Sinai–Chicago).[44] Two-thirds of persistent breeches at 37 weeks gestational age were successfully converted to a vertex presentation; approximately 50% of remaining breeches were vaginally delivered. Strict diagnostic criteria for the cesarean section indication of "fetal distress" resulted in only a 1% cesarean section rate for this diagnosis. A physician on-call system placed a board-certified obstetrician on call for hospital (labor floor) coverage at all times. The use of intrauterine pressure transducers whenever labor is augmented was encouraged. No time limits are set for the length of the second stage so long as mother and fetus are in good condition and progress in labor is observed. Epidural anesthesia was discouraged. A "homey" delivery environment and, finally, a committed physician staff were also considered key.

Table 11.2 summarizes the cesarean section experience of a suburban Chicago community hospital for the years 1988–1992. These data are presented to demonstrate the importance of the training process that then creates the mindset for practicing obstetricians. Sherman Hospital in Elgin is a small, nonteaching community hospital, which in the Chicago area has gained a reputation for having the lowest cesarean section rate among comparable institutions.

The institution was joined in 1988 by one of the authors (R.H.D.) as a private practitioner after he graduated from the residency program at Mount Sinai Hospital Medical Center in Chicago. The departmental cesarean section rate at that point was 19.9%, with R.H.D. having a 10% and the remainder of the department a 20.2% rate. During the ensuing 5 years, the R.H.D. practice grew from 53 (in 1988) to 1481 (in 1992) deliveries and from one to five physicians (plus two midwives). The cesarean section rate for this fully private group remained stable in a range between 6.3% and 11.8%, while the remainder of the department stayed in a range of 17.3%–20.2% Because the R.H.D. practice increased their

TABLE 11.2. Cesarean section utilization at Sherman Hospital, Elgin, Illinois

Year	1988	1989	1990	1991	1992
Sherman Hospital					
Total births	1788	1741	1999	2372	2877
Cesarean section rate	19.9%	17.4%	15.5%	14.5%	15.0%
Primary	11.5	10.2	10.3	8.8	9.8
Repeat	8.4	7.2	5.2	5.7	5.2
Epidural rate	<1%	1%	10%	15%	16%
RHD & Associates					
Total births	53	167	436	917	1081
Cesarean section rate	10%	8.1%	6.3%	10.2%	11.8%
All other physicians					
Total births	1735	1574	1563	1499	1396
Cesarean section rate	20.2%	18.4%	18.1%	17.3%	19.4%

contribution to the departmental patient pool from 3% to 52%, their practice pattern significantly impacted the departmental cesarean section rate, which decreased from almost 20% to 15%.

These data are presented to demonstrate that low cesarean section rates can be achieved in a variety of settings. R.H.D. successfully transferred the basics of the Mount Sinai cesarean section reduction initiative from an inner-city teaching hospital to a private, suburban community hospital. Moreover, he transferred those principles from a large teaching department, which included a residency program, to what was initially a one-person private practice. This practice grew in the second year to two physicians, when R.H.D. was joined by his wife J.B.N., another Mount Sinai graduate. As the practice continued to expand these two Mount Sinai graduates had to transmit their philosophy of practice to other physicians who came from training programs with very different approaches. In doing so successfully, they demonstrated not only that a Mount Sinai-like program approach is feasible in private practice, but that it can be taught to any obstetrician with an open mind, wherever he or she may practice. They also demonstrated that a frequently mentioned "consumer demand" for cesarean section delivery cannot be too pervasive if they succeeded in growing their practice 20 fold over 4 years.

That cesarean section review programs will affect cesarean section rates has been demonstrated in a variety of different settings. For example, Kaiser Permanente, the nation's largest managed care provider, has reduced cesarean section rates very favorably in comparison to national and regional figures.[60] New York State and the local ACOG District developed a program to enhance in-house cesarean section review processes for participating hospitals.[61] The state's cesarean rate was reversed, and reviewed hospitals demonstrated a stronger downward trend than nonreviewed institutions. Similar effects have also been reported from outside the United States.[47,57]

Selective Contributing Factors to Cesarean Section Deliveries

Previous Cesarean Section

Guidelines from a variety of authoritative national and international bodies strongly support the concept of VBAC,[1,6–8] and generally suggest a trial of labor in women with *one* previous cesarean section, a history of transverse uterine scar (with uncomplicated healing process), and no other complicating factor. In fact, one could summarize those recommendations as an endorsement of VBAC delivery with the caveat that a repeat cesarean section may still be indicated whenever "in doubt." We disagree with such an approach in that we believe that a well thought out medical approach should always demand that the more invasive, more costly procedure must establish its benefit over less invasive and less costly alternatives. Consequently, when "in doubt" a trial of vaginal delivery should be attempted.

Under such a philosophy a trial of VBAC can and should be performed not only after one previous cesarean but independent of how many have been performed,[17–19] if a breech presentation is present (assuming vaginal breech delivery criteria are met)[62] with multiple gestations,[63] if macrosomia of the fetus is suspected,[64] or even if a scar is unknown or classical.[65] Based on a meta-analysis of morbidity and mortality, Rosen et al.[16] argued in favor of more trials of labor in women with previous cesarean sections. Our own experience[44] (and Table 11.2) suggest that 80%–90% of women should undergo a trial of labor. Among those, 80%–90% can be expected to deliver successfully vaginally.

Previous Myomectomy

A previous myomectomy, which resulted in entry into the endometrial cavity, is universally considered an indication for a cesarean section. We were unable to find even a single valid study in support of such a practice pattern in the literature and can therefore *not* support the contention of an automatic cesarean section. If one were to extrapolate from the cesarean section scar experience (and why should one scar differ from another), then at least with single entry into the cavity, the experience should not differ from that in patients with unknown or classical scars. Because uterine rupture in those appears to be a rare event,[65] a trial of labor after at least single entry into the cavity appears warranted.

As in patients after previous cesarean section, labor should be monitored carefully and the longitudinal observation of intrauterine pressures may be of benefit.

Dystocia/Cephalopelvic Disproportion

The Mount Sinai–Chicago[44] and Northwestern[54] experiences have clearly confirmed that an active obstetric management approach can reduce cesarean section rates for this indication dramatically. As noted earlier, not more than 2% of deliveries should undergo a cesarean section for dystocia.[33–35] Both programs were based on simple treatment protocols that can be easily repeated in different settings.

Breech Presentation

We reviewed in 1987 the indications for cesarean sections for breech presentations in great detail.[36,37] Our conclusions then, based on an extensive review of the literature, was that statistically valid studies which demonstrate the benefit of a cesarean section over a vaginal delivery are lacking if minimal patient selection criteria were followed.[38,42] In fact, vaginal breech delivery is indicated not only for term frank breech presentation but for nonfrank breech presentation and premature fetuses as well. The same conclusion can be applied to breech presentation in multiple gestations.[43,66] Based on published data and our own experience, we therefore conclude that a safe vaginal delivery of term and preterm breech fetuses can be accomplished in approximately 50% of cases.[38,40,44]

Preterm Delivery

A number of reports have suggested an increased utilization of cesarean sections for preterm deliveries for vertex as well as breech presentations.[1,67,68] As noted previously, there appears to be no benefit from performing cesarean section deliveries in breech presentation.[40,41,66] Malloy et al.,[69] in a study from the National Institute of Child Health and Human Development Neonatal Research Network, recently concluded (after accounting for potential other contributing maternal and fetal factors) that cesarean delivery is *not* associated with a lower risk of either mortality or intraventricular hemorrhage in very low birthweight infants, independent of presentation. Similar results regarding the influence of route of delivery on intraventricular hemorrhage in preterm infants were also reported by Anderson et al.[70] Prematurity per se does therefore not constitute an indication for cesarean section delivery.

Fetal Distress

The use of fetal monitoring has been implicated as a cause of increased cesarean section rates.[71] Because this probably is the result of a misinterpretation of fetal heart rate tracings, it is only reasonable to corroborate the true presence of fetal distress situation by other means whenever possible. The Mount Sinai–Chicago program suggested a fetal scalp pH, whenever technically feasible, in confirmation of fetal acidosis before a cesarean section was performed under the presumed diagnosis of fetal distress.[44] Others have recommended fetal scalp stimulation as an alternative.[72]

Overally, cesarean sections for fetal distress should probably occur in not more than 0.6% (Porreco[34]) to 1.5% (Quilligan[45]) of all deliveries.[44]

Advanced Maternal Age

One recent study suggests that advanced maternal age alone may influence physicians in their decision on the mode of delivery and thereby place some older women at an unnecessary risk of cesarean section.[73] While older women may, in fact, demonstrate an increased incidence of high risk factors, which by themselves can predispose to cesarean sections, age alone should not. There is also no supportive evidence in the literature for the frequently heard argument that older women have less tolerance for (prolonged) labor.

The Valuable Pregnancy

One also frequently encounters in clinical practice the concept of the so-called very valuable pregnancy. This may be a first pregnancy in an older woman, a pregnancy after prolonged infertility, or a pregnancy that has reached viability after multiple earlier losses. While such pregnancies certainly are "valuable" or even "very valuable," it is difficult to perceive of a treatment approach that should differ from any other standard of care. Because the treatment of any pregnancy should be maximal, valuable or very valuable pregnancies should have no "better" options available to them than any other conception. A concept of preferred cesarean section delivery for such pregnancies therefore does not make sense unless one believes in the (incorrect) concept that cesarean section deliveries are in general safer to the fetus than vaginal deliveries.[5] One therefore has to wonder why, for example, conceptions achieved through in vitro fertilization almost universally are reported to demonstrate cesarean section rates in the 50% range.[74,75]

Genital Herpes

The current recommendation of the Infectious Disease Society for Obstetrics-Gynecology is the performance of a cesarean section delivery in the presence of active herpetic lesions at time of labor.[76] Randolph et al.[77] recently suggested, however, that the practice to perform cesarean section deliveries indiscriminately whether the patient suffers from a first or a repeat episode should be reexamined. They claimed that the automatic performance of a cesarean section in women with repeat outbreaks of herpes simplex type 2 virus infections results in more than 1580 excess cesarean deliveries performed for every poor neonatal outcome prevented, a cost of $2.5 million for every case of prevented neonatal herpes, and a cost of $203,000 per quality-adjusted life-year gained. These calculations were made based on the fact that neonatal infection in *recurrent* herpes infection is, as a result of protective antibodies, rare (1%[78]) in comparison to primary disease.

In an accompanying editorial, Gibbs et al.[79] supported the practice to perform cesarean section deliveries only in the presence of visible herpetic lesions. They noted, however, that 38% of randomly chosen obstetricians were still performing screening cultures for herpes in pregnant women with a history of genital herpes, an observation that probably suggests that many cesarean sections are still performed in only culture-positive women with no active disease. They editorial, however, also suggested that *for the present* not enough data are available to recommend vaginal delivery in the presence of active lesions even in the women with a repeat episode of genital herpes. A cesarean section delivery thus remains indicated whether an active lesions during labor is primary or secondary in nature.

Congenital Anomalies

There are very limited data in the literature on whether the performance of a cesarean section impacts on outcome of fetuses with known congenital anomalies. Luthy et al.[80] investigated this question in regards to prenatally diagnosed meningomyelocele. Delivery by cesarean section before the onset of labor may result in better subsequent motor function than vaginal delivery or delivery by cesarean section after a period of labor.

Fetuses with congenital anomalies demonstrate an increased incidence of abnormal fetal heartrate patterns,[81] which often lead to a false assumption of fetal distress and subsequent cesarean section.

Physician Factors

Physicians do not perform cesarean sections so as to make it to the golf course in time.[82] Physician factors do, however, affect cesarean section utilization rates. Berkowitz et al.,[83] for example, have demonstrated that older, more experienced physicians performed fewer cesarean sections for dystocia and a higher percentage of forceps deliveries and breech extractions. These results reemphasize the earlier discussed (in) experience factor, which by itself has become an indication for the performance of cesarean sections. In our opinion inexperience in the performance of alternative delivery methods to cesarean sections does not represent an indication for cesarean section so long as experience is available by consultation or referral. As the pendulum swings back, it is increasingly recognized that even midforceps deliveries represent safe and acceptable alternatives to cesarean sections if patients are properly selected and operators are experienced.[84]

Social Factors

The rates of cesarean section utilization vary greatly with socioeconomic status.[85] This is a potentially very disturbing finding because it suggests that different levels of obstetric care may be provided dependent on the socioeconomic status of patients. Interestingly, the provision of health insurance to previously uninsured low-income pregnant woment may not by itself improve maternal health but it certainly increases cesarean section rates.[86]

Stress

The stress of labor appears to be a contributing factor to the rate of cesarean section utilization. In a well-designed randomized study, Kennell et al.[87] demonstrated that the presence of a supportive companion (doula) during labor and delivery significantly reduced the cesarean section rate. Rosen[88] in an accompanying editorial suggested that we may benefit from some new, nontechnological approaches to modern obstetric care. It is tempting to speculate that it is at least in part this doula effect that results in lower cesarean section rates in birthing center settings and with midwife deliveries.[59] Any national health care reform may benefit from this recognition.

Conclusions

The reduction of the overall cesarean section rate to 15 or fewer per 100 deliveries and of the primary cesarean section rate to 12 or fewer per 100 deliveries has become a national health objective.[3,89] A national consensus seems finally to emerge that cesarean sections are excessively utilized. In response, multiple alternative strategies have been proposed to combat rising rates and possibly even reverse them. Stafford[2] probably summarized them best by grouping them into six distinct strategies (Table 11.3). Most of this chapter has been concerned with his first strategy, which he called *Education and Peer Evaluation* and which he described as strategies aimed at physicians reviewing the literature, the issuance of CPGs, and formal hospital programs. As he notes, these rely on implementation by obstetricians and tend to preserve physician autonomy. As physicians we are therefore well advised to support these strategies.

The alternatives will undoubtedly increasingly restrict physician independence. The *External Audit and Review* strategy, as Stafford claims, makes the assumption that without external input (or restraint) physicians are incapable of making correct, or economically sound, decisions. They therefore require external "guidance," a suggestion clearly not very popular among physicians. The *Public Dissemination of*

TABLE 11.3. Strategies for reducing cesarean section rates

Strategy	Methods of implantation	Impact on cesarean section
Education and peer evaluation	Individual obstetricians reading the medical literature	Not effective
	Clinical guidelines issued by specialty society (ACOG) or expert panels	Have failed to modify cesarean section use
	Formal programs to reduce cesarean section use in individual hospitals	In teaching hospitals: reductions of as much as 34%
External review of practices	Audit of obstetric practices by outside agency	In Scotland: single-year reductions of 13% and 27%
	Review of practices by health care payers either before or after services are provided	Impact has not been evaluated
Public dissemination of cesarean section rates	Making data accessible to consumers for use in selection of providers	Impact has not been evaluated
Modification of physician reimbursement	Equalize physician payments for cesarean and vaginal deliveries	Reductions of 0%–50% but limited implementation and evaluation
Modification of hospital reimbursement	Remove incentive for cesarean section through prospective payment for deliveries	Limited implementation has been evaluated for impact
Medical malpractice reform	Several approaches aimed at reducing the volume and cost of malpractice claims	Implemention and assessment have been limited

Information strategy is based on the assumption that the public will make correct choices if only properly informed. In that vein, the publication of cesarean section rates by physicians or hospitals has been proposed as a mandate. Whether such data by themselve scan or will be correctly interpreted by the public is at least questionable. A *Change in Physician Reimbursement* strategy has been proposed by these authors for almost a decade.[32,49] This is not based on frequently cited (though totally unsupported) allegations that physicians perform cesarean section deliveries to benefit from increased reimbursement rates (in comparison to vaginal deliveries). In fact, nothing supports such a contention. However, there can also be no doubt that the average time effort for a vaginal birth is longer than for a cesarean section delivery, especially in regards to VBAC deliveries versus repeat cesarean sections. A financial incentive structure in favor of cesarean delivery therefore does not make sense, and the insurance industry would be well advised to create *strong* financial incentives for vaginal and against cesarean section deliveries.

Such an approach was recently also supported by Spellacy.[90] A *Change in Hospital Reimbursement* strategy is then the next logical consequence. At present hospitals have no incentive to reduce cesarean section rates. In fact, payment incentives by third-party payers are probably in favor of cesarean section deliveries because they reward hospitals on a fee-for-service basis for all additional services provided. A single rate reimbursement for "delivery" would create a strong incentive to hospitals to reduce cesarean section rates because hospital costs for cesarean sections are probably considerably higher than for vaginal deliveries. Lastly, a *Medical Malpractice Reform* strategy has to be seen as an important component of any national cesarean section effort.

As we are entering the twenty-first century, the practice of medicine is undergoing revolutionary changes. Outcome assessment for any form of medical intervention will become an increasingly important tool in an ongoing process of continuous improvement in the quality of patient care. At the same time, we should learn from past mistakes and assess treatment strategies in their respective benefits before they are widely implemented into daily clinical practice. One of the lessons of our national experience with cesarean sections is that practice patterns develop a life of their own. They not only determine how we practice but, unfortunately, also how we think and how we train. As a consequence, it is much more difficult to reverse a national practice pattern than it is to implement a new one. In that sense the national cesarean section experience teaches a very important lesson to everybody who is concerned about the future of medical care in this country.

References

1. National Institutes of Health. Cesarean Childbirth: Report of a Consensus Development Conference. Bethesda, Maryland: Department of Health and Human Services, 1981. (Publication no. DHHS (NIH) 82-2067.)
2. Stafford RS. Alternative strategies for controlling rising cesarean section rates. JAMA 1990;263:683–687.
3. Anonymous. Rates of cesarean delivery—United States 1991. Morbid Mortal Wkly Rep 1993;42:285–289. JAMA 1993;269:2360.
4. Abraham L. Cesarean on demand. Am Medical News 1988 Nov 4:1830.
5. Feldman GB, Friedman JA. Prophylatic cesarean section at term? N Engl J Med 1985;312:1264–1267.
6. Guidelines for vaginal delivery after a previous cesarean birth. statement of the Committee on Obstetrics: Maternal and Fetal Medicine. Washington, DC: American College of Obstetricians and Gynecologists, 1984.
7. New guidelines for VBAC. statement of the Committee on Obstetrics: Maternal and Fetal Medicine. Washington, DC: American College of Obstetricians and Gynecologists, 1988.
8. Hannah WJ, Baskett TF, Chance GW, et al. Indications for cesarean section: final statement of the panel of the National Consensus Development Conference on Aspects of Cesarean Birth. Can Med Assoc J 1986;134: 1348–1352.
9. Marieskind HI. An evaluation of cesarean section in the United States. Washington, DC: U.S. Department of Health, Education, and Welfare, Office of the Assistant Secretary for Planning and Evaluation, 1980.
10. Rostow VP, Osterweis M, Bulger RJ. Medical professional liability and the delivery of obstetrical care. N Engl J Med 1989;321: 1057–1060.
11. Carpenter MW, Soule D, Yates WT, Meeker CL. Practice environment in association with obstetric decision making regarding abnormal labor. Obstet Gynecol 1987;70: 657–662.
12. Goyert GL, Bottoms SF, Treadwell MC, Nehra PC. The physician factor in cesarean birth rates. N Engl J Med 1989;320:706–709.
13. Newton ER, Higgins CS. Factors associated with hospital-specific cesarean birth rates. J Reprod Med 1989;34:407–411.
14. Localio AR, Lawthers AG, Bengtson JA, et al. Relationship between malpractice claims and cesarean delivery. JAMA 1993;269:366–373.
15. Rutkow IM. Obstetrics and gynecologic operations in the United States, 1979 to 1984. Obstet Gynecol 1986;67:755–759.
16. Rosen MG, Dickinson JC, Westhoff CL. Vaginal birth after cesarean: a meta-analysis of morbidity and mortality. Obstet Gynecol 1991;77:465–470.
17. Porreco R, Meier P. Trial of labor in patients with multiple cesarean sections. J Reprod Med 1983;28:770–776.
18. Novas J, Myers SA, Gleicher N. Obstetric outcome of patients with more than one previous cesarean section. Am J Obstet Gynecol 1989;160:364–367.
19. Asakura H, Myers SA. Multiple cesarean section: A five year experience with 434 patients. Am J Obstet Gynecol 1993;168:437 (abstr).
20. Ophir E, Yagoda A, Rojansky N, Oettinger M. Trial of labor following cesarean section: dilemma. Obstet Gynecol Surv 1988;44:19–24.
21. Gleicher N, Vermesh M, Rotmensch Z, Thornton J, Elrad H. Cesarean section patterns: influence of a perinatology service. Mount Sinai J Med 1985;52:100–105.
22. Laube DW, ed. Core curriculum for residents of obstetrics and gynecology. Washington,

DC: Council on Resident Education, 1992: 103.
23. Gleicher N. Breast disease programs in obstetrics and gynecology: a plea for training in mammography. Am J Obstet Gynecol 1989;161:267–270.
24. Lomas J, Enkin M, Anderson GM, Hannah WJ, Vayda E, Singer J. Opinion leaders vs. audit and feedback to implement practice guidelines. Delivery after previous cesarean section. JAMA 1991;265:2202–2207.
25. Barrett JFR, Jarvis GJ, MacDonald HN, Buchan PC, Tyrrell SN, Lilford RJ. Inconsistencies in clinical decisions in obstetrics. Lancet 1990;336:549–551.
26. Notzen FC, Placek PJ, Taffel SM. Comparison of national cesarean-section rates. N Engl J Med 1987;316:386–389.
27. Stafford RS. The impact of nonclinical factors on repeat cesarean section. JAMA 1991; 265:59–63.
28. Gleicher N. In reply. JAMA 1985;253:3248–3249.
29. Gibbs CE. Planned vaginal delivery following cesarean section. Clin Obstet Gynecol 1980; 23:507–515.
30. Martin JN Jr, Harris BA Jr, Huddelston JF, et al. Vaginal delivery following previous cesarean birth. Am J Obstet Gynecol 1983; 146:255–263.
31. Gellman E, Goldstein MS, Kaplan S, Shapiro WJ. Vaginal delivery after cesarean section experience in private practice. JAMA 1983; 249:2935–2937.
32. Gleicher N. The cesarean section epidemic. Mt Sinai J Med 1986;53:563–565.
33. Placek PJ, Taffel SM, Molen M. Cesarean rates increases in 1985. Am J Public Health 1987;77:241–242.
34. Porreco RP. High cesarean section rates: a new perspective. Obstet Gynecol 1985;65: 307–311.
35. MacDonald D, Grant A, Sheridan-Penetra M, Boylan P, Chalmers I. The Dublin randomized control trial of intrapartum fetal heart rate monitoring. Am J Obstet Gynecol 1985;152:524–539.
36. Confino E, Gleicher N, Elrad H, Ismajovich B, David MP. The breech dilemma: a review. Obstet Gynecol Surv 1985;156:6–10.
37. Myers SA, Gleicher N. Breech delivery: why the dilemma? Am J Obstet Gynecol 1987; 156:6–10.
38. Collea JV, Chein C, Quilligan EJ. The randomized management term breech presentation: a study of 208 cases. Am J Obstet Gynecol 1980;137:235–244.
39. Green JE, McLean F, Smith LP, Usher R. Has an increased cesarean section rate for term breech delivery reduced the incidence of birth asphyxia, trauma and death? Am J Obstet Gynecol 1982;142:643–648.
40. Kitchen W, Ford GW, Doyle LW, et al. Cesarean section or vaginal delivery at 24 to 28 weeks gestation: comparison of survival and neonatal and two-year morbidity. Obstet Gynecol 1985;66:149–157.
41. Woods JR. Effect of low birth weight breech delivery on neonatal mortality. Obstet Gynecol 1979;53:735–740.
42. Gimovsky ML, Wallace RL, Shifrin BS, et al. Randomized management of the nonfrank breech presentation at term. a preliminary report. Am J Obstet Gynecol 1983;146:34–40.
43. Fishman A, Grubb DK, Kovacs BW. Vaginal delivery of nonvertex second twin. Am J Obstet Gynecol 1993;168:861–864.
44. Myers SA, Gleicher N. A successful program to lower cesarean-section rates. N Engl J Med 1988;319:1511–1516.
45. Quilligan E. Cesarean section: modern perspective. In: Queenan JT, ed. Management of high-risk pregnancy, 2nd Ed. Oradell, NJ: Medical Economics, 1985:594–600.
46. Anonymous. Childbirth. Chicago Tribune 1993. Jan (sect 7) 17:2.
47. Tay SK, Tsakok FHM, Ng CSA. The use of intradepartmental audit to contain cesarean section rate. Int J Gynecol Obstet 1992;39: 99–103.
48. Socol ML, Garcia PM, Peaceman AM, Dooley SL. Reducing cesarean section births at a primarily private university hospital. Am J Obstet Gynecol 1993;168:1748–1758.
49. Gleicher N. Cesarean section rates in the United States. The short term failure of the National Consensus Development Conference in 1980. JAMA 1984;252:3273–3276.
50. Inglehart JK. Health policy report. N Engl J Med 1987;316:960–965.
51. Myers S, Gleicher N. The Mount Sinai cesarean section reduction program: automated data selection and peer review. Qual Lett 1991;3:10–17.
52. Battaglia FC. Reducing the cesarean section rate safely. N Engl J Med 1988;319:1540–1541.
53. Heynes de Regt R, Minkoff HL, Feldman J, Schwarz RH. Relation of private or clinical

care to the cesarean birth rate. N Engl J Med 1986;315:6619–6624.

54. Lopez-Zeno JA, Peaceman AM, Adasher JA, Socol ML. A controlled trial of a program for the active management of labor. N Engl J Med 1992;326:450–454.
55. Sandmire HF. In discussion of reference 48. Am J Obstet Gynecol 1993;168:1754–1756.
56. Raynor BD. The experience with vaginal birth after cesarean delivery in a small rural community practice. Am J Obstet Gynecol 1993;168:60–62.
57. Iglesias S, Burn R, Saunders LD. Reducing the cesarean rate in a rural community hospital. Can Med Assoc J 1991;145:1456–1459.
58. Peck T. Low cesarean rates possible. ObGyn News 1992;27:9.
59. Butler J, Abrams B, Parker J, Roberts JM, Laros RK Jr. Supportive nurse-midwife care is associated with a reduced incidence of cesarean section. Am J Obstet Gynecol 1993; 168:1407–1413.
60. Piturro M. Managed care obstetrical services: an option where everyone wins. Group Practice Managed Health Care News 1992 Feb: 16.
61. Dillon WP, Choate JW, Nusbaum ML, Amao-McCarthy M, McCall M, Rosen MG. Obstetric care and cesarean birth rates: a program to monitor quality of care. Obstet Gynecol 1992; 80:731–737.
62. Ophir E, Oettinger M, Yagoda A, Makovits Y, Rojansky N, Shapiro H. Breech presentation after cesarean section. Always a section? Am J Obstet Gynecol 1989;161:25–28.
63. Strong TH Jr, Phelan JP, Ahn MO, Sarno AP Jr. Vaginal birth after cesarean delivery in the twin gestation. Am J Obstet Gynecol 1989;161:29–32.
64. Flamm BL, Goings JR. Vaginal birth after cesarean section: is supected fetal macrosomia a contraindication? Obstet Gynecol 1989;74:694–698.
65. Beall M, Eglinton GS, Clark SL, Phelan JP. Vaginal delivery after cesarean section: the unknown uterine scar. J Reprod Med 1984; 29:31–36.
66. Davison L, Easterling TR, Jackson JC, Benedetti TJ. Breech extraction of low-birth-weight second twins: can cesarean section be justified? Am J Obstet Gynecol 1992;166: 497–502.
67. Taffel SM. Cesarean section in America: dramatic trends, 1970–1987. Stat Bull Metrop Life Insur Co 1989;70:2–11.
68. Williams RL, Chen PM. Identifying the sources of the recent decline in perinatal mortality rates in California. N Engl J Med 1982;306:207–214.
69. Malloy MH, Onstad L, Wright E, et al. The effect of cesarean delivery on birth outcome in very low birth weight infants. Obstet Gynecol 1991;77:498–499.
70. Anderson GD, Bada HS, Shaver DC, et al. The effect of cesarean section on intraventricular hemorrhage in the preterm infant. Am J Obstet Gynecol 1992;166:1091–1101.
71. Haverkamp AD, Orleans M, Langendoerfer S, et al. A controlled trial of the differential effects of intrapartum fetal monitoring. Am J Obstet Gynecol 1979;134:399–404.
72. Clark SL, Gimovsky ML, Miller FC. The scalp stimulation test: a clinical alternative to fetal scalp blood sampling. Am J Obstet Gynecol 1983;148:274–277.
73. Gordon D, Milberg J, Daling J, Hickok D. Advanced maternal age and risk factor for cesarean delivery. Obstet Gynecol 1991;77: 493–497.
74. Wagner MG, St Clair PA. Are in-vitro fertilization and embryo transfer of benefit to all? Lancet 1989;2:1027–1030.
75. Sauer MU, Paulson RJ, Lobo RA. Reversing the natural decline in human fertility! An extended clinical trial of oocyte donation to women of advanced reproductive age. JAMA 1992;268:1275–1279.
76. Gibbs RS, Amstey MS, Sweek RL, Mead PB, Sever JL. Management of genital herpes infection in pregnancy. Obstet Gynecol 1988; 71:779–780.
77. Randolph AG, Washington E, Prober CG. Cesarean delivery for women presenting with genital herpes lesions. JAMA 1993; 270:77–82.
78. Chuang T. Neonatal herpes: incidence, prevention and consequences. Am J Prev Med 1988;4:47–53.
79. Gibbs RS, Amstey MS, Lezotte DC. Role of cesarean delivery in preventing neonatal herpes infection. JAMA 1993;170:94–95.
80. Luthy DA, Wardinsky T, Shurtleff DB, et al. Cesarean section before the onset of labor and subsequent motor function in infants with meningomyelocele diagnosed antenatally. N Engl J Med 1991;324:662–666.

81. Garite TJ, Linzey EM, Freeman RK, et al. Fetal heart rate patterns and fetal distress in fetuses with congenital anomalies. Obstet Gynecol 1979;53:716–721.
82. Phillips RN, Thornton J, Gleicher N. Physician bias in cesarean sections. JAMA 1982; 248:1082–1084.
83. Berkowitz GS, Fiarman GS, Mojica MA, Bauman J, Haynes de Regt R. Effect of physician characteristics on the cesarean birth rate. Am J Obstet Gynecol 1989;161: 146–149.
84. Bashore RA, Phillips WH Jr, Brinkman CR III. A comparison of morbidity of midforceps and cesarean delivery. Am J Obstet Gynecol 1990;162:1428–1432.
85. Gould JB, Davey B, Stafford RS. Socioeconomic differences in rates of cesarean section. N Engl J Med 1989;321:233–239.
86. Haas JS, Udvarhelyi S, Epstein AM. The effect of health coverage for uninsured pregnant women on maternal health and the use of cesarean section. JAMA 1993;270:61–64.
87. Kennell J, Klaus M, McGrath S, Robertson S, Hinkley C. Continuous emotional support during labor in a U.S. hospital. a randomized controlled trial. JAMA 1991;265:2197–2201.
88. Rosen NG. Doula at the bedside of the patient in labor. JAMA 1991;265:2236–2237.
89. Public Health Service. Healthy people 2000: National health promotion and disease prevention objectives—full report with commentary. Washington, DC: U.S. Department of Health and Human Services, Public Health Service, 1991. (DHHS publication no. (PHS) 91-60212.)
90. Spellacy WN. Vaginal birth after cesarean: a reward/penalty system for national implementation. Obstet Gynecol 1991;78:316–317.

Appendix I. Mount Sinai–Chicago Perinatal Database/ Neonatal Outcome

Birthweight	>2500 g		1500–2499 g		<1500 g	
	Total	%	Total	%	Total	%
A. Babies delivered	2001	86.0	245	10.9	71	3.1
1. 1 min apgar < 7	171	8.5	42	16.5	40	56
2. 1 min apgar < 3	18	0.8	9	3.5	14	19
3. 5 min apgar < 7	41	2.0	16	6.2	19	26
4. 5 min apgar < 3	3	0.1	1	0.3	2	2
Babies and Morbidity Data	1808	90.3	210	82.6	35	4.9
B. Cardiorespiratory						
1. Endotracheal tube	22	1.2	22	10.4	21	60
2. Chest tube	7	0.3	7	3.3	10	28
3. Ventilator	14	0.7	22	10.4	19	54
4. Maximum F102 > 60% 02	10	0.5	7	3.3	11	31
5. Acidosis pH < 7.2	14	0.7	10	4.7	10	28
6. Pneumo th/med	2	0.1		0.0		0
7. Meconium aspiration	14	0.7		0.0		0
8. Pneumonia	1	0.0		0.0		0
C. Hematologic						
1. Bilirubin > 10 mg/dL	71	3.9	47	22.3	15	42
2. Phototherapy	73	4.0	51	24.2	22	62
3. EXC. transfusion		0.0		0.0	1	2
4. RBC transfusion	5	0.2	5	2.3	13	37

Birthweight	>2500 g		1500–2499 g		<1500 g	
	Total	%	Total	%	Total	%
D. Neurologic						
1. Seizures	3	0.1		0.0		0
2. Paralysis		0.0		0.0		0
3. CNS hemorrhage	2	0.1	2	0.9	3	8
4. Hydrocephalus	2	0.1		0.0		0
5. Other	14	0.7	2	0.9		0
E. Infection						
1. Sepsis	7	0.3	3	1.4	2	5
2. Scalp		0.0	2	0.9		0
3. Lung	2	0.1	1	0.4	3	8
4. Other	12	0.6	3	1.4	2	5
F. Miscellaneous						
1. Genetic abnormality	6	0.3	1	0.4		0
2. GI abnormality	35	1.9	15	7.1	7	20
3. Fracture	1	0.0		0.0	1	2
4. Peripheral IV	111	6.1	49	23.3	10	28
5. Central IV	113	6.2	50	2.38	13	37

Appendix II. Mount Sinai–Chicago Perinatal Database/ Obstetrical Outcome

	Total	%	1988		1987	
A. Mothers delivered	1320		2430		2301	
Babies delivered	1341		2472		2345	
1. Single pregnancy	1300	99.4	2390	98.3	2257	98.0
2. Multiple pregnancy	20	1.6	40	1.6	44	1.9
3. Nulliparas	369	27.9	761	31.3	666	28.9
4. Multiparas	951	72.9	1669	68.8	1625	71.0
5. Private	871	50.8	1471	80.5	1418	61.2
6. Faculty	649	49.1	959	39.4	851	36.7
7. Walk-ins	39	2.9	104	4.2	69	2.9
8. Home births	23	1.7	40	1.6	35	1.5
B. Vaginal deliveries	1175	89.0	2132	87.7	2035	86.4
1. NSVD	1107	83.8	1997	82.1	1895	82.3
2. Outlet forceps	28	2.1	63	2.5	79	3.0
3. Breech	34	2.5	55	2.2	43	1.6
4. Mildpelvic delivery	5	0.3	15	0.6	27	1.1
A. Low or mild forceps	3	0.2	2	0.0		0.0
B. Mild forceps	2	0.1	10	0.4	13	0.5
C. Vacuum extractions		0.0	3	0.1	14	0.6
C. Abdominal deliveries	145	10.9	298	12.2	265	11.5
1. Private	76	5.7	193	7.9	168	7.3
2. Faculty	69	5.2	105	4.3	87	4.2
3. Low transverse	131	9.9	293	12.0	205	11.5

	Total	%	1988		1987	
4. Low vertical	11	0.8	4	0.1		0.0
5. Classical	3	0.2	1	0.0		0.0
6. Second opinion	24	20.1	18	7.5		0.0
7. Primary	99	7.5	191	7.8	169	0.9
A. Dystocia	44	44.4	76	39.7	54	33.7
B. Breech	13	13.1	20	10.4	27	18.1
C. Fatal distress	23	23.2	51	26.7	46	28.7
D. Others	19	13.1	44	23.0	31	19.3
8. Repeat	48	3.4	107	4.4	165	4.5
9. Hysterectomy	1	0.0		0.0		0.0
D. Previous cesarean section	125	9.4	275	11.3	285	11.5
1. Elective	12	9.6	32	11.6	37	13.9
2. Trial of labor	113	90.4	243	88.3	228	86.0
A. Vaginal	79	69.9	167	68.7	169	70.1
B. Failed trial	34	30.0	76	31.2	68	29.8
E. Breech deliveries	54	4.0	109	4.4	113	4.8
1. Spontaneous breech	3	5.5	16	14.6	18	15.9
2. Assisted breech	27	50.0	40	36.6	23	20.3
3. Breech extraction	3	5.5	3	2.7	7	6.1
4. Cesarean section	21	38.9	49	44.9	65	57.5
A. Breech primary	15	71.4	29	59.1	38	56.4
B. Breech repeat	6	28.5	20	40.8	27	41.5
F. Multigestational deliveries						
1. Mothers	20	1.5	40	1.8	44	1.8
2. Babies	41	3.0	82	3.3	88	3.7
3. Babies, vaginal	25	1.8	52	2.1	53	2.2
4. Babies, cesarean section	16	1.1	30	1.2	35	1.4
A. Primaries	16	1.1	20	0.0	27	1.1
B. Repeats		0.0	10	0.4	8	0.3
G. Gestational hypertension	59	4.4	155	8.3	174	7.5
1. Mild preeclampsia	30	2.2	88	3.6	97	4.2
2. Severe preeclampsia	11	0.6	14	0.5	86	0.4
3. Eclampsia	1	0.0	3	0.1	2	0.0
4. Chronic hypertension	17	1.2	46	1.6	58	2.5
5. Mild preeclampsia and hypertension		0.0	2	0.0	5	0.2
6. Severe preeclampsia and hypertension		0.0	2	0.0	2	0.0
7. Eclampsia hypertension		0.0		0.0		0.0
H. Episiotomy						
1. No episiotomy	565	48.0	1240	56.0	1183	56.9
A. No episiotomy or laceration	550	97.3	1211	97.6	1147	96.6
B. No episiotomy-third degree	12	2.1	17	1.3	12	1.0
C. No episiotomy-fourth degree	3	0.5	12	0.9	4	0.3
2. Episiotomy	811	61.9	897	41.9	879	43.0
A. Midline	256	41.8	441	49.1	504	57.5
1. Midline no laceration	237	92.5	374	94.5	438	86.7
2. Midline third degree	14	5.4	42	9.5	35	0.9
3. Midline fourth degree	5	1.9	25	5.6	32	0.3
B. Mediolateral	355	58.1	456	50.6	373	42.4
1. Mediolateral laceration	346	97.4	433	94.9	362	97.0
2. Mediolateral–third degree	7	1.9	23	5.0	7	1.8
3. Mediolateral–fourth degree	2	0.5		0.0	4	1.0
3. Cervical laceration	10	0.0	7	0.3		9.0
4. Vaginal laceration	8	0.0	4	0.1		8.0

	Total	%	1988		1987	
I. Miscellaneous complications						
1. Premature labor	268	20.3	474	19.5	464	20.1
2. Induced labor	113	8.5	325	13.3	333	14.4
3. Previa abruption	15	1.1	27	1.1	22	0.9
4. Srom before labor	210	15.9	480	19.7	412	17.9
5. Drug abuse total	96	7.2	123	5.0	126	5.2
A. Suspected	15	1.1	11	0.4	12	0.5
B. Admitted	81	8.1	112	4.6	106	4.6
6. 1 min apgar <7	99	7.3	257	10.3	253	10.7
7. 1 min apgar <3	7	0.5	44	1.7	41	1.7
8. 1 min apgar <7	16	1.1	70	2.0	76	3.2
9. 1 min apgar <3	2	0.1	12	0.4	6	8.2
10. Stillbirths >500 g	26	1.9	35	1.4	21	0.0

APPENDIX III. Mount Sinai–Chicago CPGs for the Most Frequent Cesarean Section Indications

1. Previous Cesarean Section

The department recognized in principle that vaginal delivery was preferred for all patients who had previously undergone a cesarean section. This included patients with vertical scars in the lower segment of the uterus, those who had undergone classic cesarean sections or myomectomies in which the uterine cavity had been entered, and those with previous uterine incisions of unknown type. Patients were usually advised during a counseling session at their first prenatal visit that a trial of labor was anticipated.

2. Dystocia/Cephalopevic Disproportion

The diagnosis of dystocia was accepted as an indication for a cesarean delivery only after no progress of labor was observed for more than 2 h of regular uterine contractions of appropriate strength, as evaluated by intrauterine pressure catheters. A contraction frequency of once every 2 to 3 min and a contraction intensity of 50 torr in early labor and 70 torr in late labor were expected. Oxytocin was administered freely to achieve this goal.

3. Fetal Distress

A diagnosis of fetal distress, based on monitoring of the fetal heartrate, had to be corroborated by sampling of blood from the fetal scalp whenever technically feasible. A maternal-fetal pH difference of 0.20 to 0.28 was required for confirmation, depending on the maternal pH. The pH of the unbilical cord blood had to be obtained immediately after the birth of every infant delivered by cesarean section because of fetal distress. The result was used in the subsequent case review.

4. Breech Presentation

Vaginal delivery was recommended for all breech fetuses with the exception of those with true hyperextension of the cervical spine or macrosomia. Macrosomia was defined as an estimated fetal weight of more than 4300 g, determined by ultrasound with use of standard nomograms for fetal biparietal diameter and abdominal circumference. The parity of the patient and type of breech presentation were not considered in the decision-making process.

12 Cesarean Delivery: A Medical-Legal Perspective

JEFFREY P. PHELAN

Since the NIH consensus conference on cesarean birth in 1980,[1] the cesarean delivery rate has continued to rise.[2,3] Today, cesarean delivery continues to be the number one hospital-based operative procedure in the United States. This rise has continued despite the increased availability of cesarean alternatives such as vaginal birth after cesarean[4–6] and external cephalic version.[7–10]

In fact, the rise in cesarean births appears to be caused by factors that do not readily lend themselves to statistical analysis. For example, the incidence of cesarean births has paralleled the rise in malpractice claims. During the 1980s, the average number of paid claims rose 450%.[11] Currently, more than 80% of the fellows of the American College of Obstetricians and Gynecologists (ACOG) have been sued at least once[12] and many members more than once. Of greater concern to the practicing obstetrician is the fact that the rise in malpractice claims has been accompanied by an increase in the number of awards in excess of $1 million. Since 1975, these awards have risen 232 fold.[13] By 1994, the midpoint brain-injured baby jury award was estimated to be $2.5 million.

In obstetrics, the cesarean problem is complicated by parental expectations of a "perfect child" and the belief that cesarean delivery can provide that child. Notwithstanding, and unlike other medical specialists, the obstetric health care provider must simultaneously balance the interests of two and sometimes more individuals. When it comes to cesarean delivery, the balancing continues. Should a cesarean delivery be performed and thus expose the mother to a greater risk of morbidity and mortality? Or should vaginal delivery be allowed because it reduces these maternal risks, yet exposes the fetus to a greater one? It may be that the circumstances of the case make cesarean delivery the only apparent option. Under different circumstances, vaginal rather than cesarean delivery would have been the best choice.

It is these complexities that have frequently contributed to the rise in malpractice claims and lawsuits. For these reasons, this chapter is devoted to the medical-legal issues associated with cesarean birth, and reviews the basic concepts of the duty to refer, the doctrine of informed consent, timely cesarean delivery, and foreign bodies left in the abdomen.

A detailed analysis of each of these issues is not within the scope of this chapter. First, the laws vary from state to state. Second, the individual circumstances of each case vary. Third, a purpose of this chapter is to provide the reader with an overview of the medical-legal issues that could arise from the performance of a cesarean. If a more detailed analysis with applicable state law is desired, the reader is referred to an attorney in the reader's state or jurisdiction. Finally, this chapter is intended to be used for educational purposes and is not designed to provide legal advice.

Referral to a Specialist

The technological advances and the information explosion of the past 20 years have made it harder for the practicing physician to keep pace with the changes in medicine. Subspecialization has been a necessary outgrowth of these advances. The net effect of subspecialization has been to make America the number one health care system in the world. Obstetrics and gynecology has also followed the subspecialization pathway. During the past 20 years, for example, fetal monitoring and ultrasonography have been introduced into the contemporary practice of obstetrics. In many respects, the three obstetrics and gynecology subspecialty areas—gynecologic oncology, reproductive endocrinology, and maternal-fetal medicine—are a necessary outgrowth of the technological and information explosion of the 1970s and 1980s. Initially, these specializing physicians limited their practices to university hospitals. It is estimated, however, that, in the past 5 years the number of maternal-fetal subspecialists practicing in community hospitals has risen exponentially.

Nevertheless, a nonspecialist, general obstetrician-gynecologist, appears to have a duty to refer a patient to a specialist or to request the assistance of a specialist whenever circumstances warrant it. For example, with the increased availability of maternal-fetal medicine subspecialists, an increasing number of high-risk patients could be referred to these subspecialists, not necessarily for primary prenatal care, but for consultation. If, for example, an obstetrician-gynecologist fails to refer a high-risk obstetrical patient to a perinatologist, that physician could be held to the standard of care of the perinatologist (Table 12.1). This means a higher standard of care could be applied to the general ob-gyn.

TABLE 12.1. Commonly used bases for the possibility of a failure to refer the patient to a specialist

A careful obstetrician/gynecologist would have recognized a problem existed
The obstetrician/gynecologist was not capable of handling the problem
A timely referral was not made

TABLE 12.2. Clinical situations in which the failure to consult a specialist at the time of cesarean could arise

Intraoperative complications, e.g., bladder, bowel, or ureteral injury
Intensive care unit admissions
Postpartum hemorrhage with or without hysterectomy
Hypogastric or uterine artery ligation

For example (Table 12.2), assume a bladder laceration occurs during the performance of a cesarean. Under most circumstances, a fully trained specialist in obstetrics and gynecology can repair this laceration without calling in a consultant. But, a more extensive laceration of the bladder, bowel, or ureter, might, depending on the learning, skill, and experience of the obstetrician and the circumstances of the case, may require a physician more highly skilled in the repair of these complications. This could include, but is not limited to, a urologist, a gynecologic oncologist, or a general surgeon. If the repair is undertaken, however, by a nonspecialist and the patient sustains a causally related injury, the nonspecialist could be held to the standard of care of one of these specialists. Thus, the underlying rationale for this principle is to encourage frequent consultation in an effort to improve the quality of care. After all, more frequent consultation provides a direct benefit to the patient and an indirect one to her primary obstetrician.

The primary issue in today's obstetric practice, however, is whether the specialist referral will be authorized? During the past decade, we have witnessed an expansion in health maintenance organizations with their nonphysician gate-keeper control of medical decision making. Not infrequently, referrals to specialists are not authorized and the burden of care will ultimately rest with the general ob-gyn. How does one in this practice setting reconcile the catch-22 of pericesarean complications? The cornerstone of referrals is an awareness of one's limitations. This does not mean that the general ob-gyn is not competent to repair a bladder laceration. The issue is twofold: one, our limited experience in handling selected complications; and two, our goal is to provide quality care to our obstetric patients. This necessarily means that frequent consultation benefits the patient and improves patient care.

Informed Consent

As patients have sought greater control of their health care, informed consent has assumed a prominent role in the practice of medicine today. One concern among practitioners is not whether informed consent was given but whether sufficient information has been provided to the patient to permit an informed choice.[14] In obstetrics, the adequacy of informed consent becomes complicated by the pain of labor, analgesics that can affect the patient's mental state, and the expanding rights of the fetus.[15] It is these complexities that make each patient dependent on her physician to provide sufficient information before undertaking the proposed operation, such as a cesarean.

What constitutes adequate patient informed consent varies from state to state and according to the circumstances of the case.[16] But, there appears to be a common thread. For example, the doctrine of informed consent focuses on the characteristics of the physician–patient relationship.[17] First, knowledge of medicine between the physician and the patient is not the same. Second, a competent adult has the right to decide whether to agree to lawful medical treatment. Third, before submitting to the treatment, the patient has the right to be informed. Fourth, the patient is unlearned in the medical sciences and, as a result, is dependent on her physician to provide her with sufficient information to make an informed choice.

The underlying rationale for the doctrine of informed consent is illustrated in *Cobbs v. Grant*, 8 Cal 3d 229, 104 Cal Rptr 505, 502 P2d 1 (1972). In that case, the patient agreed to undergo surgery for a duodenal ulcer. However, the inherent risks associated with the surgical procedure were not discussed with the patient. Following the surgery, the patient sustained internal bleeding and required a splenectomy. A month after the second surgery, the patient was readmitted for a gastric ulcer and underwent a partial gastrectomy. Subsequently, the patient was readmitted for abdominal bleeding because of an abdominal suture dissolving prematurely. At issue in this case was whether the patient had been adequately informed of the risks of surgery.

What must be disclosed? In many respects, the information to be disclosed is that which you or I would want to know if we were considering a proposed medical treatment or surgical procedure. This would include sufficient information for the patient to make an informed decision. Frequently, the difficult part is deciding what is considered material. This decision often relates to the informed consent requirements in your state. In a patient-oriented standard (minority rule), the obstetrician would have to consider what a reasonable patient would consider significant.[14] In contrast, the professional standard (majority rule) is directed at what a reasonable obstetrician would disclose under the same or similar circumstances.[14]

Under a patient-oriented standard, the patient is entitled to know about any known risk of death or serious illness.[14] At the same time, the disclosure or explanation to the patient should be given in lay terms or at the patient's level of understanding. What if there are lesser risks associated with a proposed procedure or treatment? Under these circumstances, disclosure of lesser risks may depend on whether physicians practicing under the same or similar circumstances would be required to disclose those risks. But where the procedure is simple and the danger is remote and commonly appreciated to be remote, such as drawing blood or the patient's request not to be informed, there may be no duty to disclose the risks of that procedure.

With respect to cesarean delivery, the patient is generally entitled to be informed of the reason for the cesarean, the risks and potential complications of the procedure, alternative treatment approaches such as external cephalic version or vaginal birth after cesarean, and the consequences if the procedure is not performed and the problems of recuperation. Equally important is the fact that the performance of a cesarean without the patient's consent[18] or where the consent is deemed invalid[19] could be considered battery.

Even though informed consent appears to be an infrequently litigated issue today,[16] it is important to understand how these cases might be approached. As noted in Table 12.3, the basic elements of negligence are also involved. But, once a doctor–patient relationship is established, the question becomes whether the physician breached her duty by failing to give sufficient information so that a proposed treatment could be chosen intelligently. This again will depend on the applicable informed consent standard, whether patient oriented or professional. This will frequently focus on whether the physician informed the patient of alternative treatments, the reasonably foreseeable risks of each alternative, the consequences of no treatment, and items of a personal interest. Additionally, if the material risk of an alternative treatment has been disclosed, the patient, it seems, could have chosen either no treatment or a different one. Finally, the patient would have to show that she was injured as a result of submitting to treatment. What consequences will follow if the patient has consented to a proposed treatment or operation and the physician has failed to inform the patient adequately? The failure to obtain "such consent is negligence and renders the physician or surgeon subject to liability for any injury resulting from the [treatment or operation] if a reasonably prudent person in the patient's position would not have consented to the [treatment or operation] if he had been adequately informed of all the significant perils."[20]

Informed refusal or the consequences of nontreatment was established in a California case, *Truman v. Thomas*.[21] In that case, Mrs. Truman saw her physician on several occasions between 1964 and 1969. Her physician, however, never obtained a Pap smear, nor did he specifically inform her of the risks involved in a failure to perform this test. He said "You should have a Pap smear," but this was not supported by the medical records. Subsequently, she was diagnosed as having inoperable carcinoma of the cervix and died at age 30. Suit was brought by her dependent children. The finding of the court was that the patient should have been given adequate information regarding the consequences of refusing to undergo a Pap smear.

The failure to inform the patient of alternative treatment approaches may also serve as a basis for medical malpractice claims. Common clinical examples are demonstrated in Table 12.4. To illustrate, the recent resurgence of vaginal birth after cesarean where it has been shown that maternal morbidity and mortality can be significantly reduced has made the vaginal birth option available to prior cesarean patients. For example, the performance of an elective repeat cesarean delivery without first disclosing to the patient the trial-of-labor alternative could be an example of inadequate informed consent. This means that the physician could be held liable even if a re-

TABLE 12.3. Basic elements of adequate informed consent for a cesarean birth

Indications for procedure
Alternative treatments
Material risks of each option
Problems of recuperation
Consequences of nontreatment
Items of personal interest

TABLE 12.4. Selected clinical examples and the options available in each one

Example	Options
Breech presentation	Selected vaginal delivery External cephalic version Cesarean delivery
Prior cesarean	Vaginal birth after cesarean Repeat cesarean
Postdates	Induction of labor with or without cervical ripening Fetal surveillance testing Cesarean delivery
Transverse lie	External version with attempted vaginal delivery Cesarean delivery

peat cesarean is performed nonnegligently.

During the last decade, informed consent requirements have changed dramatically. For example, consent guarantees have recently been recommended in a North Carolina case.[22] In this case, the hospital was found liable for failure to assure a patient with a breech presentation that her obstetrician had properly informed her of the risks of vaginal birth. In that case, Mrs. Campbell was permitted to deliver her known breech infant vaginally. During the course of delivery, a prolapsed cord occurred, and the infant became severely asphyxiated and is now neurologically impaired. In holding against the nurse-hospital, the Campbell court stated that "... explaining the risk of alternative procedures [is] the responsibility of the physician ... assuring that the patient has had an explanation [is] the responsibility of the nurse."[22]

Recently, items unrelated to a patient's health but of a personal interest to a physician must be disclosed. In a 1990 California case,[23] a physician must disclose "... personal interests unrelated to a patient's health whether research or economic, that may affect the physician's judgment." In that case, a University of California at Los Angeles (UCLA) physician had substantial propriety interest in the selling of a patient's blood without first disclosing that interest to the patient.

Failure to Perform a Cesarean in a Timely Manner

"Cases frequently are generated by the failure of physicians or staff to recognize and manage immediate or imminent emergencies. Failure to properly assess and react is fertile ground for a negligence claim."[24] But, when is it a failure to perform a cesarean in a timely manner? According to the most recent guidelines of the ACOG,[25] any facility that provides obstetric care should have, as a minimum, a "cesarean delivery capability within 30 minutes." This would imply, therefore, that a decision–incision interval in excess of 30 min could constitute a breach in the standard of care. If the decision–incision interval does exceed 30 min and the infant is born with a neurologic handicap, should the physician be held liable for those injuries? For this reason, there may be a "fertile ground for a negligence claim," but it does not necessarily follow that the obstetrician-gynecologist proximately caused the injuries. Circumstances may be such that cesarean delivery may not have been technically feasible within the "allotted time."

Moreover, there is a popular misconception that neonatal brain damage is proximately related to the events of labor and is caused by not the natural course of events of pregnany but the negligent conduct of the attendant physician, nursing staff, or hospital. As sugested by Perkins,[26] "the number of infants injured before labor is highly underestimated [and those infants injured] during labor is highly overestimated." In support of Perkins' assertion, Paul and associates[27] have demonstrated that fetal injuries such as intracranial hemorrhage, myocardial infarction, and meconium aspiration can and do occur before the onset of labor.

What then is the role of electronic fetal monitoring? Although "fetal heart rate changes can provide an indicator of hypoxia that precedes neurologic damage,"[28] the fetal heart rate pattern at the onset of labor can be indicative of fetal prelabor injury.[29] This suggests, therefore, that "intrapartum asphyxia may be the result rather than the cause of neurologic abnormality."[30] Moreover, neonatal compromise may represent a continuum rather than a single isolated event that "is not . . . influenced markedly by the circumstances of labor and delivery."[26] This is not to say that an adverse fetal outcome cannot be related to the events of labor or to the conduct of the obstetrician or nurse, but rather that the failure to perform a cesarean within a specified period of time is not necessarily the cause of the neonate's injuries.

Nevertheless, the physician should make reasonable efforts either to evaluate a patient [*Thomas v. Ellis*, 329 Mass 93, 106 NE 2d 687 (1952)] or to begin a cesarean in a timely manner.[31] For example, in *Thomas v. Ellis*, the pregnant woman's husband first notified her physician around 11:30 P.M. that she was bleeding heavily from her vagina. At that time, the husband was advised to keep her at home and observe her for approximately 1 h. About a half-hour later, the husband called the physician again because there was "more blood all over the blanket, a lot of blood" and the patient had pain. The physician told the husband that it sou¢ded like an abruption. He advised them to go directly to the hospital and stated that he would meet them there. The couple arrived at around 12:30 A.M. The record was unclear as to when the physician arrived, but a cesarean was done about 11:00 A.M. At the time of delivery, an abruption was confirmed and the fetus was dead.

This case illustrates the importance of a timely evaluation and delivery. An earlier evaluation and delivery might have provided a more favorable outcome for the fetus. But because that evaluation and delivery were not undertaken earlier, we will never know whether it would have made a difference in this case. However, a timely evaluation is the first step in minimizing the risk of harm.

In the event of acute fetal distress such as an acute, prolonged fetal heartrate deceleration, the timeliness of the cesarean may protect the fetus from neurologic injury or death. With in-house anesthesia, cesarean delivery can readily be achieved. But in-house anesthesia is not always encountered where obstetric services are provided. Under these circumstances, cesarean delivery would be necessarily delayed while waiting for anesthesia personnel to come to the hospital.

If anesthesia personnel are not in the hospital, several options are available. Two options are available to correct the fetal heartrate pattern. One is to attempt to arrest uterine activity with a beta-mimetic such as terbutaline.[32] The other option is to reestablish the amniotic fluid volume with saline amnioinfusion.[33] This would allow intrauterine resuscitation of the fetus and provide additional time for anesthesia support personnel to arrive. If the beta-mimetic is contraindicated, technically not feasible, or unsuccessful, cesarean delivery under local anesthesia is a reasonable consideration.[34]

In summary, reasonable efforts to evaluate a patient or to perform a cesarean in a timely manner would appear prudent. Whether this will have an effect on the incidence of neonatal injury or death remains to be proven.

Failure to Perform a Cesarean Properly

The failure to perform a procedure, such as the administration of rhogam to an Rh-negative gravida, may give rise to a claim against the physician for maternal or neonatal injuries in a subsequent pregnancy.[35] This concept has been extended to include a host of cases in which physicians have been held liable for a child's injuries that resulted from negligence toward the child's mother in a prior pregnancy.[36] This extension includes the failure to perform a cesarean properly during an earlier pregnancy. In *Bergstreser v. Mitchell*, 577 F2d22 (CA8 MO, 1978), the plaintiff had undergone a cesarean in a previous pregnancy. The prior incision, however, was not described in the legal summary. At approximately 30

TABLE 12.5. Documentation considerations in the patient with a classical or upper-segment cesarean incision

The type of incision
The patient has been told of the incision
The patient has been advised that attempted vaginal delivery in a future pregnancy is not recommended
The risk of uterine rupture in a subsequent pregnancy

weeks gestation in a subsequent pregnancy, she sustained an uterine rupture that resulted in an exploratory laparotomy and delivery of the infant. The child suffered brain damage. The plaintiff alleged that the injuries sustained by the infant resulted from the negligent performance of the prior cesarean and the failure to inform her of the prior uterine incision.

In this era, where vaginal birth after cesarean is an acceptable alternative,[4–6] it is important to inform the patient of the type of uterine incision performed and its consequences (Table 12.5). In addition, reasonable efforts to avoid a cesarean birth or to reduce the necessity for an upper-segment incision with external cephalic version[7–10] would appear prudent. If this is done, the likelihood of liability for a subsequent pregnancy might be lessened.

Foreign Bodies Left in the Patient Following a Cesarean

Throughout surgical history, numerous objects have been left in the abdomen following surgery. These have included beetles, needles, drains, sponges, surgical instruments, forceps, and a cloth sac.[37] Although foreign bodies may be more frequently left behind during extensive surgery, a needle, sponge, and instrument count is designed to protect against this event. The failure to follow such a procedure could be evidence of negligence [*Bowers v. Olch*, 120 Cal App 2d 108, 260 P2d 997 (1953)].

Before the performance of a cesarean, a needle and sponge count is done by the surgical scrub nurse and the circulating nurse. Once completed, the results of the count are given aubibly. Before the end of the cesarean, none of the sponges, needles, or instruments should leave the operating room. Before and after the closure of the peritoneum, the count is done again and the results are given audibly to the surgeons and scrub nurse. The numbers must match.

If the count indicates a missing sponge, needle, or instrument, a search for the missing item is appropriate. The failure to search for a missing item could be evidence of negligence [*Smedra v. Stanek*, 187, F2d 892 (CA 10 COLO 1951)]. Fortunately, most items are radiopaque or, in the case of sponges, contain a radiopaque tape. Thus, if manual and visual exploration is unsuccessful, an abdominal x-ray may help locate the missing item.

Reliance on the needle and sponge count alone may not support the claim that a physician exercised due care.[37] For example, in the case of *Key v. Caldwell* [139 Cal App 2d 698, 104 P2d 87 (1940)], a laparotomy sponge was left in a patient's abdomen following a cesarean delivery. The needle and sponge count was done and was said to be correct. Testimony demonstrated that the surgeon in this case had the practice of placing a laparotomy sponge in the same location in the abdominal cavity at each cesarean. Althogh the physician attempted to rebut the allegations, the sponge had nevertheless not been removed.

The Key case suggests that the surgeon has an independent duty to determine whether a sponge, for example, has been left behind. Thus, in addition to an accurate needle, sponge, and instrument count, manual and visual exploration of the abdomen would also appear prudent.

When an instrument has been broken[38] during a cesarean, reasonable efforts to locate the missing object would also be appropriate. These efforts may be sufficient if the item is not located [*Roark v. Peters*, 162 La 111, 110 So 106 (1926)], but the patient is entitled to be informed of the missing object unless there are medical reasons for not informing her.[37]

TABLE 12.6. At the time of a cesarean, the count: what you should include[39,40]

Needle
Sponge
Instrument
Fetal scalp electrode

If an item is discovered to be missing after a cesarean, is consent required to remove it? According to Jones,[37] the operating surgeon can perform a second operative procedure for the purpose of removing a foreign body immediately following the primary procedure without the consent of the patient. If, however, a sponge is found to be missing some time after the cesarean, full disclosure and consent would be appropriate[37] [*Delahunt v. Fenton*, 244 Mich 226, 221 NW 168 (1928)].

In summary, needle, sponge, and instrument counts are an important part of a cesarean delivery. With a cesarean, the fetal scalp electrode should also be counted (Table 12.6). In addition, a manual and visual exploration of the abdomen before closure would appear prudent. Consideration should be given to document the manual and visual exploration in the medical records. If it is subsequently determined that an item is missing, the physician should tell the patient and obtain her consent to remove it.

References

1. Cesarean Childbirth Report of a Consensus Development. (DHHS publication no. 82-2076.) Conference sponsored by the National Institute of Child and Human Development, Washington, DC, 1980.
2. Taffell SM, Placek PJ, Liss T. Trends in the USA cesarean section rate and reasons for the 1980–1985 rise. Am J Public Health 1987;77(8):955–959.
3. Barr R. C-section under fire. OBG Management 1990;(October):18–25.
4. Phelan JP, Clark SL, Diaz F, et al. Vaginal birth after cesarean. Am J Obstet Gynecol 1987;157:1510–1516.
5. Phelan JP, Ahn MO, Diaz F, et al. Twice a cesarean, always a cesarean? Obstet Gynecol 1989;73:161–165.
6. Flamm BL, Newman LA, Thomas SJ, Fallon D, Yashida MM. Vaginal birth after cesarean delivery: results of a 5-year multicenter collaborative study. Obstet Gynecol Surv 1991; 46:360–365.
7. Phelan JP, Stine LE, Edwards NB, et al. The role of external cephalic version in the intrapartum management of the transverse lie presentation. Am J Obstet Gynecol 1985; 151:724–726.
8. Dyson DC, Ferguson JE, Hensleigh P. Antepartum external cephalic version under tocolysis. Obstet Gynecol 1986;67:63–67.
9. Morrison JC, Myatt RE, Martin JN, et al. External cephalic version of the breech presentatoin under tocolysis. Am J Obstet Gynecol 1986;154:900–905.
10. Stine LE, Phelan JP, Wallace RL, et al. Update on external cephalic versoin performed at term. Obstet Gynecol 1985;65:642–646.
11. Jacobson PD. Medical malpractice and the tort system. JAMA 1989;262:3320–3327.
12. Swan JE. Your views on the crisis. Contemp Ob/Gyn 1989;33:13–24.
13. Harrison TF. The top ten jury awards of 1990. *Lawyers Alert* 1991;11:4–16.
14. Appelbaum PS, Lidz CW, Meisel A. The legal requirements for disclosure and consent: history and current status. In: Informed Consent: Legal Theory and Clinical Practice. New York: Oxford University Press, 1987: 35–65.
15. Phelan JP. The maternal abdominal wall: a fortress against fetal health care? South Cal Law Rev 1991;65:461–490.
16. Appelbaum PS, Lidz CW, Meisel A. Informed Consent: Legal Theory and Clinical Practice. New York: Oxford University Press, 1987.
17. *Cobbs v. Grant*, 8 Cal 3d 229, 104 Cal Rptr 505, 502 P2d 1 (1972).
18. *Mohr v. Williams*, 95 Minn 261, 104 NW 2d 523 (1905).
19. *Rainer v. Community Memorial Hosp*, 18 CA 3d 240, 95 Cal Rptr 901 (1971).
20. Book of Approved Jury Instructions. St. Paul: West, 1986.
21. *Truman v. Thomas*, 27 Cal 3d 285, 165 Cal Rptr 308, 611 P2d 902 (1980).
22. *Campbell v. Pitt County Hosp*, 84 NC App 314, 352 SE 2d 902 (1987) [decertified in later decision].
23. *Moore v. Regents of California*, 90 Daily Journal D.A.R. 8010 (1990).

24. Olender JH. Obstetric negligence. Trial 1984;May:52.
25. Standards for Obstetric-Gynecologic Services, 7th Ed. Washington, DC: American College of Obstetricians and Gynecologists, 1989: 33–39.
26. Perkins RP. Perspective on perinatal brain damage. Obstet Gynecol 1987;69:807.
27. Paul RH, Yonekura ML, Cantrell CJ, et al. Fetal injury prior to labor: does it happen? Am J Obstet Gynecol 1986;154:1187.
28. Adamson SK, Myers RE. Late decelerations and brain tolerance of the fetal monkey to asphyxia. Am J Obstet Gynecol 1977; 128:893.
29. Phelan JP, Ahn MO. Perinatal observations in 48 neurologically impaired term infants. Am J Obstet Gynecol 1994;171:424–431.
30. Niswander K. Asphyxia in the fetus and cerebral palsy. In: Pitkin RM, Zlatnik FJ, eds. 1983 Year Book of Obstetrics and Gynecology. Chicago: Year Book Medical, 1983:107–125.
31. Gee DR. Physician's failure to perform timely cesarean. 19 POF 2d 285 (1979).
32. Patriarco MS, Viechnicki BM, Hutchinson TA, et al. A study on intrauterine fetal resuscitation with terbutaline. Am J Obstet Gynecol 1987;154:384.
33. Strong TJ, Phelan JP. Amnioinfusion: an intrapartum technique for the '90's. Contemp Ob/Gyn 1991;36(5):15–24.
34. Gilstrap LC, Hankins GDV. The uncomplicated patient. In: Phelan JP, Clark SL, eds. Cesarean Delivery. New York: Elsevier, 1987:139–154.
35. *Renslow v. Mennonite Hosp*, 67 I11 2d 348, 10 I11 Dec 484, 367 NE 2d 1250 (1977).
36. Liability for child's personal injuries or death resulting from tort committed against child's mother before child was conceived. 91 ALR 3d 316 (1979).
37. Jones JL. Malpractice: liability of physician, surgeon, anesthetist, or dentist for injury resulting from foreign object left in patient. 10 ALR 3d 9 (1966).
38. Zitter JM. Medical malpractice: instruments breaking in course of surgery or treatment. 20 ALR 45th 1179 (1986).
39. Thoulon JM, Gonnet C. Les accidents de l'electrocardiographic foetale directe (a propos d'une de l'electrode de scalp). J Gynecol Obstet Biol Reprod 1978;7:1257.
40. Nieburg P, Gross SJ. Breakage of a fetal scalp blade with retention of fragments in the infant's scalp. Am J Obstet Gynecol 1987;157:441.

13 Economic Considerations in Cesarean Section Use

LAURA B. GARDNER

Cesarean section was originally indicated to preserve the life of a mother with obstructed labor or to deliver a viable infant from a mother who was dying.[1] Although now in widespread use in less dire situations, the operative approach to delivery is associated with significantly greater economic costs compared with vaginal delivery. The increased cost of cesarean section is related both to higher delivery charges and the costs of maternal and neonatal morbidity and mortality. In a health care policy environment that increasingly emphasizes cost-effectiveness and cost containment, high rates of use of expensive health care technologies such as cesarean section must be justified by improvements in outcome that are worth their cost.

This chapter describes the nationwide economic cost of current levels of cesarean section use, outlines the economic factors that influence and contribute to cesarean section use, and reviews the clinical and economic issues to be considered in attempting to model the optimal choice of delivery modality using decision analysis theory.

Economic Impact of Cesarean Section Use

In a 1987 nationwide sample of U.S. hospitals, the average total hospital charge across all payers for hospital discharges with cesarean section as the primary procedure was $3386.[2] Total hospital charges for the 934,000 cesarean sections performed in the United States in 1987, which comprised 24.4% of all deliveries,[3] amounted therefore to more than $3.1 billion in 1987 dollars. In comparison, the average charge for all other hospital discharges with an obstetric procedure as the principal procedure, exclusive of normal deliveries, was $1587, or 53% lower than the charges for cesarean section.[2] For discharges with a normal delivery, the average hospital charge was $1396, almost 59% lower than for cesarean section.

These data suggest that decreasing the overall cesarean section rate would lead to a substantial reduction in expenditures for obstetric hospitalizations nationwide. Assuming conservatively that for each delivery in which a cesarean section was avoided there was instead a vaginal delivery, even one with one or more reportable complications, the annual savings associated with each percentage point reduction in the cesarean section rate would amount to more than $68 million (1987 dollars). This estimate of savings does not include the additional savings that would result from lower physician fees and avoided cesarean section-related morbidity and mortality.

Experts have estimated that the ideal cesarean section rate lies in the range of 10% to 18%.[4–6] On the basis of these estimates, if the cesarean section rate were reduced nationwide from 24% to 18% (the upper end of the estimated ideal range), the expected reduction in expenditures for hospital services alone would surpass $435 million per year in 1987 dollars.

Cesarean section-related morbidity and mortality has significant social and economic implications. The risk of maternal death after cesarean section has been reported to be approximately 2 deaths per 1000 operations,[1] or between 2 and 26 times that of vaginal delivery.[7–11]

Maternal cesarean-related morbidity, which includes cerebrovascular events, endometritis, hemorrhage, thrombophlebitis, wound and urinary tract infections, pulmonary embolism, and intestinal obstruction, occurs in 25%–50% of patients.[1] Postpartum infection occurs in as many as 20% of women delivering by cesarean section.[12] Overall, maternal cesarean section morbidity is associated both with higher direct costs of care and with indirect costs such as lost productivity, delayed recovery, and impaired maternal–infant bonding.[13]

Elective cesarean section occurring before labor begins also carries the risk of iatrogenic (physician-induced) prematurity and respiratory distress syndrome in the infant.[14] Pulmonary complications are more likely to occur with cesarean section than with vaginal delivery, because optimal infant lung function depends on the physical pressures exerted on the fetal thorax during vaginal delivery.[15] A study of neonatal outcomes reported by Martin et al.[16] confirmed that, controlling for the mother's personal and clinical characteristics, babies delivered by cesarean section were much more likely to have health problems, many of which could be attributed to the use of cesarean section.

Analysis of California hospital discharge abstract data (California Office of Statewide Health Planning and Development) confirms that in 1991 the risk of sudden maternal death was 2.2 fold higher for patients receiving cesarean section than for those who had a nonoperative delivery. The combined risk of the five most serious postpartum complications other than maternal death (cerebrovascular disorder, pulmonary embolism, postpartum infection, thrombophlebitis, wound dehiscence or infection, and pyrexia) was more than 6 fold higher for patients receiving cesarean section than for those who had a nonoperative delivery. The presence of one or more of these postpartum complications increased the mean inpatient cost of hospital discharges with cesarean section as the principal procedure between 38% and 250%.

Although use of cesarean section in appropriate clinical situations (e.g., clearcut fetal distress) can dramatically reduce neonatal morbidity, it has been shown that higher cesarean section rates do not explain the trend toward decreased perinatal mortality.[4,8,17] A recent study of very low birthweight infants reported that there was no improvement in the survival of such infants despite an increasing rate of both cesarean section and active resuscitation.[18] If cesarean section rates are higher than clinically necessary, decreasing the number of cesarean sections performed could be cost-saving without adversely affecting quality of care.

Economic Factors Influencing Use of Cesarean Section

There has been much research to indicate that economic and other nonclinical factors contribute to the treatment decisions that physicians make in addition to clinical considerations. In the case of obstetric decision making, the choice of cesarean section frequently has been shown to be associated with factors other than the objective clinical risks and benefits of cesarean delivery. Nonclinical and economic factors that have been shown to be associated with the rate of use of cesarean section include patient preferences, payment incentives, physician and hospital characteristics, and the organization of medical care delivery.[19–22]

Economic theory holds that self-interested suppliers of services, in furtherance of their own economic interests, will seek to maximize both the profitability of their services and their personal satisfaction.[23–25] In discussing the influence of economic incentives in health care, Luft[26] suggested that "the rapid growth in the use of medical technologies may stem not just from clinical efficacy but also from the high returns physicians receive by using such technology" (p. 505). The following discussion focuses on physician- and hospital-level economic factors that have been shown to influence the use of cesarean section.

Physician-Level Economic Incentives

The most direct economic incentive that may influence a physician's decision making in a fee-for-service system is the relative reimbursement amount. In particular, the difference in reimbursement between cesarean section and vaginal delivery has been cited as a contributor to the high rate of cesarean deliveries. Tussing and Wojtowycz[27] reported a positive and significant association between the ratio of physician charges for cesarean section to physician charges for vaginal delivery and the probability of a cesarean section.

The reimbursement disparity between cesarean section and vaginal delivery is considerable. Data from a private payer in Northern California indicate that physician fees for deliveries in 1989 differed by 42% for the two procedures, with an average fee of $1377 for a vaginal delivery and $1951 for a cesarean section, both including prenatal care[28] (and Robinson et al., unpublished data). A recent survey of doctor charges indicated that nationwide the median physician fee for a cesarean section in 1993 was $2400, or 20% higher than the $2000 median physician fee for vaginal delivery.[29]

In addition to the influence of reimbursement differentials, there are indirect economic incentives that influence physician behavior. When the supplier of a service is able to decrease the costs of production, he is able to provide the service at a greater profit. One of the main production costs a physician faces is the time spent to provide patient care. The economic value of time, known as the "opportunity cost," is defined as the value of the next best use that could have been made of that time. Luft[26] noted that technology can often substitute for physicians' time and effort, allowing physicians to use that time elsewhere to their net advantage.

Cesarean section deliveries require much less total physician time than do vaginal deliveries. Physicians acknowledge that they often feel pressured to make "efficient" use of their time when attending a delivery. Thus, in a fee-for-service setting the incentive to minimize time spent attending a prolonged delivery is twofold: (1) the physician's current income is directly dependent on the total number of patients seen, and (2) the viability of the physician's practice and thus its future income are affected by the attractiveness of the practice to new patients, who do not generally appreciate having to wait in the office while the physician is in the hospital. Tussing and Wojtowycz[27] reported that weekend deliveries (a proxy for deliveries less apt to be affected by a physician's need to return to patients in the office) were significantly less likely to be by cesarean section.

Conversely, in systems in which physicians are prepaid, capitated, or salaried, economic factors may be a disincentive to the use of higher cost treatment alternatives.[22,30] In numerous analyses, health maintenance organization (HMO) patients have been shown to have lower cesarean section rates than fee-for-service patients. For example, Wilner et al.[31] observed higher primary cesarean section rates for physicians paid on a fee-for-service (FFS) basis (14%) than for HMO physicians in the same hospital (11%), with apparently equal quality of outcome after risk adjustment. Wright et al.[32] found that although patients were being delivered by the same physicians, FFS patients had a primary cesarean rate of 15%, compared with 8% for HMO patients, again with no apparent difference in outcome.

McCloskey et al.[33] reported that a woman with an HMO physician was 78% as likely to have a cesarean as a woman delivered by a physician in private practice. Stafford et al.[34] reported that in California, Kaiser HMO patients treated in Kaiser hospitals, which pay physicians by salary, were observed to have a lower primary cesarean section rate than patients of non-Kaiser HMOs, which pay physicians on a negotiated fee-for-service basis (18.1% compared with 26.5% in 1989). The cesarean section rate in non-Kaiser HMO patients was closer to the 28.7% rate seen in patients with private fee-for-service insurance. The authors noted that the difference in cesarean section rate between members of Kaiser and non-Kaiser HMOs lends support to the hypothesis that both the method of physician payment and the organization of medical care delivery affects cesarean section rates.

The presence of a reimbursement differential effect on the cesarean section rate is a characteristic of a fee-for-service system. However, other economic incentives that favor cesarean section, such as physician convenience, are relevant in both fee-for-service and capitated or salary-based systems. Fraser et al.[21] reported a small but significant trend toward higher rates of cesarean section during evening hours, consistent with the hypothesis of physician convenience. Also consistent is the finding that physicians who shared night call with other colleagues had lower cesarean section rates than those who were responsible for their own night coverage.[20]

Finally, another key factor in the degree of a physician's susceptibility to nonclinical influences on decision making is the degree of clinical uncertainty. Hillman[35] has noted that physicians will generally act in the patient's best interest whenever the medical decision is clear cut, but that financial considerations may come into play when the "correct" decision is uncertain. Uncertainty also predisposes the physician to be more sensitive to factors such as patient preference for cesarean section, which has been reported to be a factor in the historical increase in cesarean section rates, especially in the rate of repeat cesarean section,[36,37] and to the risk of malpractice liability.[38] Thus, physician coverage, convenience, and clinical uncertainty are factors with economic implications that contribute, along with the reimbursement differential effect, to increasing the use of cesarean section.

The theoretical basis for the reimbursement differential effect is grounded in the principal agent economics concept of "supplier-induced demand." This term generally has negative connotations because it implies that physicians are attempting to provide a level of care that is beyond what is in the best interests of patients. However, as Evans[23] (p. 164) has pointed out, "Such a reaction [i.e., attempting to increase the level of care provided] by physicians should *not* be interpreted as the deliberate provision of unnecessary care. If physicians believe that the public 'needs' more care [for clinical reasons] than it now receives... then they may well [attempt] to take better care of their patients through more frequent recalls and follow-ups, more extensive testing, consultations, and more services generally."

With regard to physician convenience, McCullough[39] argued that physicians have a legitimate self-interest that is compatible with their ethical obligations to their patients. McCullough's concept of legitimate self-interest suggests that physician satisfaction depends not just on economic rewards but also on having both the opportunity for activities that result in improved patient care, such as "the time to study, reflect, and learn and the time to rest and to maintain an alert mind" (p. 12), and to spend the time necessary to fulfill obligations to persons and responsibilities outside the medical network, such as family and community.

Physicians who are responsible for attending their patients' deliveries during the same hours that they are scheduled to see patients in the office feel the greatest pressure to speed a delivery and return to the office. It is reasonable to suggest that the organization of medical care delivery could be improved so as to address issues such as coverage and the conflicting demands on physicians' time. In fact, many HMOs are arranged so as to provide continuous coverage, which allows physicians to leave at the end of their "shift." Obstetricians, for example, would be relieved from being responsible for both delivering patients at the hospital and seeing patients in the office during the same time period. This type of arrangement neutralizes the economic pressure on obstetricians to choose cesarean section when faced with a prolonged delivery. The strong organizational culture in certain HMOs such as Kaiser promotes peer interaction and standard setting as well, thus providing a measure of peer support in situations of clinical uncertainty.

Hospital-Level Economic Incentives

Various hospital characteristics have been found to be associated with a hospital's cesarean section rate. These include hospital ownership and economic organization, teaching status, size, location, and technological sophistication (such as having a neonatal intensive care unit). For example, a lower hospital occupancy rate has been shown to be associated with a higher cesarean section rate.[22] This is consistent with the expectation that the lower the hospital's occupancy rate, the greater the incentive to use a treatment choice that entails a longer average length of stay.

As with physician fees, hospital inpatient charges for cesarean section deliveries are significantly greater than charges for vaginal deliveries. The average hospital charge for uncomplicated cesarean sections in California in 1991 ($5877) was 2.4 times that for uncomplicated vaginal deliveries ($2404), according to hospital discharge data for all payers.

Elaborating the influence of hospital-level economic incentives on use of medical care services is made difficult by the increasingly complex interdependence between hospitals and physicians.[22] Not uncommonly, physicians and hospitals face conflicting economic incentives.[40] For example, HMOs and preferred provider organizations (PPOs) often use different methods to reimburse physicians and hospitals; that is, hospitals may be paid a fixed amount per admission while physicians are paid on a fee-for-service basis. This dichotomy places a hospital's fixed reimbursement incentive (to manage the cost of the admission) in conflict with the physician's fee-for-service incentive (to enhance reimbursement by increasing the number or the intensity of services provided). Conversely, if the hospital is paid fee-for-service or per diem, the hospital's incentive favors clinical choices that prolong the patient's length of stay. However, physicians would benefit more from higher charges than from longer patient stays because relatively fewer physician services are needed during the later part (recovery portion) of the patient's admission.

Hospitals that respond successfully to economic incentives are more likely than those that do not so respond to have a process by which hospital administrators monitor and attempt to influence physician decision making. Gardner[22] reported that the hospital–physician relationship in hospitals of different ownership types can be characterized with respect to two specific features: (1) the degree to which the hospital and the physicians share a common set of values and goals in response to existing economic incentives, and (2) the presence of salaried, vertically integrated, or exclusive arrangements between the hospital and its physicians by which the hospital monitors, controls, or coordinates physician behavior. Together these features comprise a concept termed "close linkage." Gardner reported that the degree of close linkage in the hospital–physician relationship, as represented by the hospital's ownership category, appeared to be significantly related to the hospital's cesarean section rate.

Why is the rate of cesarean section use as high as it is across all hospital ownership categories? A pertinent explanation for this phenomenon was offered recently in a discussion of the discrepancy between physicians' decisions for individual patients versus those for groups. Redelmeier and Tversky[41] observed that physicians gave more weight to patients' personal concerns when they viewed the patient as an individual, and more weight to general criteria of effectiveness when they were making decisions for a group of patients (e.g., establishing clinical policies). The authors further observed that physicians dealing with individual patients were more likely to recommend a therapy with a high probability of success but the chance of an adverse outcome. Both these situations pertain to the physician's decision to perform a cesarean section. Although the authors stated that the observed differences in physician behavior were not related to economic incentives, it may be that physicians are more sensitive to economic incentives when dealing with patients individually than when they are asked to form general opinions as to the optimal treatment approach in a particular clinical situation.

Using Decision Analysis Theory to Model the Choice of Delivery Modality

Analysis that explore the influence of economic factors on medical care decisions are increasingly common, although clinicians still may be generally unfamiliar or uncomfortable with the economic perspective. However, economic analyses are not independent of clinical considerations; rather, they require and build on existing clinical knowledge about the nature of a particular disease, its management, and its outcome. Understanding the full economic value of a treatment entails analyzing the clinical effectiveness of the treatment in the context of both the costs of the treatment and the clinical, economic, and psychosocial outcomes that result. A valid and meaningful economic analysis of a particular treatment option thus will reflect a thorough understanding of the following:

1. The clinical course, expression, and outcomes of the condition being treated
2. The baseline economic impact of the condition
3. The relevant incremental costs and benefits that derive from existing clinical management alternatives
4. The expected costs and benefits of the treatment alternative being studied

A rigorous prospective economic analysis enables the body of knowledge on which subsequent clinical decision making is based to be more complete, less driven by implicit value judgements, and more responsive to consideration of limited economic resources.

When prospective economic analyses are infeasible, as is the case with medical technologies that are already accepted as the standard of care, decision analysis theory can be used to model the expected costs and benefits of the treatment and its relevant comparators; thus the economic and clinical consequences can be explored explicitly and systematically. Decision analysis is also particularly appropriate for analyses of treatment choices made under conditions of clinical uncertainty.

Decision analysis theory has been applied previously in obstetrics to model the appropriate rate of repeat cesarean section,[42,43] and the economic effects of alternative methods of inducing labor.[44] However, to date there have been no modeling analyses focusing on the use of primary cesarean section.

Lowering the cesarean section rate among patients with a previous cesarean section is a high priority because repeat cesarean sections account for 35% of all U.S. cesarean sections,[3] and 9.5% of deliveries involve women with previous cesarean section. The use of routine repeat cesarean section evolved as a response to the need to minimize the risk of uterine rupture. However, changes in operative procedure decreased the incidence of uterine rupture and brought into question the assumption that repeat cesarean section should be performed routinely.

The professional organization of obstetricians, the American College of Obstetrics and Gynecology, now recommends a trial of labor in most patients with a previous cesarean section.[45] However, uncertainty exists as to which patients will successfully deliver vaginally after having had a previous cesarean section. In 1987, only 9.8% of women with previous cesareans delivered vaginally.[3] Physicians report that they remain concerned about the potential risks to mother and child posed by the trial of labor and by the subsequent need, if the trial is unsuccessful, to perform an emergency cesarean section, which is not as safe (or as convenient) as a planned cesarean section.

In the two studies addressing repeat cesarean section,[42,43] the decision that was analyzed was whether to schedule a routine operative delivery or a trial of labor in women with previous cesarean section. Because the clinical issues were so closely focused on one single outcome, uterine rupture, the decision analytic models in the studies of repeat cesarean section could be limited to the incidence and expected outcomes associated with uterine rupture compared with the expected outcomes associated with a trial of labor. Modeling the costs and benefits for primary cesarean section is much more complicated than for repeat cesarean section because of the wide range of clinical entities that comprise the indications for primary cesarean section. The model must incorporate the epidemiology of these clinical entities, the ambiguities inherent in their clinical presentation, the complex assortment of clinical management options, and the full range of possible outcomes in each case given the clinical management pathway.

Currently, three of the clinical entities that are associated with high primary cesarean section rates—breech presentation, fetal distress, and dystocia—are conditions for which there is a great deal of clinical uncertainty. This uncertainty relates to when and whether to use an operative approach and to the extent of the improvement in clinical results to be expected from performing a cesarean section rather than a vaginal delivery. Decision analysis is a useful tool for modeling the costs and benefits of alternative delivery options for such conditions.

Constructing the Decision Model

The first step in the analysis of a decision problem is to define the events in the clinical pathway and their sequence. Events in the clinical pathway include (1) conditions that are present by chance, such as the occurrence of specific indications for intervention; (2) the set of clinical decisions that must be made, such as the choice between a vaginal or cesarean delivery; and (3) the set of all possible outcomes and consequences that might result from each intervention. The probability of occurrence of chance events, the interventions, and their sequence make up the decision flow diagram, or decision tree. A decision tree should include all relevant alternative courses of action and outcomes for a specific clinical situation.

Determining Probabilities

The second step in a decision analysis is to ascertain or estimate the probabilities of occurrence of the chance events. In medical decision analyses, the probabilities are generally estimated from epidemiologic studies or clinical trials. In the absence of rigorous published data, the decision theorist may have to rely on expert opinion or case studies.

When a decision analysis is used to determine the preferred intervention for an individual patient, each treatment decision point is evaluated as a yes/no choice by the model. However, decision trees can also be used to evaluate aggregate costs and benefits using the experience of a group of people. For this purpose, each decision point is modeled as a chance event, and the treatment alternatives are each assigned a probability equal to the observed frequency of use of that treatment option in that population.

Assigning Values to Outcomes

The third step is to assign relative values to each of the possible outcomes. One can evaluate single outcomes such as costs or mortality, or dual outcomes such as cost per year of life saved or cost per survivor. The assigned value of an outcome multiplied by the probability of that outcome results in a score, known as the expected value, for each choice option. Calculating expected values allows the treatment options to be quantitatively compared to eath other. This process is known as "folding back" the decision tree.

Sensitivity Analysis

The results of a decision analysis are only as valid as the information on which the analysis is based. If there is uncertainty regarding the probabilities or the values that are assigned to outcomes, it is advisable to perform a sensitivity analysis using a range of estimates. Sensitivity analysis indicates how robust the results of the decision analysis are to changes in a key variable.[46]

Applying the Decision Model

Patients with no prior history of cesarean section can be categorized into five groups based on indication for cesarean section: (1) patients with breech presentation, (2) patients with fetal distress, (3) patients with dystocia, (4) patients with other indications for cesarean section, and (5) patients without an indication for cesarean section. The following discussion uses the group of patients with breech presentation to illustrate two ways that decision analysis theory can be applied to analyze the choice of delivery modality.

First, using reported rates of cesarean section for each patient subgroup, we evaluate the decision to attempt an external cephalic version, one of the first clinical decisions physicians face when discovering, before delivery, that a patient has a breech presentation. This decision is evaluated from two perspectives: (1) a cost analysis of the decision to attempt external cephalic version in patients who are eligible, and (2) the sensitivity of the cesarean section rate among all patients with breech presentation to the rate at which version is attempted, which allows us to draw conclusions about how changes in the management of breech presentation might be expected to affect the overall cesarean section rate.

Second, we explore the impact of various cost-effectiveness scenarios on the yes/no decision regarding cesarean section for any individual patient. We also test the sensitivity of this decision to certain assumptions. Either of these applications of decision theory similarly could be applied to patients with other presenting diagnoses.

The Decision to Attempt Version

Patients with breech presentation comprise approximately 4% of all patients at term. In 1987, 84.4% of all women who had breech presentation were delivered by cesarean section.[3] However, studies of breech deliveries have reported that the increase in cesarean section rate for term breech deliveries to nearly universal use did not appear to significantly reduce unfavorable outcomes.[18,47,48] For this reason it is now recommended that external cephalic version be attempted in patients with term breech presentation before the onset of labor, except in women with ruptured membranes, a previous cesarean section, an engaged presenting part, or an estimated fetal weight of 4000 g or greater.[49]

Using data from the published literature, we first construct the decision model by outlining the clinical pathway for breech presentation and inserting reported probabilities for each chance event. Descriptions of the clinical pathway indicate that patients with breech presentation may or may not be eligible for external cephalic version. Of those who are eligible, a subset will undergo a version attempt and the remainder will not.* Among those who undergo an attempt, a percentage will be successful. There will also be a spontaneous version rate among women who are not eligible for assisted version or who are eligible but who do not undergo a version attempt. The frequency of use of each delivery approach must then be determined for each subgroup of patients.

Data on total delivery costs were obtained from 1991 California hospital discharge abstracts (California Office of Statewide Health Planning and Development). The California hospital data indicate that among patients with breech presentation, the average total hospital cost is $3186 for a vaginal delivery after a successful version attempt, $3519 for a vaginal delivery without version, $5840 for a cesarean delivery after a successful version attempt, and $5876 for a cesarean delivery without version.

Data regarding maternal and neonatal survival rates were obtained from published clinical literature. Based on numerous reports, it is reasonable to use 0.2% for the maternal mortality rate associated with cesarean section and 0.04% for the maternal mortality rate associated with vaginal delivery.[8,43,50] For neonatal mortality, we initially use rates of 1% for cesarean section deliveries and 2% for vaginal deliveries. Subsequently we evaluate a range of neonatal mortality rates as part of the sensitivity analysis. Mortality statistics and costs are summarized in Table, 13.1.

Figure 13.1 depicts the initial decision tree using expected probabilities and rates of cesarean section for each patient subgroup from various sources. Figure 13.2 demonstrates that the decision to attempt

TABLE 13.1. Cost and mortality statistics used in the decision analyses

Clinical presentation	Cost	Maternal mortality rate (%)	Neonatal mortality rate (%)
Vaginal delivery after successful version	$3186	0.04	2
Vaginal breech delivery	$3519	0.04	2
Cesarean delivery after successful version	$5840	0.2	1
Cesarean delivery without version	$5876	0.2	1

version is preferred on the basis of the cost analysis. Figure 13.2 also allows us to see the cost per live outcome and percentage of patients in each of the patient subgroups.

We can evaluate the sensitivity of the overall cesarean section rate to the rate of attempted version by converting the decision node at which version is attempted into a chance node, inserting various rates of attempted version, and assessing the expected rate of cesarean section summed over each patient subgroup. The impact on the cesarean section rate of three alternative scenarios for the rate of attempted version is assessed: a 50% attempt rate, a 75% attempt rate, and a 95% attempt rate. Table 13.2 presents the results, which indicate that simply by increasing the rate of attempted version among eligible patients from 50% to 95%, the cesarean section rate among all patients with breech presentation decreases from 62.0% to 42.4%, a relative decrease of 32%. The impact on the overall cesarean section rate (all presentations) is an absolute decrease of more than 0.75%.

The Impact of Cost-Effectiveness Analysis on the Decision to Use Cesarean Section

The preceding analysis indicates how changes in the management of patients with breech presentation before delivery can have a substantial impact on the overall cesarean section rate, regardless of the rela-

*Patients who are eligible for a version attempt but do not receive one are either: (1) patients in whom the breech is not detected before the onset of labor, or (2) patients who do not have access to a physician trained in external version or whose physician chooses not to attempt version for reasons other than a clinical contraindication.

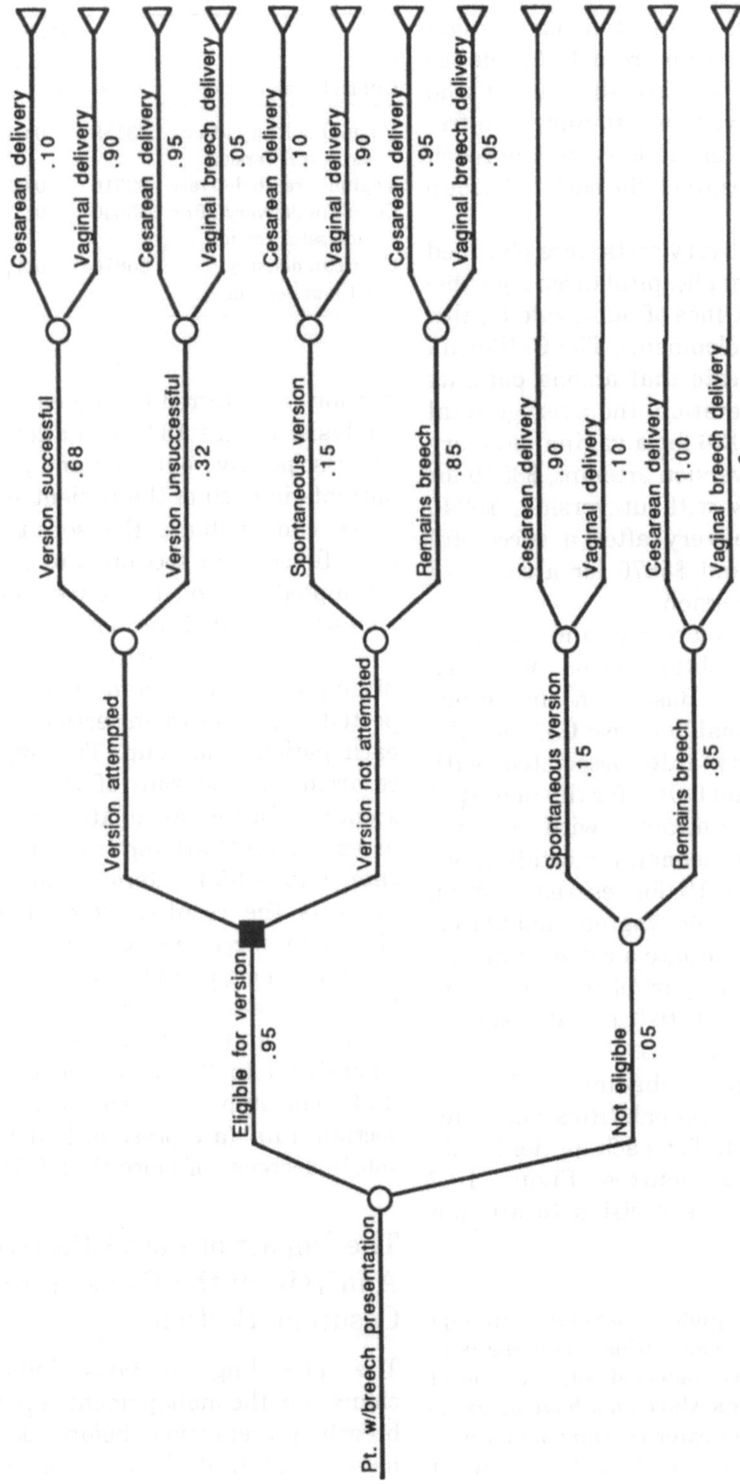

FIGURE 13.1. Initial decision tree for patients with breech presentation.

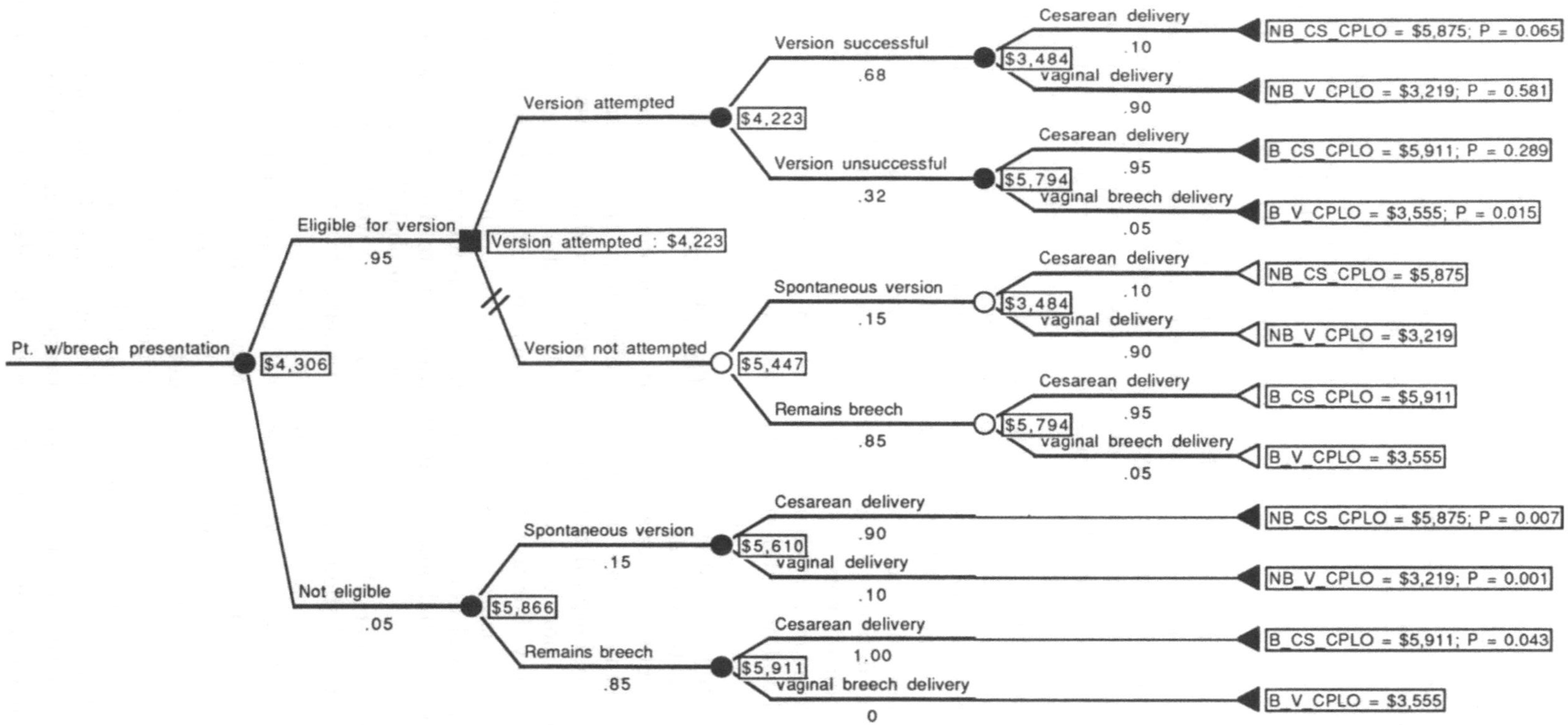

FIGURE 13.2. Expected costs and probabilities for decision analysis of attempted version (NB = not breech; B = breech; CS = cesarean section; V = vaginal delivery; CPLO = cost per live outcome; P = percentage of total patients in selected pathway given use of preferred alternative at the decision node).

TABLE 13.2. Sensitivity of the overall cesarean section rate to the rate of attempted version

Rate of attempted version (%)	Cesarean section rate among patients with breech presentation (%)	Overall breech-related cesarean section rate (%)
50	62.0	2.47
75	51.0	2.04
95	42.4	1.70

tive costs or outcomes of cesarean section compared with vaginal delivery. Another important application of decision analysis theory is to explore which method of delivery is preferred if we consider both costs and clinical outcomes together.

Cost alone is not a useful outcome to consider in this decision analysis because, in current practice, cesarean section delivery always costs more than vaginal delivery and therefore a decision based on cost alone will always favor vaginal delivery. Clearly, the key issue is the trade-off between the increased costs of cesarean section and the impact on maternal and neonatal mortality and morbidity. For this reason, we need to evaluate both costs and clinical outcomes.

Using hospital data to evaluate delivery charges, we can simplify the analysis by focusing on costs and mortality, under the assumption that actual hospital charges include the impact of morbidity occurring within the hospital stay. For this analysis, the cost-effectiveness measure that is used is the ratio of average total delivery costs to average combined maternal and neonatal survival rate.

It is relevant to consider whether maternal and neonatal survival should be valued equally in calculating the cost-effectiveness ratio. This issue can be addressed by performing a separate sensitivity analysis on the relative utility, or value, that is associated with maternal survival compared with neonatal survival. We will assume initially that maternal and neonatal survival are given equal utility.

Figure 13.3 depicts the decision tree for breech presentation using the mortality and cost data presented in Table 13.1. Because we are now considering the optimum decision for any one patient individually, only one of the delivery decision nodes will be relevant. Each patient subgroup should be considered independently of the others. The preferred delivery method is determined by comparing the outcome measure we have chosen (the cost per live outcome) for the two delivery modalities for that patient subgroup.

As can be seen in Figure 13.3, the expected value calculation indicates that the cost per live outcome for cesarean section exceeds the cost per live outcome of vaginal delivery in every patient subgroup, and that therefore the preferred delivery modality for all patients is vaginal delivery. From this we conclude that, given the previous assumptions about cost, morbidity, and mortality rates in our simplified model, the difference in neonatal outcome between cesarean and vaginal delivery is not of sufficient magnitude to counterbalance the incremental cost and maternal morbidity of cesarean delivery.

Threshold Analyses

It is reasonable to suggest that, with a sufficiently great disparity in neonatal mortality between cesarean and vaginal delivery, the improvement in outcome associated with cesarean delivery would become worth the increase in cost. Threshold analysis is a type of sensitivity analysis that allows us to determine the level of a particular probability at which the decision maker would consider the alternative strategies to be equivalent. Generally, only one probability is allowed to vary at a time. All other probabilities and assumptions remain constant.

We first perform a threshold analysis to determine the neonatal survival rate at which a cesarean delivery would become worth its cost. Table 13.3 shows the cost per live outcome of a vaginal delivery over a range of neonatal mortality rates. This table was generated by substituting vaginal de-

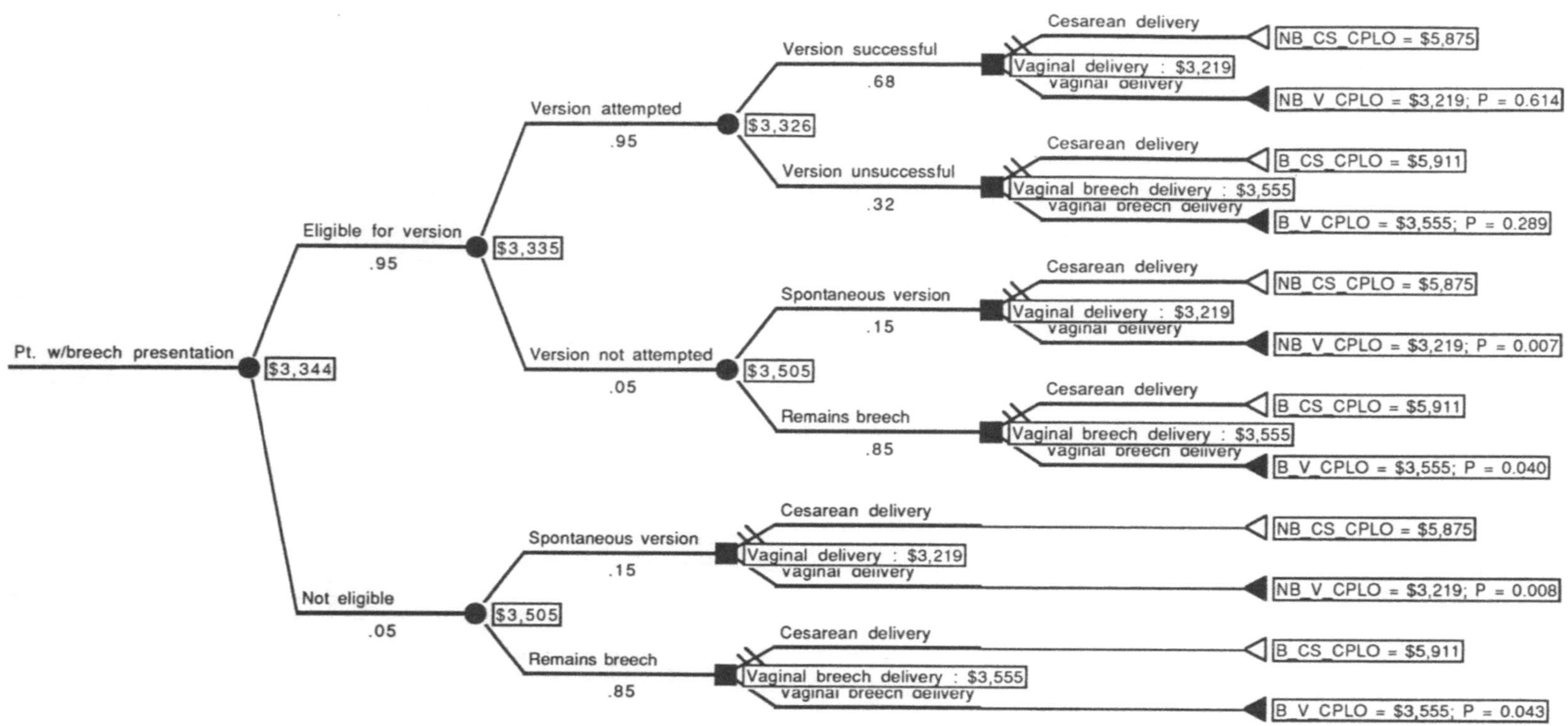

FIGURE 13.3. Cost-effectiveness decision tree for choice of delivery modality in breech presentation (NB = not breech; B = breech; CS = cesarean section; V = vaginal delivery; CPLO = cost per live outcome; P = percentage of total patients in selected pathway given use of preferred alternative at the decision nodes).

TABLE 13.3. Cost per live outcome of vaginal delivery over a range of values for neonatal mortality rate

Neonatal mortality rate	Cost per live outcome ($)
0.00	3187
0.05	3268
0.10	3354
0.15	3445
0.20	3541
0.25	3642
0.30	3749
0.35	3863
0.40	3983
0.45	4112
0.50	4249
0.55	4396
0.60	4553
0.65	4721
0.70	4903
0.75	5099
0.80	5312
0.85	5543
0.90	5795
0.95	**6071**
1.00	**6375**

TABLE 13.4. Threshold analysis of neonatal mortality rate and relative utility

Ratio of maternal to neonatal utility	0.001	0.01	0.1	1.0	10.0
Neonatal mortality rate	Cost per live outcome of cesarean section ($)				
0.01	5899	5900	5904	5929	5954
	Cost per live outcome of vaginal delivery ($)				
0.00	3,186	3,186	3,186	3,187	3,187
0.10	3,540	3,536	3,505	3,354	3,216
0.15	3,748	3,742	3,689	3,445	3,231
0.20	3,982	3,973	3,894	3,541	3,246
0.25	4,247	4,234	4,123	3,642	3,261
0.30	4,549	4,532	4,381	3,749	3,277
0.35	4,899	4,876	4,673	3,863	3,292
0.40	5,306	5,275	5,007	3,983	3,307
0.45	5,788	5,746	5,392	4,112	3,323
0.50	**6,366**	**6,310**	5,841	4,249	3,339
0.55	**7,071**	**6,995**	**6,372**	4,396	3,355
0.60	**7,953**	**7,849**	**7,010**	4,553	3,371
0.65	**9,086**	**8,939**	**7,789**	4,721	3,387
0.70	**10,595**	**10,380**	**8,762**	4,903	3,404
0.75	**12,706**	**12,377**	**10,014**	5,099	3,420
0.80	**15,867**	**15,323**	**11,684**	5,312	3,437
0.85	**21,120**	**20,112**	**14,021**	5,543	3,454
0.90	**31,576**	**29,254**	**17,527**	5,795	3,471
0.95	**62,534**	**53,635**	**23,370**	**6,071**	3,489
1.00	**3,190,462**	**321,915**	**35,060**	**6,375**	3,506

Portion of table not in boldface, cost per live outcome for cesarean section exceeds that for vaginal delivery; boldfaced portion, cost per live outcome for cesarean section is less than or equal to that for vaginal delivery.

livery neonatal mortality rates ranging from 0 to 100% into the cost-effectiveness ratio used in the attempted version arm of the decision tree depicted in Figure 13.3, then recomputing the cost per live outcome. Given that the cost per live outcome of a cesarean section in this case is $5929, the results indicate that, in this simplified model, cesarean delivery would not be the preferred method unless the vaginal delivery neonatal mortality rate was greater than 90%.

This conclusion is based on the assumption that maternal and neonatal survival should be considered to have equal value when calculating the cost-effectiveness ratio. The next threshold analysis explores the two-way impact of changes in both the vaginal delivery neonatal mortality rate and the relative utility of maternal survival compared with neonatal survival.

Table 13.4 presents the cost per live outcome of vaginal delivery over a range of values for neonatal mortality rate and relative utility, compared with the cost per live outcome of cesarean section delivery at a neonatal mortality rate of 1%. It can be seen from Table 13.4 that if the utility of maternal survival were one-tenth that of neonatal survival, cesarean section delivery would be worth its cost at vaginal delivery neonatal mortality rates above 50%, compared with the threshold of greater than 90% that is seen at equal utilities. In this analysis, the threshold value for neonatal mortality does not appear to drop below 45%, even when the relative utility of maternal survival is as little as one-thousandth that of neonatal survival. Furthermore, if the relative utility of maternal survival is increased to 10 fold that of neonatal survival, then cesarean section would not be worth its cost even at a neonatal mortality rate of 100%.

In truth, actual vaginal delivery neonatal mortality rates are likely to be far smaller than the threshold values calculated by these sensitivity analyses, and actual utility for maternal survival is likely to at least equal if not surpass that for neonatal survival. We must conclude that if hospital charges and mortality rates were indeed the only outcomes to be considered, cesarean section would never be the preferred delivery modality for breech presentation.

Discussion and Conclusions

The importance of economic analyses of medical interventions is that they enable those paying for an intervention to understand the conditions under which the intervention is worth the cost. Including economic considerations explicitly in studies of choice among alternative therapeutic options enables the economic issues to be analyzed in conjunction with the clinical considerations.

Our decision analysis of cesarean section versus vaginal delivery for women with breech presentation indicates that under the given assumptions, vaginal delivery is always the preferred method of delivery. However, this conclusion is subject to change as a function of variations in any of the underlying probabilities or assumptions that comprise the model. Particularly important to consider is the way the results of the decision analysis would change if outcomes other than maternal and neonatal mortality, and costs other than delivery charges, were included.

In many areas of medical care, prices are not an accurate guide to the true cost of a procedure. Finkler and Wirtschafter,[17] in a study of one HMO, reported that cesarean section rates were unrelated to direct costs. Table 13.5 shows how, if the sensitivity analysis in Table 13.4 were repeated us-

TABLE 13.5. Threshold analysis of neonatal mortality rate and relative utility with equal charges for cesarean section and vaginal delivery

Ratio of maternal to neonatal utility	0.001	0.01	0.1	1.0	10.0
Neonatal mortality rate	Cost per live outcome of cesarean section ($)				
0.01	3218	3219	3221	3235	3248
	Cost per live outcome of vaginal delivery ($)				
0.00	3,186	3,186	3,186	3,187	3,187
0.01	**3,218**	3,218	3,215	3,203	3,190
0.02	**3.251**	**3,250**	**3,245**	3,219	3,193
0.03	**3,284**	**3,284**	**3,275**	**3,235**	3,196
0.04	**3,319**	**3,317**	**3,306**	**3,252**	3,199
0.05	**3,354**	**3,352**	**3,338**	**3,268**	3,202
0.10	**3,540**	**3,536**	**3,505**	**3,354**	3,216
0.15	**3,748**	**3,742**	**3,689**	**3,445**	3,231
0.20	**3,982**	**3,973**	**3,894**	**3,541**	3,246
0.25	**4,247**	**4,234**	**4,123**	**3,642**	**3,261**
0.30	**4,549**	**4,532**	**4,381**	**3,749**	**3,277**
0.35	**4,899**	**4,876**	**4,673**	**3,863**	**3,292**
0.40	**5,306**	**5,275**	**5,007**	**3,983**	**3,307**
0.50	**6,366**	**6,310**	**5,841**	**4,249**	**3,339**
0.60	**7,953**	**7,849**	**7,010**	**4,553**	**3,371**
0.70	**10,595**	**10,380**	**8,762**	**4,903**	**3,404**
0.80	**15,867**	**15,323**	**11,684**	**5,312**	**3,437**
0.90	**31,576**	**29,254**	**17,527**	**5,795**	**3,471**
1.00	**3,190,462**	**321,915**	**35,060**	**6,375**	**3,506**

Portion of table not in boldface, cost per live outcome for cesarean section exceeds that for vaginal delivery; boldfaced portion, cost per live outcome for cesarean section is less than or equal to that for vaginal delivery.

ing the same charge for both cesarean and vaginal delivery (set for this illustration at $3186), then cesarean section delivery would be worth its cost at much more realistic values for neonatal mortality rate (e.g., more than 2% at equal utility for maternal and neonatal survival).

Furthermore, including the probability and costs of maternal and neonatal *morbidity* will dramatically influence the decision in favor of cesarean section, because neonatal morbidity related to vaginal breech delivery (e.g., asphyxia and trauma), are undesirable outcomes that also can be extremely expensive. Including the probabilities and costs of malpractice liability following each of the two methods of delivery similarly is likely to exert an effect that would favor the decision to choose cesarean section.

Decision analysis is generally a normative tool when used in ongoing decision making regarding the care of a particular patient. However, the clinical decision that feels "right" may not always mirror the results of the decision analysis. In such circumstances, comparing the results of the decision analysis with the clinical decision that is actually made can make it clear that there are additional values and judgments that are implicitly influencing the clinical decision. Whether these values and judgments ought to have a place in the decision making process is the purview, not of decision theorists, but of clinical leaders, policy makers, and society.

References

1. Dunn LJ. Cesarean section and other obstetric operations. In: Scott JR, DiSaia PJ, Hammond CB, Spellacy WN, eds. Danforth's Obstetrics and Gynecology, 6th Ed. Philadelphia: Lippincott, 1990:639–658.
2. Elixhauser A, Andrews RM, Fox S. Clinical classifications for health policy research: discharge statistics by principal diagnosis and procedure. AHCPR Publication No. 93-0043. Division of Provider Studies Research Note 17, Agency for Health Care Policy and Research. Rockville, Maryland: Public Health Service, 1993.
3. Taffel SM, Placek PJ. An overview of recent patterns in cesarean delivery and where we stand today. Presented at the 116th Annual Meeting of the American Public Health Association, November 13–17, 1988, Boston, Massachusetts.
4. Porreco RP. High cesarean section rate: a new perspective. Obstet Gynecol 1985;65(3): 307–311.
5. de Regt RH, Minkoff HL, Feldman J, Schwarz RH. Relation of private or clinical care to cesarean birth rate. N Engl J Med 1986;315:619–624.
6. Petitti DB. The ideal cesarean section rate. In: Parer JT, ed. Antepartum and intrapartum management. Philadelphia: Lea & Febiger, 1989:207–215.
7. Evrard JR, Gold EM. Cesarean section and maternal mortality in Rhode Island. Obstet Gynecol 1977;50:594.
8. Minkoff HL, Schwarz RH. The rising cesarean section rate: can it safely be reversed? J Am Coll Obstet Gynecol 1980;56(2):135–143.
9. Rubin GL, Peterson HB, Rochat RW, et al. Maternal death after cesarean section in Georgia. Am J Obstet Gynecol 1981;139:681.
10. Petitti DB, Cefalo RC, Shapiro S, et al. In-hospital maternal mortality in the United States: time trends and relation to method of delivery. Obstet Gynecol 1982;59:6–12.
11. Rochat RW, Koonin LM, Atrash HK, Jewett JF. Maternal mortality in the United States: report from the Maternal Mortality Collaborative. Obstet Gynecol 1988;72(1):91–97.
12. Guldholt I, Espersen T. Maternal febrile morbidity after cesarean section. Acta Obstet Gynecol Scand 1987;66:675–679.
13. Tulman L, Fawcett J. Return of functional ability after childbirth. Nurs Res 1988; 37:77–81.
14. NIH. Cesarean childbirth: report of the NICHD task force on cesarean childbirth. Bethesda, MD: National Institutes of Health, 1981. (DHHS publication no. (NIH) 82-2067.)
15. Bland R. Pathogenesis of pulmonary edema after premature birth. Adv Pediatr 1987; 34:175–221.
16. Martin DP, Fadden MK, Holt VL, et al. The relationship among delivery method, quality and outcomes of obstetrical care. Presented at the annual meeting of the Association for Health Services Research, 1992, Chicago, Illinois.
17. Finkler MD, Wirtschafter DD. Why pay extra for cesarean deliveries? Inquiry 1993; 30(2):208–215.
18. Hack M, Fanaroff AA. Outcomes of extremely-low-birth-weight infants between 1982 and 1988. N Engl J Med 1989;321(24):1642–1647.
19. Anderson GM, Lomas J. Explaining variations in cesarean section rates: patients, facilities or policies? Can Med Assoc J 1985; 132:253–259.
20. Carpenter MW, Soule D, Yates WT, Meeker CI. Practice environment is associated with obstetric decision making regarding abnormal labor. Obstet Gynecol 1987;70:657–662.

21. Fraser W, Usher RH, McLean FH, et al. Temporal variation in rates of cesarean section for dystocia: does "convenience" play a role? Am J Obstet Gynecol 1987; 156:300–304.
22. Gardner LB. Hospital-physician relationships and cesarean section rate variations. Doctoral dissertation, University of California at Berkeley.
23. Evans R. Supplier-induced demand: some empirical evidence and implications. In: Perlman M, ed. The economics of health and medical care. New York: Wiley, 1974.
24. Anderson R, House D, Ormiston M. A theory of physician behavior with supplier-induced demand. South Econ J 1981;July:124–133.
25. Wilensky GR, Rossiter LF. The relative importance of physician-induced demand in the demand for medical care. Milbank Mem Fund Q 1983;61(2):252–277.
26. Luft HS. Economic incentives and constraints in clinical practice. In: Aiken LH, Mechanic D, eds. Applications of social science to clinical medicine and social policy. New Brunswick: Rutgers University Press, 1986:500–518.
27. Tussing AD, Wojtowycz MA. The cesarean decision in New York State, 1986. Med Care 1992;30(6):529–540.
28. Robinson JC, Gardner LB, Luft HS. Health plan switching in anticipation of increased medical care utilization. Med Care 1992; 31(1):43–51.
29. Murray D. Do you know how much to charge? Medical Economics 1993 October 11:106–133.
30. Pauly MV, Hillman AL, Kerstein J. Managing physician incentives in managed care. Med Care 1990;28:1013–1024.
31. Wilner S, Schoenbaum SC, Monson RR, Winickoff RN. A comparison of the quality of maternity care between a health maintenance organization and fee for service practices. N Engl J Med 1981;304:784–787.
32. Wright CH, Gardin TH, Wright CL. Obstetric care in a health maintenance organization and a private fee-for-service practice: a comparative analysis. Am J Obstet Gynecol 1984;149:848–856.
33. McCloskey L, Petitti DB, Hobel CJ. Variations in the use of cesarean delivery for dystocia. Lessons about the source of care. Med Care 1992;30:126–135.
34. Stafford RS, Sullivan SD, Gardner LB. Trends in cesarean section use in California, 1983–90. Am J Obstet Gynecol 1993;168(4): 1297–1302.
35. Hillman AL. Health maintenance organizations, financial incentives, and physicians' judgments. (Editorial.) Ann Intern Med 1990;112(12):891–893.
36. Feldman GB, Freiman JA. Prophylactic cesarean section at term? N Engl J Med 1985;312(19):1264–1267.
37. Cepicky P, Stembra Z, Zeman J, Lomickova T, Mandys F. When is it possible to meet the wish of a woman to terminate her labour by cesarean section? Eur J Obstet Gynecol Reprod Biol 1990;38:109–112.
38. Rock SM. Malpractice premiums and primary cesarean section rates in New York and Illinois. Public Health Rep 1988;103: 459–463.
39. McCullough LB. The physician's virtues and legitimate self-interest in the patient-physician contract. Mt Sinai J Med 1993; 60(1):11–14.
40. Feldstein PJ. Health care economics, 3rd Ed. Albany: Delmar, 1988.
41. Redelmeier DA, Tversky A. Discrepancy between medical decisions. for individual patients and for groups. N Engl J Med 1990; 322(16):1162–1164.
42. Shy KK, LoGerfo JP, Karp LE. Evaluation of elective repeat cesarean section as a standard of care: an application of decision analysis. Am J Obstet Gynecol 1981;139(2): 123–129.
43. Silver RK, Minogue J. When does a statistical fact become an ethical imperative? Am J Obstet Gynecol 1987;157(2):229–233.
44. Davies LM, Drummond MF. Management of labour: Consumer choice and cost implications. Journal of Obstetrics and Gynecology 1991;11[Suppl. 1]:S28–S33.
45. American College of Obstetrics and Gynecology. Guidelines for Vaginal Delivery after a Previous Cesarean Birth. Washington, DC: Committee on Obstetrics: Maternal and Fetal Medicine, 1988.
46. Luce BR, Elixhauser A. Standards for the socioeconomic evaluation of health care services. Berlin: Springer-Verlag, 1990.
47. Green JE, McLean F, Smith LP, Usher R. Has an increased cesarean section rate for term breech delivery reduced the incidence of

birth asphyxia, trauma, and death? Am J Obstet Gynecol 1982;142:643–649.

48. Myers SA, Gleicher N. Breech delivery: why the dilemma? Am J Obstet Gynecol 1987; 156(1):6–10.
49. Cruikshank DP. Malpresentations and umbilical cord complications. In: Scott JR, DiSaia PJ, Hammond CB, Spellacy WN, eds. Danforth's Obstetrics and Gynecology, 6th Ed. Philadelphia: Lippincott, 1990: 567–584.
50. Danforth DN. Cesarean section. JAMA 1985;253(6):811–818.

14 Ethical Issues in the Utilization of Cesarean Section

THOMAS E. ELKINS AND DOUGLAS BROWN

This chapter explores the ethical issues surrounding the utilization of cesarean section for delivery. A few definitions are in order.

Ethics has to do with the decisions about what ought to be done, all things considered.[1–5] This definition distinguishes what could be or is done from what decision makers decide ought to be done. This definition calls for thorough consideration of nonclinical as well as clinical factors. Ethical discourse concentrates on who decides and the means by which decisions are made. Participants attempt to interpret immediate experiences against the backdrop of societal wisdom about human behavior.

Medical ethics focuses on decision making in situations involving health care. From a societal perspective, ethical reflection on health care issues concentrates on individual human and civil rights and on a fair distribution of resources. From a professional perspective, ethical reflection orients clinical decisions toward the goal of enhancing a patient's well-being or easing a patient's decline. From a patient perspective, ethical reflection measures clinical choices by the impact on one's life story, which has been disrupted by circumstances such as accidents or illnesses.

Until the 1960s, the Hippocratic tradition provided historical context for ethical reflection about medicine in the United States.[6,7] Technological advances had not yet given patients sufficient cause to presume that physicians could significantly alter the course of a disease or the consequences of a serious accident. Minorities did not yet fully benefit from the democratic promotion of individual rights. Physicians were measured by the standard of compassion and were trusted to "do no harm" when they paternalistically made beneficent decisions "in the patient's best interests."

Technological and cultural revolutions eventually combined to restructure radically the discussion of medical ethics. Patient expectations shifted from care to cure, from acquiescence to self-determination. Ethical principles (with autonomy and justice added to nonmaleficence and beneficence) formed a core vocabulary for physicians, philosophers, and theologians.[8] The application of these principles in clinical settings was seen as the "end" by some (i.e., deontologists), and as the "means" by others (i.e., utilitarians).[9,10] (Their ethical conclusions were not invariably at odds.)

Each of these pillar principles initially held prima facie importance. However, given the affirmation of cultural diversity and individual rights, the principle of patient autonomy rose to "trump" status in societal, and eventually in medical-ethical, discourse.[11,12] Case-by-case decisions were thought to originate from within the privacy of individual or familial traditions. Attention on patient values increased.[13,14]

Most recently, perceived deficiencies have been identified amid the strengths of autonomy-based ethics, giving impetus to revised affirmations of physicians' autonomy, accountability, and character relative to the decision-making process.[15–17] The patient–physician relationship is now being framed more as an overlapping of patient and physician "stories."[18] Also, the range of participants in medical-ethical deliberations, frequently organized as ethics committees, has expanded to include lawyers, economists, minority advocates, politicians, and "lay" persons. The American College of Obstetricians and Gynecologists (ACOG) was among the first medical associations to establish a committee on ethics with the responsibility to issue guidance papers.

Reproductive ethics narrows the focus to decision making associated with procreative physiology. The clinical situations cover a wide range, for example, gynecologic care, infertility, birth control, pregnancy termination, prenatal care, fetal tissue research, labor and delivery serivice, high-risk pregnancy management, and fetal therapy. Promotion of a woman's procreative well-being and control provides the primary ethical common ground for these varied interests. Given advances in fetal diagnosis and therapy as well as in neonatal intensive care, it should be noted that physicians involved in decisions to utilize cesarean section are trained as surgeons[19] and have distinguishable specialties within the field of maternal-fetal health care.

Obstetric ethics (as distinguished from gynecologic ethics) concentrates on pregnancy management decisions.[20–23] The presence of two mutually dependent beings—the pregnant woman and the fetus—is the most obvious justification for isolating the issues of this area of health care. The goal of obstetric ethics is the delivery of a healthy term baby without any compromise to maternal well-being.

Ethical discussions and the associated literature often leave the impression that obstetric ethics has to do with dilemmas and tragic choices at the commencement of life. Instead, in the majority of pregnancies, commitments to respect for new life, to confidentiality, to maternal and fetal health, to mutual trust, and to the naturalness of procreation flow without ambiguity or conflict.

It should be noted that obstetricians and other professionals comprising perinatal teams are hardly neutral participants in cases in which cesarean section delivery is under consideration. Clinicians hold no uniform value system. For example, one may link the value of an endangered fetus to racial, economic, or cognitive development status. Another may view every fetus as inherently valuable.

Clinicians are likely to see situations from the perspective(s) established by their specialties. Clinicians have unique patchworks of experience that shape their reaction to each situation. For example, one professional in the midst of a malpractice suit may intend to avoid the risk of another suit. That professional may have a colleague who recently became the mother of a healthy baby after fetal therapy. That colleague may at least temporarily fail to grasp the grounds for indecision about pursuing intervention.

Clinicians have opinions about numerous related issues (e.g., elective abortion, selective nontreatment of critically ill newborns, the right of a fetus to the status of "person," a patient's right to refuse treatment, and wrongful life/wrongful death suits). These opinions clarify their responses to clinical conflicts over recommended but refused interventions.

Anatomy of a Recommendation

Many issues that surround the use of cesarean section have broad ethical implications. In developing countries where the number of trained physicians is extremely low, the question is debated as to who should be trained to perform cesarean sections—nurses, midwives, traditional birth attendants. Cesarean section is the pivotal procedure that influences funding decisions for vast projects aimed at reducing maternal mortality rates (which reach 10 deaths per 1000 births).

The soaring cesarean section rate in the Unites States has raised a separate set of ethical concerns. Issues of increased physician fees for operative deliveries, the use of cesarean sections to avoid long hours on weekends of busy physicians, and a concern for medical decisions being made as a result of legal fears have all been explored in recent years.

If put in international perspective, an ethics chapter on this relatively simple operative procedure could obviously become quite broad. Our discussion, accordingly, puts primary attention on decisions about the utilization of cesarean section in American settings. However, the significance of international considerations should not be forgotten.

Questions about cesarean section delivery are at times raised in situations that lack compelling medical need.

Case One

A 15-year-old primigravida woman, progressing in labor and having received an epidural anesthetic as well as narcotics, screams for a cesarean section to end her labor pain.

Case Two

A 39-year-old woman requests a cesarean section at 37 weeks so as to be able to make a trip 1 month later.

Every resident physician in obstetrics has been confronted by the distraught patient in labor who is demanding as loudly as possible for someone to "Take my baby!" The anguish of labor does preoccupy decision-making considerations on occasion. A cesarean section is a welcome relief for many patients. Much less frequently, a woman may request delivery by cesarean section to facilitate the timing of events relative to career plans or even vacation plans. Such extreme appeals to autonomy are, fortunately, uncommon.

Cesarean sections in these situations would be disturbing evidence that the cesarean section rate in the United States is not simply a result of decision making based on medical factors and is strikingly influenced by physician "style" in handling nonclinical considerations.[24-27] We doubt that even physicians for whom a pregnant woman's self-determination is the trump ethical consideration would often acquiesce to such requests.[28] It should, nonetheless, be noted that refusing such patient requests does imply that patient autonomy is a conditional, rather than absolute, right.

Responses to requests for cesarean sections when compelling medical need is lacking are value-laden responses.[29] Still, an argument based on the absence of compelling medical need may satisfy physicians wary of expanding ethical reflection to non-clinical considerations. However, a more thorough exploration of nonclinical considerations is necessary in situations complicated by possible or evident risk to the pregnant woman or to the fetus.

Case Three

A 23-year-old primigravida woman presents at 29 weeks gestation in active labor with twins in A breech, vertex presentation. A cesarean section is recommended.

Case Four

A 37-year-old primigravida woman presents in labor at 41 weeks gestation with a footling breech presentation. Delivery by cesarean section is recommended.

Case Five

A 24-year-old woman undergoes ultrasonography at 28 weeks gestation and fetal hydrocephaly is identified. The patient asks about routes of delivery and her obstetrician recommends cesarean delivery.

Case Six

A 19-year-old woman presents at 27 weeks gestation in active labor. Fetal monitor tracings reveal severe bradycardia with late decelerations. The obstetrician recommends delivery by cesarean section for fetal distress.

In some developing countries where cesarean section is rarely done for fetal well-being, all these cases might "properly" be managed by vaginal delivery. In situations in which antenatal assessment, fetal monitoring in labor, neonatal intensive care, and even routine pediatric follow-up (and malpractice attorneys) do not exist, vaginal deliveries are preferred. But is this ethically acceptable?

Ethical arguments take on new foci in such countries. For example, in rural Africa, should a cesarean section be done to deliver a fetus barely maintaining a heartbeat after 3 days of obstructed labor and sepsis? Or should a destructive procedure with vaginal delivery be done for the mother? Geography, technology, and culture may alter traditional decisions normally made so easily in a technologically advanced country. However, to ignore the viable fetus as a patient today, in America, is to be truly noncontemporary.

The dilemma in the foregoing cases, in which only residue of a noneventful pregnancy remain, has to do with the task of making choices that benefit one patient at some expense to the other. Multiple experiences—for example, with pregnancy, medical education, litigation, choice of specialty, and indigent patients—significantly influence the sort of obstetrician a professional becomes.

A pregnant woman in America today who invites an obstetrician into a chapter of her "story," has, in turn, made the decision to become or remain pregnant, has adopted a perception of childbearing, has formed an opinion about the place of prenatal care, and has some opinion about the relation of her interests to the well-being of her fetus.

Because our society remains grounded in the presumption that its members can be trusted to exercise their freedom of choice in such a way that others will not be harmed and the ideals of a just society will be approximated, the fundamental role of medicine is to avoid harming patients in the effort to maximize the health they seek so as to continue their pursuit of a complete life. For a fiduciary relationship to unite mother and obstetrician in the task of responding to any circumstances that may disrupt the flow of the mother's childbearing, the mother needs information about the values of the hospital and the obstetrician sufficient in detail to establish confidence. The obstetrician involved needs information about the mother that is adequate to plan an initial strategy that integrates his or her values and medical skills with the mother's values and aspirations.[30]

We propose that the relationship formed between the obstetrician and the pregnant woman necessitates compatibility on at least four matters:

1. Aim. Is there agreement about the place of fetal outcome in the envisioned pregnancy?
2. Perception of childbearing.[31] Is childbearing seen as a natural process? Has the pregnant woman assumed responsibilities related to fetal outcome? What role does the obstetrician, trained to intervene to correct deviations in the process of childbearing, have?
3. Decision-making process. Has the pregnant woman clarified her interest to be involved in decision making? Has she introduced other participants (e.g., father of the fetus, parents) to the physician? Is the physician's role clear?
4. Perception of the fetus.[31] Is there compatibility in the moral status assigned to the fetus?

So long as the prenatal period is uneventful, the obstetrician should remain inconspicuous but alert as the pregnant woman focuses on the mystery of bringing a new life to birth. A threshold is crossed when the obstetrician perceives a risk or risks to the pregnant woman or the fetus. The obstetrician, reacting with a heightened sense of responsibility, becomes more conspicuous in the case. A "pause" to consider the ethical implications in this shift is warranted, because drawing the pregnant woman's attention to risks in the pregnancy changes her childbearing experience. The obstetrician's role becomes more aggressive. The privacy of the physician–patient relationship[32] is diminished when consultation is required.

The obstetrician faces another threshold if the pregnant woman resists or rejects attempts to reframe her situation as having significant risk. She may dispute the interpretation. She may refuse recommended interventions designed to stand between ("obstetrics" is a Latin combination, meaning "to stand between") her and the proposed threat to her well-being or that of her fetus.

Another "pause" is critical here, in view of the ethical decisions created by such refusal. The obstetrician, whose clinical instinct to intervene often jumps ahead of reflection, may find this pause most difficult. The prevailing assumption is that recommended medical or surgical treatment is beneficial.[33] Rhoden[34] has argued, on the basis of extensive clinical exposure, that an obstetrician's clinical intuition to intervene is so ingrained that a pregnant woman's refusal is often virtually unthinkable. The ethical questions facing the obstetrician, which must be resolved with "exteme caution,"[35] have been variously worded:

When are the risks of fetal intervention to the woman sufficiently great to allow her to maintain autonomy and when are the benefits to the fetus sufficiently great to justify overriding her autonomy?[36]

Where risk of serious injury or death to the mother is minimal and the possibility of benefit to the fetus is considerable, when should the physician intervene on behalf of the unborn child?[37]

When, if ever, is it ethical to provide medical treatment against a patient's will?[33]

Which patient—the pregnant woman or the fetus—is the primary patient?[36]

Which patient choices ought to be respected?[38]

Does force ever have a place in medicine?[39]

What Has Been Done

The Kolder et al. Review

Exact determination of the number of cesarean sections that have been done without the agreement of the pregnant woman is not possible, in part because the lines that distinguish recommendation, persuasion, and coercion are blurred. We agree with the observation by Brody[40] that power pervades the physician–patient relationship and is disproportionately weighted toward the physician. Brody's assessment means that every communication from a physician, in both what is said and how it is said, puts leverage on a patient.

What can be documented are cases that have passed from recommendation and persuasion to court-ordered intervention. Kolder et al.[41] conducted a review of cases, through 1986, in which court orders had been sought to sanction interventions over maternal refusal. They found 21 cases involving cesarean sections, hospital detentions, or intrauterine transfusions. Fifteen court orders for cesarean sections had been sought in 11 states. Thirteen orders had been obtained (2 of which were not used when maternal agreement subsequently occurred). The women were all Afro-American or Asian; English was not the primary language of 1 in 4 of these women. Seven of the 15 court orders were sought earlier than 37 weeks of gestation. The most common diagnoses were fetal distress, previous cesarean section, and placenta previa. All cases had occurred within teaching hospitals. No case had resulted in maternal morbidity or motality. In 1 case, a Nigerian father, strongly opposed to the ordered intervention, committed suicide a few months after his wife's cesarean section. Two infants experienced significant morbidity. No fetal deaths occurred.

Kolder et al. also sought opinions from the heads of maternal-fetal medicine programs. Nearly half the respondents could conceive of situations in which women whose refusal endangered thel life of the fetus should be detained. A similar number thought sanctioned interventions should include procedures other than cesarean sections. A quarter of the respondents advocated state surveillance.

These investigators regarded the central question to be whether a patient's decision-making rights should ever be usurped by a physician or by the government to advance the therapeutic interests of a second patient.[41]

> Acceptance of forced cesarean sections, hospital detentions, and intrauterine transfusions may trigger demands for court-ordered prenatal screening, fetal surgery, and restrictions on the diet, work, athletic activity, and sexual activity of pregnant women. (p. 1195)

An accompanying editorial argued that, without the pregnant woman's informed consent, fetal access is only possible "by treating her as a fetal container, a nonperson without rights to bodily integrity."[42] The article and editorial drew several critical letters to the editor.

Elkins et al.[43] studied one subject group related to the Kolder et al. investigation. This study reported the perceptions of 31 resident and faculty physicians in obstetrics and gynecology in two residency programs in Michigan, a state to which the Kolder et al. article referred several times. The respondents showed no consensus of opinion on the management of the following case presentation:

> A competent 27-year-old white female presented to Planned Parenthood at 34 weeks gestation, requesting an abortion. She was advised against pregnancy termination. Because her blood pressure was 180/110, she was transferred to a tertiary care hospital, where a physical examination revealed that she had severe preeclampsia. Monitoring, which the patient did not refuse, revealed serious fetal distress. Physicians informed the patient of the fetal distress and advised her that she should be delivered immediately by cesarean section. She refused the operation, stating that she did not want the baby. Various labor and delivery personnel, as well as family members, tried to persuade her to have the surgical delivery. Following the patient's initial refusal, the hospital attorney, at the request of the delivery team, spoke with three local judges, each of whom refused to hear the case. Eventually, a stillborn female fetus weighing 2140 g was delivered spontaneously. Although her hypertension and proteinuria had not resolved, on her third postpartum day the woman signed out of the hospital against medical advice.

In an effort to assess the willingness of the respondents to appeal for judicial leverage against pregnant women who refuse their physicians' recommendations, the study posed nine situations not involving cesarean section. The respondents' willingness to force intervention in circumstances not involving delivery increased as fetal risk at term increased. The circumstances for which the respondents were most willing to force intervention—insulin treatment and excessive use of alcohol—would requie monitoring the pregnant woman's life-style, beginning in the early weeks of pregnancy.

To determine the respondents' support of court-ordered cesarean sections, the study posed 10 situations. The respondents indicated little compulsion to seek a court order if fetal complications had not yet occurred. Their readiness increased as fetal risk grew to the point at which fetal health was at grave risk without cesarean delivery (as in cases of well-documented fetal distess near term) or the future health of the newborn was in jeopardy (as in the refusal to deliver by cesarean section a breech presentation with hyperextension of the head), or when the health of both mother and fetus were at grave risk (as in refusal of cesarean section delivery in spite of documented placenta previa at term). The majority rooted their clinical responses in medical reasonableness (62%), ethical principles (62%), and malpractice concerns (54%).

These data confirm the identification by Kolder et al.[41] of a willingness on the part of physicians practicing in a teaching hospital setting to employ court-ordered interventions. However, these data did not support the impression that physicians are determined to maximize control of pregnant women or the conclusion that court-ordered interventions must be categorically rejected so as to avoid abuse.

In Re: Angela Carder[44,45]

Reaction to the decision by a District of Columbia judge on June 16, 1987, to sanction a cesarean delivery has catapulted a tragic story into "a paradigm case that has a secure niche in the annals of law and ethics."[46] An explanation and defense of the decision was filed November 10, 1987. Two weeks later, 39 organizations, including the American Medical Association (AMA) and the American College of Obstetricians and Gynecologists (ACOG), submitted requests for the Court of Appeals to reverse the ruling. The Court of Appeals decided on April 26, 1990, to reverse the judge's decision, with one judge dissenting in part.

Attempts to reconstruct any case, including cases devoid of controversy, fall short of the reality. The privacy, intensity, spontaneity, and emotion unique to every case cannot be recovered. Looking back introduces self-examination and reflection into the recollection. This limitation was certainly true with the reports of the Angela Carder case. However, initial descriptions of the Angela Carder case were further compromised by the selection of information and the interspersed biases of authors who appearred to be drawn to the case for its political capital.[47–49]

In spite of these limitations, the undisputed prominence of this case makes an attempt to tell the story necessary. One of the authors of this chapter (T.E.) was asked by defense attorneys, apparently looking for a witness to support the court order, to review the testimony and related literature.[46,50,51] The review led to this reconstruction of the Angela Carder case.

Angela Carder was a 27-year-old high-risk patient of the obstetric service of George Washington University Hospital. Diagnosed at age 13 with cancer, she had a long history of bone cancer and had undergone the amputation of one leg. After years of courtship, she had married a man who accepted her even with her major disabilities. This couple shared a deep desire to have a child together. When she became pregnant, she had not undergone chemotherapy for more than a year. She was approximately 25 weeks pregnant when a large tumor in one of her lungs, detected during her June 9 prenatal visit, confirmed that the cancer was no longer in remission and had in fact spread markedly. She was admitted to the hospital on June 11. She was informed on June 15 that the tumor was inoperable. With her condition terminal, the couple's desire to leave a child behind as a symbol of their relationship was underscored when they reaffirmed their June 12 decision to administer palliative treatment with the goal of extending her pregnancy to 28 weeks (thus significantly improving the baby's chances) and then delivering the baby by cesarean section.

Discussion of potential infant morbidity made Angela's mother, herself confined to a wheelchair, extremely anxious. She feared that, with Angela's death, she would be responsible for raising a child that might be severely handicapped after birth. The father of the fetus, emotionally overwhelmed, added little to subsequent discussion. Angela's mother became increasingly the dominant figure. The attending physician, a maternal-fetal medicine fellow, supported the plan, but apparently failed to explore with Angela her preferences should her condition dete iorate more rapidly. The oncology resident on call on June 15 asked, during a chart review, what was to be done should Angela die during the night. The attending physician, convinced that Angela would not have chosen to deliver a potentially severely impaired child, communicated that no cesarean delivery would be attempted. Thus, the initial question had to do with a postmortem procedure.

By early morning, Angela's condition had deteriorated more rapidly than expected. The family had been called. They had last rites administered by 8:00 A.M. The attending physician met with the family. The family opposed a postmortem delivery. Angela's husband expressed doubt that he could face the child without Angela. Her mother argued that Angela had gotten pregnant with the confidence that she would be able to raise the baby. The attending physician was convinced that the family had Angela's wishes sufficiently in mind to make their guidance reasonable. This management plan was communicated on June 16 to the department head, who happened to be in a meeting with a hospital administrator at the time. The administrator, made aware of the case, asked whether the baby was viable. He was told viability was probable. Without further inquiry into the details of the case, he called the case to the attention of the hospital's senior legal counsel, who advised that the case should be treated as an emergency and that physician opinion should be followed.

Pediatric opinion suggested a high probability of neonatal survival if the fetus was delivered.

Hospital administrators or attorneys called a Superior Court judge. Legal representation for Angela, the baby, the hospital, and the District of Columbia was arranged.

Angela was no longer alert or capable of communication by the time the judge and the various lawyers met near Angela's room with the medical personnel and the family. The initial postmortem question by now had evolved into a premortem question; that is, given that Angela would soon die, should a cesarean section be immediately performed. Angela's lawyer argued against such a surgery, because consent could not be obtained and discussion of a cesarean section earlier than 28 weeks gestation had not been previously conducted with Angela. The baby's lawyer argued for surgery as the baby's only hope, maternal interests being greatly diminished by Angela's imminent death. Angela's mother strenuously opposed the surgical delivery, expressing her opposition in terms of her interests and the potential burden to her. Angela's husband contributed little to the discussion. The attending physician opposed the intervention. However, other physicians present, including the pediatrician, were not of one opinion. The hospital lawyers backed the attending physician.

On the basis of on Angela's evident willingness to put herself at risk to have a child and on the lack of a surrogate decision maker, the judge concluded that the balance was weighted in favor of the baby and that the state had sufficiently compelling interest to order cesarean section delivery. Entering Angela's room to prepare her for surgery, the attending physician and a colleague found the patient roused. He presented the judge's decision as a fiat accompli. She agreed. When Angela's alertness and response were reported to the judge, her tearful husband and her mother, along with her attending physician and a senior colleague, went to her room for confirmation. The mother urged her daughter not to attempt delivery and instead carry the baby with her to heaven. Angela wanted to know who would be doing the surgery. Her attending physician told her he would if she authorized the surgery. Finally she mouthed the words, "I don't want it done." It was not clear to the physicians in the room whether Angela's response could be considered an informed response. The reported equivocation did not change the mind of the judge, who had made his decision on grounds other than the presence or absence of maternal competence. A local appeals court chose not to issue a stay. The cesarean section was performed at 6:30 P.M. The baby, born at approximately 26 weeks gestation, survived only a couple of hours. Angela's husband did not want her informed about the death. However, the attending physician informed Angela. Angela died 2 days later. The surgery was listed on the death certificate as a contributing cause. Lindsay Marie was buried with her mother.

The Court of Appeals reheard the case in September 1988 and issued a reversal in April 1990.[45] The Court of Appeals:

1. Rejected the judge's attempt to balance interests
2. Argued that the right of bodily integrity protected competent and incompetent individuals from having to submit to any invasive procedure for the benefit of another
3. Stressed that cesarean section is major surgery, thus necessitating "an extraordinary case" before court-ordered intervention could be justified
4. Concluded that the state's interest in preserving the Carder baby's life was not truly compelling
5. Criticized the judge for not seeking a substituted judgment as to Angela's preferences
6. Noted the inability of the judicial system to properly function in such circumstances

The Angela Carder case has been classified as a paradigm case for management of pregnancies complicated by disagreement over the utilization of cesarean delivery or other prenatal interventions.[46] (A paradigm is a way of seeing and interpreting data in the effort to make sense of life experiences. A paradigm case would be a precedent-setting case which, as a frame or point of reference, provides clarification and guidance for decision making in a present case.[52]) We counterpropose that the Angela Carder case is a paradigm case for *mis*managing pregnancies complicated by disagreement over the utilization of cesarean delivery or other prenatal interventions. Every facet of the decision-making process seems to have failed.

1. The patient failed to prepare a functional process of substituted judgment.
2. The patient's husband failed to interpret clearly his understanding of his wife's perceptions or his own perceptions.
3. The patient's mother disqualified herself as a surrogate decision maker by appealing to burdens she might face personally.
4. The attending physician left questions undiscussed that should have been anti--cipated in the event the mother deteriorated more rapidly than expected.
5. The attending physician's senior colleague brought the case to the attention of an administrator without strong reason.
6. The alerted administrator failed to gather sufficient information before bringing in legal counsel.
7. The hospital's ethics committee, ironically meeting just down the hall at crucial points on the day of decision, was never consulted.
8. The judiciary, minus the time needed to function appropriately, failed to evaluate the mother's competency and failed to establish a decision-making process.
9. All parties involved failed to enable Angela's husband to take a significant position in the process as the legitimate surrogate decision maker.
10. Commentators on the case, both in defense of and in opposition to the court-ordered surgery, failed, with few exceptions, to show respect for the profoundly human and tragic dimensions of this case in their grasp for political gain.

Far from closing discussion about these cases, the significance of the Court of Appeals' ruling has been widely disputed. Some saw in the decision a sweeping policy that "turned the law in the proper direction."[53] Others felt that the decision did not "provide any practical assistance to doctors, hospital administrators, lawyers, or judges faced with these difficult cases."[54]

We view this case as one tragedy (i.e., the patient story) overlaid by another tragedy (i.e., the social and political response to the story). The limitations embedded in this case leave the clinician with the question: "What should be done?"

What Should Be Done?

The AMA Board of Trustees Report: Legal Interventions During Pregnancy

A few months after the Angela Carder decision was reversed, the Board of Trustees for the American Medical Association pub lished a report in which their legal and policy concerns about court-ordered medical treatments were discussed.[55] The report argued that "a woman who chooses to carry her pregnancy to term has a moral responsibility to make reasonable efforts toward preserving fetal health," but drew a clear distinction between moral and legal responsibilities. The report advised physicians that "the duty to protect the health of both the pregnant woman and the fetus precludes balancing one against the other" and that the principle of informed consent "indicates that a pregnant woman's refusal of treatment should not be overridden for the benefit of the fetus." The report adopted a "rule" utilitarian position (rather than an "act" utilitarian position, in which the consequences taken into consideration are limited to the case at hand)[36] in reasoning that "while the health of a few infants may be preserved by overriding a pregnant woman's decision, the health of a great many more may be sacrificed."

The report concluded that physicians, whose duty is to ensure that the pregnant woman makes an informed and thoughtful decision, "should refrain from using the courts to impose personal value judgments on a pregnant woman who refused medical advice." However, the report did leave room for "an exceptional circumstance . . . in which a medical treatment poses an insignificant—or no—health risk to the woman, entails a minimal invasion of her bodily integrity, and would clearly prevent substantial and irreversible harm to her fetus." In such a situation, "it might be appropriate for a physician to seek judicial intervention."

This report attempts to provide guidance for cases such as the Angela Carder case, thus agreeing with the Court of Appeals judge who, in his dissent in part, argued that these cases are medically and ethically complex.[45] By contrast, one commentator concluded that there are no medical and ethical complexities to resolve "if the only relevant issue is what the woman wants done."[39] He further exempted differences of judgment from these cases by dismissing footnotes in a Court's opinion as space reserved "for losers."[39] We find these remarks difficult to apply to clinical realities and in conflict with the recent revision by the ACOG Committee on Ethics of its position on informed consent[56] (in which the committee affirmed that patient will is not the only factor to consider in medical decisions).

However, the Board of Trustees Report and the commentator do encourage one critical question: Is the obstetrician's role simply a matter of complying with maternal preference or referring to someone who will?

As noted at the beginning of the chapter, "obstetrics ethics" does not occur in isolation from the many other clinical situations faced in the full range of obstetric and gynecologic care. "Comply or refer" may be good law, but it is not good medicine.

Complying with an informed patient's preference or referring the patient to a physician who will comply has been accepted in our society as both legally and ethically appropriate management in cases involving pregnancy termination or prenatal genetic counseling. However, continuing that process past the mother's decision to remain pregnant and into the late stages of gestation is seriously debated among equally thoughtful clinicians. The difference lies in the judgment about the medical, legal, and social status of the viable fetus. By the late stages of gestation, the burden of proof sits with decision makers who propose actions that diminish or eliminate the well-being of the fetus. This follows because by that time the fetus can be treated as a patient, has certain legal rights, and is being incorporated into the moral community.

A comply-or-refer approach to decision making is clearly inappropriate if carried beyond pregnancy into routine gynecologic care. For instance, the ACOG committee on ethics debated and finally rejected a comply-or-refer approach to requests for sterilization for women with mental retardation.[57] In a similar way, a comply-or-refer approach collapses when applied to requests for tubal ligation reversal from a patient with a history of child abuse and manic-depressive disorder.

Clinicians can easily multiply these illustrations. The point is that patients should rarely be in a position to dictate medical decision making. Patient and physician have critical and mutually dependent roles. Physicians, although deeply respectful of patient judgment, will on occasion find themselves with no ethical alternative but to hold to their judgment. Decisions about the utilization of cesarean section for delivery fall between abortion decisions and prenatal genetic counseling on the one hand and sterilization decisions on the other hand.

Ethics Statements by the American College of Obstetricians and Gynecologists and the American Academy of Pediatrics on Conflicts in Maternal-Fetal Care

The ACOG committee on ethics and the American Academy of Pediatrics (AAP) issued opinions (in October 1987 and June 1988, respectively) about the ethical implications of decisions involving conflict over the use of invasive interventions such as cesarean section.[13,58] Both statements recognized that new technology has penetrated the mystery that had surrounded the fetus. AAP differentiated the various procedures as "standard practices of proven efficacy," "routine practices," and practices founded on "a more ambiguous data base." The ACOG, with a tenuous view of all new techniques, focused on the limitations and fallibility of the technology.

Both statements expected these multifaceted conflicts to be rare. However, they differed sharply on two points in their description of these cases. First, the AAP called attention to the attending physician's "dilemma" when forced to decide which patient—the mother or the unborn child—will receive primary attention. Although concern for the well-being of the mother and the fetus was expressed, the ACOG assumed that the relationship between the pregnant woman and the obstetrician is the primary relationship. Second, the AAP included the father and extended family in the decision-making process. The ACOG made no mention of either the father or the extended family.

Neither statement commented on the moral or legal status of the viable fetus. Only the AAP referred to the pregnant woman as "mother" or "parent." Both statements expected physicians to respect the rights of maternal autonomy and bodily integrity. Both statements expected the attending physician to be medically informed, to formulate treatment plans designed to "foster the greatest benefit with the least risk" (ACOG), to educate the pregnant woman, and to make recommendations.

The ACOG left a surprisingly paternalistic picture of a conflict-free pregnancy. Accordingly, the obstetrician should assess risks and benefits, decide on a course of action, and ask the pregnant woman to consent. If she fails to "cooperate" (a term ACOG used three times), the obstetrician then attempts to "convey the reasons for the current recommendations to the pregnant woman." However, ACOG also implied that the obstetrician's strong leadership in decision making, when confronted with persistent maternal refusal, should yield to maternal choice, which "must not be violated."

In response to maternal refusal, the ACOG advised obstetricians to make the case for the recommended procedure, to encourage "responsible behavior," and to urge the pregnant woman "to seek consultation and counseling." When the recommended procedure is "of proven efficacy and has concomitant low maternal risk," the AAP encouraged even more pressure by sanctioning efforts to make parents "feel obligated to participate." Accordingly, the attending physician should stress "the responsibility of the mother to accept some personal risk for the potential benefit of her fetus."

Both statements discouraged "actions of coercion to obtain consent or force a course of action." However, they took contrasting positions about managing an impasse that cannot be resolved in a "forum of discussion" (ACOG). The AAP listed three criteria: "substantial likelihood" of fetal morbidity or mortality, a clearly appropriate intervention procedure, and low maternal risk. These criteria are intended to identify those "unusual cases" in which the physician's lack of sympathy with the treatment refusal warrants "actively opposing the woman's choice." In regard to those cases, the AAP sounded confrontational (e.g., allowing the attending physician to inform the mother that he or she "finds her refusal extremely unreasonable" and to warn her "that the hospital ethics committee might be brought into the case"). The AAP cautiously approved, as "a last resort," an appeal to the courts.

The ACOG clearly opposed such a response. Although stopping short of categorically rejecting court-ordered interventions, the ACOG offered no criteria for identifying the implied exceptional cases (as did the Court of Appeals in the Angela Carder case). The ACOG did make a four-point case against "the use of judicial authority to implement treatment regimens in order to protect the fetus": (1) benefit to the fetus cannot be "guaranteed absolutely"; (2) court orders for fetal intervention often undercut the reliability of the judicial system by requiring a quick decision; (3) such management threatens or even destroys the physician's relationship with the pregnant woman; and (4) "undesirable societal consequences, such as the criminalization of noncompliance with medical recommendations," may follow.

Suggested Criteria

Given the potential for disagreement over how to respond to maternal refusal of a recommended cesarean section, it should be noted that both the ACOG and AAP encouraged consensus by affirming a pregnant woman's right to privacy, defining an ideal outcome in terms of maternal and fetal well-being, and expressing skepticism about appeals to the judicial system.

However, we find that, when adapted to clinical situations, both statements fall short. The criteria in the AAP statement are too elastic to keep the cases "extremely rare." The ACOG statement, by failing to offer criteria by which to identify truly exceptional cases, leaves the reader unprepared to detect or respond thoughtfully to such cases.

The literature addresses the need for criteria that are clinically realistic, that take seriously the complexities of these cases, and that keep the cases exceptional.[33,59–66] We propose that actively opposing informed maternal refusal of cesarean section delivery should only be considered:

1. When maternal opposition occurs at or near term[67]
2. When the procedure puts the pregnant woman at essentially no greater risk than nonintervention does
3. When the morbidity/mortality risk to the fetus and/or newborn is virtually certain and very significant[68]
4. When the fetus' condition is clearly dissociated from neonatal reference points such as anencephaly, Potter's syndrome, extreme osteoporosis, and other lethal or near-lethal anomalies
5. When it is clear that the refusal would be contested without regard to the woman's social privilege or minority status
6. When plans are in place to diligently seek ways to repair any damage to the pregnant woman's "life story"

We are persuaded that active opposition should be exercised when, given the fore going set of circumstances, maternal intent is to harm or kill the fetus at term (i.e., truly maternal-fetal conflict). In such truly exceptional situations, the conflict is no longer between patient and physician over method of delivery with both hoping for a "good outcome."

Conclusion

These cases, although very rare, lend credence to the judgment that "obstetrics and gynecology are being devoured by ethical issues."[5] In this chapter, we have distinguished obstetric ethics within the larger study of "ethics," looked through clinician eyes at the dynamics surrounding the decision to recommend cesarean section delivery, placed a literature context around cases in which the pregnant woman refuses such recommendation, reviewed professional guidelines, and offered criteria for conflict management.

The minimal attention to perinatal ethics that has traditionally characterized residency training needs to be radically revised, if physicians attending childbirth are to participate responsibly in cases involving complex ethical decisions. It is false to assume that technical authority in matters of medical indications implies ethical authority in matters of values and beliefs.[69]

References

1. The APGO Task force on Medical Ethics. Exploring issues in obstetrics and gynecologic medical ethics. Washington, DC: The Association of Professors of Gynecology and Obstetrics, 1990.

2. Frankena W. Ethics, 2nd Ed. Englewood Cliffs, New Jersey: Prentice-Hall, 1983.
3. Ryan K. Ethics in obstetrics and gynecology. Am J Obstet Gynecol 1985;151:840–843.
4. Hill E. Your morality or mine? an inquiry into the ethics of human reproduction. Am J Obstet Gynecol 1986;154:1173–1180.
5. Park R. Old bedfellows: ethics and obstetrics and gynecology. Obstet Gynecol 1989;73: 1–3.
6. Moffic S, Coverdale J, Bayer T. The Hippocratic oath and clinical ethics. J Clin Ethics 1990;1:287–292.
7. Bulger R, ed. In search of the modern Hippocrates. Iowa City: University of Iowa Press, 1987.
8. Beauchamp T, Childress J. Principles of biomedical ethics, 3rd Ed. New York: Oxford University Press, 1989.
9. Elkins T. Introductory course in biomedical ethics in the obstetrics-gynecology residency. J Med Educ 1988;63:294–300.
10. Silver RK, Minogue J. When does a statistical fact become an ethical imperative? Am J Obstet Gynecol 1987;157:229–233.
11. MacIntyre A. After virtue. Notre Dame: University of Notre Dame Press, 1983.
12. Engelhardt T. The foundations of bioethics. New York: Oxford University Press, 1986.
13. The American College of Obstetricians and Gynecologists Committee on Ethics. Patient choice: maternal-fetal conflict. Washington, DC: American College of Obstetricians and Gynecologists, 1987.
14. Reich W. Caring for life in the first of it: moral paradigms for perinatal and neonatal ethics. Semin Perinatol 1987;11:279–287.
15. Pellegrino E, Thomasma D. For the patient's good: the restoration of beneficence in health care. New York: Oxford University Press, 1988.
16. Dunstan G, Shinebourne E, eds. Doctor's decisions: ethical conflicts in medical practice. New York: Oxford University Press, 1989.
17. Drane J. Becoming a good doctor: the place of virtue and character in medical ethics. Kansas City: Sheed and Ward, 1988.
18. Brody H. My story is broken; can you help me fix it? Medical Ethics and the joint construction of narrative. Lit and Med Spring 1994; 13:79–92.
19. Guillemin J. Babies by cesarean: who chooses, who controls? Hastings Center Rep 1981;11:15–18.
20. Pellegrino E. The anatomy of clinical-ethical judgments in perinatology and neonatology: a substantive and procedural framework. Semin Perinatol 1987;11:202–209.
21. Bosk C. Social, medical, and ethical dilemmas in fetal medicine. In: Milunsky A, Annas GJ, eds. Genetics and the law, Vol. III. New York: Plenum, 1985.
22. Chervenak FA, McCullough LB. What is obstetric ethics? Clin Obstet Gynecol 1992; 35:709–719.
23. Nelson LJ, Milliken N. Compelled medical treatment of pregnant women: life, liberty, and law in conflict. JAMA 1988;259:1056–1066.
24. Price MR, Broomberg J. The impact of the fee-for-service reimbursement system on the utilization of health services. Part III: a comparison of caesarean section rates in white nulliparous women in the private and public sectors. S Afr Med J 1990;78:136–138.
25. Goyert GL, Bottoms SF, Treadwell MC, Nehra PC. The physician factor in cesarean birth rates. N Engl J Med 1989;320:706–709.
26. Stafford RS. The impact of nonclinical factors on repeat cesarean section. JAMA 1991; 265:9–63.
27. Kirk EP, Doyle KA, Leigh J, Garrard ML. Vaginal birth after cesarean or repeat cesarean section: medical risks or social realities? Am J Obstet Gynecol 1990;160:1398–1403.
28. Evans MI, Richardson DA, Sholl JS, Johnson BA. Cesarean section: assessment of the convenience factor. J Reprod Med 1984; 29:670–676.
29. Johnson SR, Elkins TE, Strong C, Phelan JP. Obstetric decision-making: responses to patients who request cesarean delivery. Obstet Gynecol 1986;69:847–850.
30. Morreim EH. Ethical considerations in obstetric management. In: Evans MI, et al., eds. Fetal diagnosis and therapy: science, ethics, and the law. Philadelphia: Lippincott, 1989: 233–242.
31. Brown D, Elkins TE. Ethical issues in obstetric cases involving prematurity. Clin Perinatol 1992;19:469–481.
32. Annas GJ, Glantz LH, Mariner WK. The right of privacy protects the doctor-patient relationship. JAMA 1990;263:858–861.
33. Jonsen AR. Involuntary treatment in medicine. Annu Rev Med 1986;37:41–48.
34. Rhoden NK. Informed consent in obstetrics: some special problems. West Engl Law Rev 1987;9:67–88.

35. Obade CC. In re A.C. reversed: judicial recognition of the rights of pregnant women. J Clin Ethics 1990;1:251.
36. Fleischeman AR, Macklin R. Fetal therapy: ethical considerations, potential conflicts. In: Weil WB, Benjamin M, eds. Ethical issues at the outset of life. Boston: Blackwell, 1987:121–148.
37. Kaufmann CL, Williams PR. Fetal surgery: the social implications of medical and surgical treatment of the unborn child. Women Health 1985;10:25–37.
38. Shultz M. Informed consent: a symbol analyzed. Hastings Center Rep 1987;17:24–25.
39. Annas GJ. Foreclosing the use of force: A.C. reversed. Hastings Center Rep 1990;20:27–29.
40. Brody H. The healer's power. New Haven, Connecticut: Yale University Press, 1992.
41. Kolder VE, Gallagher J, Parsons MT. Court-ordered obstetrical interventions. N Engl J Med 1987;316:1192–1196.
42. Annas GJ. Protecting the liberty of pregnant patients. N Engl J Med 1987;316:1213–1214.
43. Elkins TE, Brown D, Barclay M, Andersen HF. Maternal-fetal conflict: a study of physician concerns in court-ordered cesarean sections. J Clin Ethics 1990;1:316–319.
44. In re A.C. No. 87-609, District of Columbia. Atlantic Reporter, 2nd series. November 10, 1987 (date of decision); 533:611–617.
45. In re A.C. No. 87-609, District of Columbia. Court of Appeals, en banc. Atlantic Reporter, 2nd series. April 26, 1990 (date of decision); 573:1235–1264.
46. DS Davis. Reflections on A.C. BioLaw 1990; (special section);448–451.
47. Annas GJ. She's going to die: the case of Angela C. Hastings Center Rep 1988;18:23–25.
48. Mishkin B. [Letter to the editor.] Hastings Center Rep 1988;18:40–41.
49. Annas GJ. [Reply]. Hastings Center Rep 1988;18:41–42.
50. White ML. Reflections on In re A.C.: a tragedy of decision making. BioLaw 1990;(special section):433–442.
51. Allen AE. In re A.C.—an affirmation of ACOG committee opinion number 55: maternal-fetal conflict. Women's Health Issues 1990;1:37–40.
52. Childress JF. Analogical reasoning: organ/tissue donation and caesarean sections. BioLaw 1990;(special section):443–447.
53. Curran WJ. Court-ordered cesarean sections receive judicial defeat. N Engl J Med 1990; 23:489–492.
54. Pacheco DA. Court Ordered Cesarean Sections. [Letter to the editor.] N Engl J Med 1991;324:272–273.
55. Law and Medicine/Board of Trustees Report. Legal interventions during pregnancy: court-ordered medical treatments and legal penalties for potentially harmful behavior by pregnant women. JAMA 1990;264:2663–2670.
56. The American College of Obstetricians and Gynecologists Committee on Ethics. Informed Consent. Washington, DC: American College of Obstetricians and Gynecologists, 1992.
57. The American College of Obstetricians and Gynecologists Committee on Ethics. Sterilization of women who are mentally handicaped. Washington, DC: American College of Obstetricians and Gynecologists, 1988.
58. American Academy of Pediatrics Committee on Bioethics. Fetal therapy: ethical considerations. Pediatrics 1988;81:898–899.
59. Elkins TE, Andersen HF, Barclay M, et al. Court-ordered cesarean section: an analysis of ethical concerns in compelling cases. Am J Obstet Gynecol 1989;161:150–154.
60. Elkins TE. The case for a middle ground. *Women's Health Issues* 1990;1:34–36.
61. Rosner F, Bennett AJ, Cassell EJ, et al. Fetal therapy and surgery: fetal rights versus maternal obligations. N Y State J Med 1989; 89:80–84.
62. Fletcher JC. Drawing moral lines in fetal therapy. Clin Obstet Gynecol 1986;29:595–601.
63. Chervenak FA, McCullough LB. Clinical guides to preventing ethical conflicts between pregnant women and their physicians. Am J Obstet Gynecol 1990;162:303–307.
64. Kluge EW. When cesarean section operations imposed by a court are justified. J Med Ethics 1988;14:206–211.
65. Chervenak FA, McCullough LB. Perinatal ethics: a practical method of analysis of obligations to mother and fetus. Obstet Gynecol 1985;66:442–446.
66. Strong C. Ethical conflicts between mother and fetus in obstetrics. Clin Perinatol 1987; 14:313–328.
67. Anderson G, Strong C. The premature breech: cesarean section or trial of labor? J Med Ethics 1988;14:18–24.

68. Chervenak FA, McCullough LB. Ethical analysis of the intrapartum management of pregnancy complicated by fetal hydrocephalus with macrocephaly. Obstet Gynecol 1986;68:720–724.
69. Veatch R. Medical authority and professional medical authority: the nature of authority in medicine for decisions by lay persons and professionals. In: Agich G, ed. Responsibility in health care. Dordrecht: Reidel, 1982: 127–139.

15 The Patient Who Demands Cesarean Delivery

Bruce L. Flamm

The Cesarean Decision Spectrum

Although very little research has been done on the subject of cesarean decisions, it is probably safe to say that in most cases the patient and her physician are in full agreement about the need (or lack of need) for cesarean delivery. The doctor–patient interaction would thus fall near the middle of what I have termed the "cesarean decision spectrum" (Fig. 15.1). However, there are many cases in which at least some degree of conflict exists and the interaction would fall slightly on either side of the middle of the spectrum. For example, position 4 on the spectrum would represent a woman who would prefer not to have a cesarean section but certainly would not refuse one if it was recommended by her physician. Position 6 on the scale represents a woman who would prefer a cesarean delivery but would not demand one if her physician explained that it was not medically indicated. Position number 1 at the far left end of the spectrum represents the unusual case in which the patient absolutely refuses a cesarean in spite of her doctor's insistence that the operation is urgently needed. Finally, number 10 at the far right of the spectrum represents the case in which the patient absolutely demands a cesarean operation although her doctor has explained it is clearly not needed. The representation of the distribution of doctor–patient interactions as a fairly symmetrical bell-shaped curve centered around the point of total agreement is hypothetical but does serve to illustrate two very important points.

First, the shape of the curve would almost certainly change depending on whether the interaction was measured before or during labor. While the majority of patients and their physicians remain in agreement about the need (or lack of need) for cesarean throughout the entire antepartum and intrapartum interval, a significant number of women request or even demand medically unindicated cesarean delivery in the later stage of labor. This might be illustrated as a shift of the curve to the right with the majority of doctor–patient interactions still centered near the line of total agreement (Fig. 15.2). The secondary peak at the far right of the curve represents those women who strongly request or even emphatically demand a cesarean at some point during their labor. Note that the numbers at the bottom of the spectrum signify the strength of agreement or disagreement between the patient and her doctor, *not* cervical dilatation. If dilatation was plotted against patient demand for cesarean the resulting graph might look something like Figure 15.3.

The second point that is easy to illustrate on the cesarean decision spectrum is that while almost all discussions of cesarean decision conflicts in the medical, ethical, and legal literature have focused on the cases

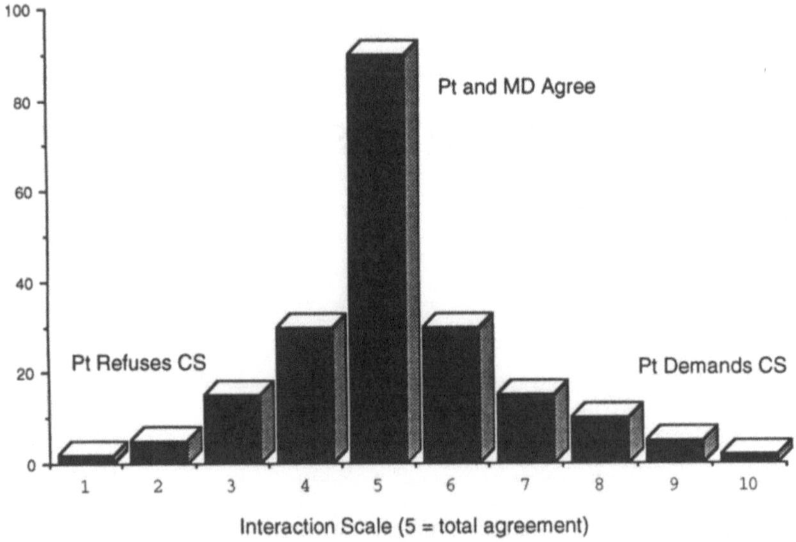

FIGURE 15.1. The cesarean decision spectrum. Pt, patient; MD, physician; CS, cesarean section.

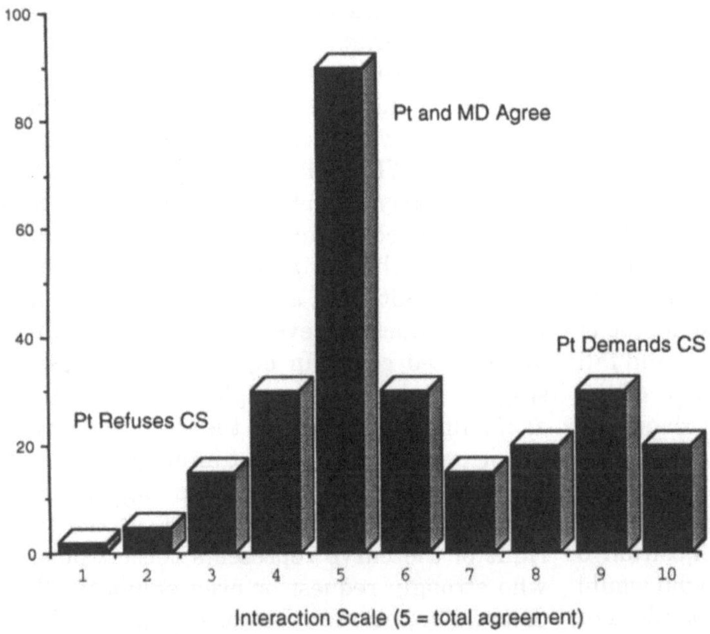

FIGURE 15.2. The cesarean decision spectrum in active labor.

that fall at the extreme left side of the spectrum (coerced or court-ordered cesareans), such cases are extremely rare (Fig. 15.4). Although they often make national headlines, only a few dozen such cases have ever been documented.[1–5] This is not meant to be critical of the amount of attention that has been devoted to court-ordered cesareans but only to point out that issues at opposite end of the spectrum have been virtually ignored. This is unfortunate because conflicts at the right of the spectrum occur far more frequently in the day-to-day practice of obstetrics.

Although obstetricians must frequently deal with requests or demands for medically unindicated cesareans, methods of handling such situation are not even mentioned

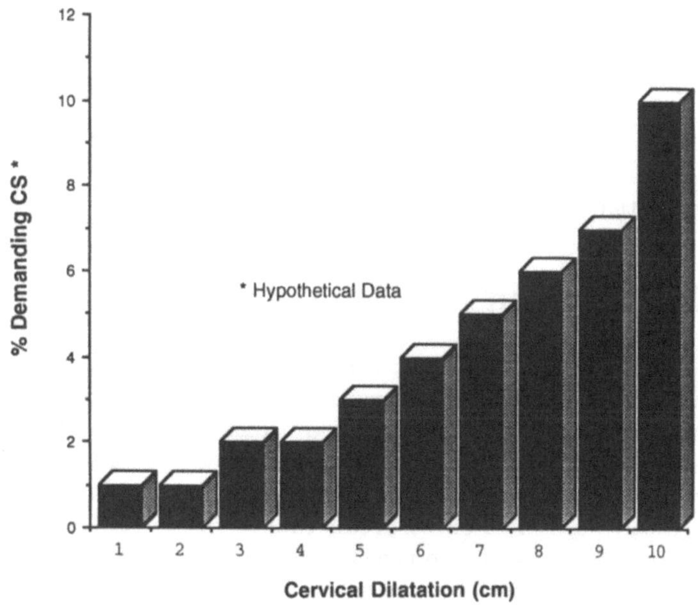

FIGURE 15.3. Opinions about cesarean change during labor.

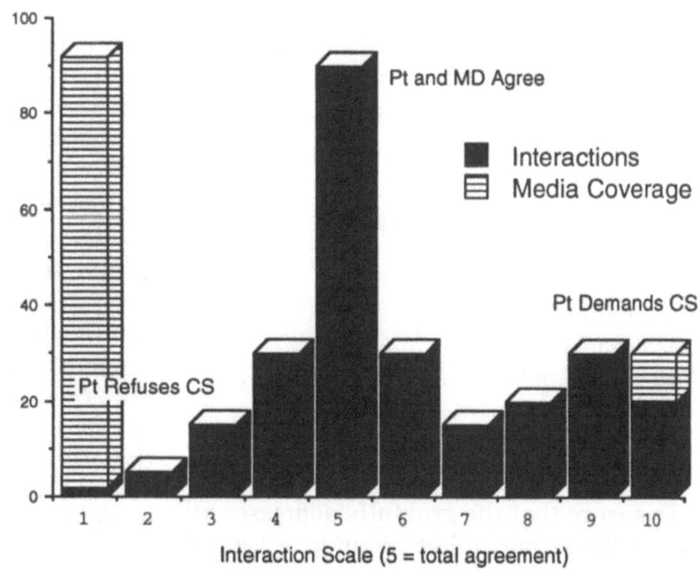

FIGURE 15.4. The media has focused on rare, court-ordered cesareans.

in current obstetric textbooks. *Williams Obstetrics* does not discuss the issue but does point out that, "In modern obstetrical practice, there are virtually no contraindications to cesarean section."[6] Perhaps even more surprisingly, a computerized search of 10 years of medical literature revealed only one paper devoted specifically to the subject of women who demand cesarean operations.[7] When the search was broadened to include "request cesarean" rather than "demand cesarean" (moving from the right end of the decision spectrum toward the middle), only two additional papers were found.[8,9] The ethics chapter (Chapter 14) in this volume delves deeply into the issue of the court-ordered cesarean (left end of spectrum) but only briefly mentions patient-demanded cesarean. Such a focus was not requested by the editors; it simply reflects the current emphasis on court-ordered cesarean in the field of medical ethics. In contrast, this chapter focuses on conflicts that arise on the right side of the spectrum: patient-requested and patient-demanded cesarean.

The Cesarean Decision Grid

Before going on to discuss patient-requested and patient-demanded cesareans, it is useful to introduce another straighforward concept that I call the cesarean decision grid. Conflicts that arise when a patient demands a medically *un*indicated cesarean can be divided into four separate types, as shown in the following grid. Because special circumstances arise in each of these cases that do not apply to the others it is useful to discuss them separately. The only information needed for the classification is whether the patient is in labor and whether she has had a previous cesarean operation.

	No prior cesarean	Prior cesarean
Patient demands cesarean section before labor	A	B
Patient demands cesarean section during labor	C	D

To ensure that the grid fully addresses all the conflicts that might occur in a typical obstetric practice, it is useful to take the process one step further and subdivide the grid into situations in which there is *definitely no medical indication* for a cesarean (designated "1") and cases in which there is a *possible but not definite medical indication* for the operation (designated "2"). The final grid then looks like the following:

	No prior cesarean	Prior cesarean
Patient demands cesarean section before labor	A1, A2	B1, B2
Patient demands cesarean section during labor	C1, C2	D1, D2

Again, "1" is designated as definitely *no* medical indication for a cesarean, and "2" is designated as a possible (but not definite) medical indication for a cesarean.

Almost every possible cesarean decision conflict that could arise will fit into one of the eight divisions of this grid. Examples of each of these eight scenarios are now discussed in detail. Note that although the distinction between patients who request, strongly request, or actually demand a cesarean is not always clear, the following discussions would apply fairly well to all these scenarios. It should be again emphasized that all the cases discussed here reflect situations in which some degree of conflict exists. Decisions that fall in the middle of the cesarean decision spectrum imply that the patient and her physician agree about the need (or lack of need) for cesarean delivery. This encompasses, by far, the majority of cases. The eight divisions of the grid could also be used to categorize and evaluate conflicts that arise at the left end of the cesarean decision spectrum (court-ordered cesarean, etc.); however, that is beyond the scope of this chapter.

Type "A" Cases

Case A: The patient has no prior cesarean and demands a cesarean before labor (i.e., *an elective primary cesarean*).

These cases are by far the rarest of the possible scenarios. Recall that we are not discussing primary cesareans in which no conflict exists, such as the case of a woman who has premature labor and breech presentation along with ruptured membranes and requests a cesarean.

Case A1: The patient has no prior cesarean and demands a cesarean before labor but there is no medical indication (i.e., *an elective primary cesarean with no medical indication*).

The prime example of this type of case is the woman who has no prenatal problems whatsoever but decides, at some point during her pregnancy, that she wants a cesarean delivery before labor begins. Such a woman may have a great fear of labor pain or perhaps a concern that labor is risky for her baby and hence demands to be delivered by cesarean for personal reasons. Although this scenario may seem ludicrous to many physicians (and nonphysicians), a scholarly argument supporting a woman's right to make such a decision was made by Drs. Feldman and Freiman in an article published in the prestigious *New England Journal of Medicine* in 1985.[9] The authors became interested in the subject after reviewing a lawsuit involving a placental abruption that occurred during labor at term. The doctor had apparently acted quickly and did all that was medically possible under the circumstances, but the baby had permanent neurologic damage. Although the malpractice charge was eventually dismissed after a lengthy trial, the authors were haunted by a novel approach taken by the plaintiff's attorney. The attorney had asked the question, "Would this baby be alive, healthy, and undamaged today if it had been delivered by cesarean section one week earlier?" The attorney went on to ask, "Don't you think a mother has the right to assume the extra risks of cesarean section for the sake of her unborn child if she wishes to?"

In an attempt to answer this question, Drs. Feldman and Freiman set up a hypothetical model. They assumed that between 0.2% and 2% of fetuses reach maturity (37 weeks) and are subsequently involved in some type of catastrophe resulting in severe neurologic damage or fetal death. They plugged this information into an equation along with data involving the risks of cesarean and reached some very interesting conclusions. For one thing, they concluded that both societal and legal factors and, "our own data all seem to support the notion that patients ought to be given a definitive opportunity to consider electing prophylactic cesarean section, especially since personal and subjective value judgments are as important to the decision as uncertain quantitative estimates of morbidity, mortality, and cost." They went on to say, "If an informed patient opts for prophylactic cesarean section at term, can it be denied?" Recall that they were not discussing women with prior cesareans or high-risk patients with special risk factors. Their conclusions were meant to apply to normal low-risk pregnant women.

One of the letters to the editor prompted by this article pointed out that, "requiring that healthy women be informed of a surgical alternative to a normal physiologic process" would probably not represent reasonable legal doctrine or public policy.[10] The letter went on to say, "As malpractice suits based on unnecessary cesarean sections become more common, might not a court someday rule that a physician should have informed a pregnant woman of the availability of midwife-attended, out-of-hospital birth as a prophylactic measure against an unwanted cesarean section?" Although the Feldman and Freiman article was initially considered by many readers (including myself) to be some type of joke or parody, it was not. It did, in fact, raise several important issues that will no doubt be debated further in coming years. In the meantime, where do we stand with the issue of elective primary cesarean on demand?

With regard to the scientific (medical) aspects of this question, it would seem that most obstetricians currently believe that the cesarean operation has not become so safe that we can offer it to anyone who desires it. Although there are apparently a few obstetricians who do perform elective primary cesareans (with no medical indication) on request, they are currently in the distinct minority. With regard to the legal aspects of this question, the fact that the doctor prevailed in the lawsuit in the case that spawned the Feldman and Freiman article is reassuring. However, if the judicial system allows the same argument ("Would this baby be alive, healthy, and undamaged today if it had been delivered by cesarean section one week earlier?") to be used in similar cases, other physicians may not fare so well.

Perhaps the key element that must be proven for a plaintiff's attorney to win a malpractice case is that the doctor breached a standard of care. If it is ever decided that one such "standard of care" is that all pregnant women should be informed of the option of elective primary cesarean before labor (for any reason they desire), then any physician who failed to do so would risk losing a lawsuit any time a labor did not have a perfect outcome. This would be true even if the quality of care during labor was not contested. In other words, the doctor could lose the case even if he or she did absolutely nothing wrong and had not breached any other standards of care.

The A1 type of case, elective primary cesarean with no medical indication, was also discussed in a Swedish study in which 10 of 33 women who demanded cesarean before labor did so solely because of fear of the pain of childbirth.[7] Many of the women had feared childbirth from an early age, and several reported that they had been brought up by mothers who told them horrendous childbirth stories. The publication pointed out that, "There are no private obstetric hospitals in Sweden. Thus, a Swedish woman cannot buy a cesarean." However, the women's concerns were taken seriously and each patient was seen from 1 to 13 times during her pregnancy for counseling and short-term psychotherapy. Of the 33 women who demanded cesarean, 14 went on to have uncomplicated vaginal births and 11 of these described their births as good or glorious experiences. Some of the women were believed to have "substantial psychiatric problems," and several did have elective primary cesareans (some under general anesthesia). In cases in which fear becomes pathologic and does not respond to counseling or psychotherapy, the line between type A1 (no medical indications) and A2 (possible medical indications) begins to blur.

Case A2: The patient has no prior cesarean and demands a cesarean before labor but there is no definite medical indication (*an elective primary cesarean with possible medical indications*).

An example of the A2 type of case is a woman with a difficult prior vaginal birth that resulted in a fourth-degree laceration (tearing into the rectum). If the tear was repaired without lasting sequelae, most obstetricians would certainly not think that subsequent vaginal birth was contraindicated. But if the patient went on to develop a rectovaginal fistula that was successfully repaired, the decision becomes more difficult. It certainly possible that the previous repair could break down during childbirth. In a survey that addressed this very situation (although the demand for cesarean was made during labor), fully 55% of obstetricians said that they would agree to perform a cesarean.[8]

Because type A (and type B) cases occur, by definition, before labor, there are some options available. First, there is time to discuss the situation at length with the patient and her partner. With time and understanding, a consensus can often be reached. It may also be wise to obtain a second opinion from another physician or perhaps even present the case at a departmental meeting to obtain the opinions of several colleagues. If the doctor and patient still cannot reach an agreement, there may also be time for the patient to seek the care of another physician. Of course, these options do not exist when conflicts arise during labor (see case types C and D).

Type "B" Cases

Case B: The patient has had a prior cesarean and demands a cesarean before labor (*an elective repeat cesarean*).

Type B cases represent some of the most common and controversial dilemmas faced by contemporary obstetricians. While obstetric textbooks have yet to discuss this situation, such cases are currently being debated on a daily basis at obstetric staff meetings all over the United States and in many other countries around the world.

Case B1: The patient has had a prior cesarean and demands a cesarean before labor but there is no medical indication (*an elective repeat cesarean with no medical indications*).

An example of a B1 type of case is the patient who had a primary low transverse cesarean several years ago for breech presentation and refuses a subsequent trial of labor. A multitude of data show that such a patient has a very high (≥75% probability) of vaginal birth and that the risk of uterine rupture is very low (<1%).[11–13]

During the past 20 years, dozens of studies have addressed technical issues related to trial of labor. Several of these issues are discussed in detail in another chapter of this book. However, one perplexing issue has not been resolved. Who decides if a given patient will have an elective repeat cesarean or a trial of labor? This question has already surfaced as a major issue in malpractice litigation. Some doctors fear that if a uterine rupture (or any significant problem) occurs during a trial of labor, an attorney will always be able to argue that the doctor should have "just" done an elective repeat cesarean. This chain of events has now taken place. Similar to the case presented by Drs. Feldman and Freiman in their *New England Journal of Medicine* article discussed earlier, this case involved a neurologically impaired infant. However, this case involved a prior cesarean and a ruptured uterus. The tragic outcome occurred in spite of the rapid performance of a cesarean section at the first signs of fetal distress. The physician had documented the prior uterine scar to be low transverse and had managed the labor appropriately. This was reaffirmed by the fact that the plaintiff's attorney was unable to show that the obstetrician had breached any standard of care.

Just as it appeared that the case was over and that the defendant physician was not guilty of any negligence or malpractice, the plaintiff's attorney raised the issue of informed consent. He argued that his client had wanted an elective repeat cesarean and that if this "simple" procedure had been done, the baby would not have suffered any injury. Although the patient claimed to have insisted on an elective repeat cesarean long before the onset of labor, the fact that she presented at term in labor argued against that contention. If the doctor had indeed refused to go along with her request she would have had ample time to seek the care of, or at least the opinion of, another physician. This had not been done. But the plantiff's attorney continued to argue that it is the right of the patient, not a physician, to decide if an operation will be performed. He went on to forcefully point out to the jury that doctors are *not* "gods".

Perhaps not coincidentally, just before the case went to trial, a movie entitled "Malice" had shown all over America. Alec Baldwin starred as a brilliant but arrogant young surgeon who, in a chilling courtroom scene relating to the need for informed consent, essentially stated that he was indeed a god. Undoubtedly some of the jurors had recently seen the movie. In spite of the plaintiff attorney's cleverly timed innuendos, the jury sided with the defendant doctor and agreed that the trial of labor had been reasonable and the uterine rupture had been unavoidable.

The outcome of this case should not be misconstrued as signifying that poor outcomes relating to uterine rupture carry little medical-legal risk. For example, a Utah obsetrician recently lost a case in which a trial of labor had ended with a uterine rupture and a brain-damaged baby. The plaintiff's attorney argued that trial of labor had been instituted as a cost-containment measure. The judge apparently directed a defense verdict on the cost-containment issue because this charge was not proven by the evidence, but the jury came back with a guilty verdict and awarded the plaintiff $8 million. Whether juries composed of lay individuals have the capacity to render reasonable verdicts in cases involving extremely complex medical issues is debatable.

Sensational trials are even shedding doubt on the ability of the jury system to deal with cases that do not require an advanced degree of medial sophistication. For example, in Los Angeles, separate trials of Lyle and Eric Menendez both concluded with mistrials because neither jury could decide whether the verdict should be murder or manslaughter. This was true in spite of the fact that the brothers, both of whom were adults, had already confessed to killing their wealthy parents with repetitive shotgun blasts.[14] After pumping several rounds into each of their parents, the brothers admitted to going outside to reload and then returning to the house to finish the job. Although both brothers claimed to have been previously abused by their father they were now adults and could have moved away from home at any time. Retrial, if it occurs, may take more than a year and cost taxpayers additional millions of dollars. If the current jury system cannot handle cases where defendants openly confess to homicide, how can the same system be expected to deal with complex issues such as deciding exactly what is the proper dose of oxytocin to utilize during labor after a prior low transverse cesarean or exactly what electronic fetal heart monitor patterns indicate that a cesarean operation should have been performed?

Patients with prior cesareans represent a significant and ever-increasing portion of the pregnant population. This brings us back to the initial question. Who decides if a given patient will have an elective repeat cesarean or a trial of labor? The most recent American College of Obstetricians and Gynecologists (ACOG) Guidelines for Vaginal Delivery after Previous Cesarean Birth recommend that "...a woman with one previous cesarean delivery with a low transverse incision should be *counseled and encouraged* to attempt labor in her current pregnancy."[15] But how strongly should she be encouraged? What if, after counseling, she is not interested in a trial of labor? These are very difficult questions to answer.

Clearly, the way trial of labor is presented to a patient will have a profound effect on whether or not she is willing to give it a try. A doctor could say, "If you opt for a trial of labor you must fully understand that your uterus could rupture and your baby could die." This is certainly true but does not put the risks into proper context. Few women given such a morbid description would even ponder an attempt at labor. Another doctor's version of informed consent might be, "If you opt for a trial of labor, many large studies have shown that you will probably be able to avoid a repeat cesarean and that the risk of your uterus rupturing is less than 1%. In the small number of cases where the uterus does rupture, the baby will often do well. But in some of these cases the baby may have problems and in very rare cases the baby could die."

Some doctors state that few if any of their patients agree to trial of labor. This may have a lot to do with exactly how the options are explained. A recent publication reported a trial of labor rate of only 19% (81% elective repeat cesarean sections) and concluded that because of both medical and patient-choice factors in favor of cesarean, "it will be difficult to substantially decrease the present rate of repeat cesarean births."[16] However, 1 year later a prospective study of 7229 patients with prior cesareans reported that fully 70% opted for trial of labor.[17] Rates of trial of labor varying from 19% to 70% cannot be explained by medical factors or demographics and must therefore reflect the way that the options are being presented to patients.

This is not to say that the typical patient will jump for joy at the prospect of going through labor if her physician has an optimistic attitude. Rather it implies that the majority of women will at least be willing to give it a try. Nevertheless, some reports contend that it is often reluctance of the patient, not her obstetrician, that prevents greater utilization of vaginal birth after cesarean section (VBAC). One study found that 50% of patients who were encouraged by their obstetrician toward VBAC opted for elective repeat cesarean without trial of labor.[18] Another study found that 40% of patients demanded elective repeat cesarean although they fulfilled the criteria for eligibility in a VBAC program.[19] A reason frequently given by these patients was that they wished to know their dates of admission, delivery, and discharge months in advance in order to arrange their schedules accordingly. Obviously, planned elective repeat cesarean provides similar scheduling advantages for the physician and hospital staff.

It is my opinion that the issue of who decides if a given patient will have an elective repeat cesarean or a trial of labor has not been resolved. To many attorneys and perhaps to many jury members, it may seem obvious that the decision to have (or not to have) surgery is always an uninfringable right of a mentally competent patient. However, things are not quite so simple. Does any women who is concerned about the risk of uterine cancer have the right to demand an elective hysterectomy? Can a physician be compelled to perform an operation that he or she feels to be unnecessary? If society agrees that the current cesarean rate is too high, then tort reform must emphasize the fact that a bad outcome is not synonymous with malpractice.

A logical system would mandate an initial review which would screen out frivolous lawsuits. In cases in which a poor outcome occurs in spite of good obstetric care with adherence to basic guidelines, an arbitration panel should rule that no lawsuit is warranted. As is discussed later in this volume (Chapter 17, Cesarean Projects at the State and National Level), such a system is now being tested in the state of Maine. The current hostile environment, in which any doctor can be dragged through a horrendous lawsuit simply because a patient "wanted a cesarean and didn't get it," has led many American doctors to conclude that the only defense is to quickly comply with every patient's request for cesarean delivery.

It should also be pointed out, however, that in the current state of affairs doing a repeat cesarean does not free an obstetrician from potential liability. In a recent malpractice case in Massachusetts a woman claimed that her doctor had agreed to letting her have a trial of labor. However, 2 weeks before her due date the doctor told her he was going on vacation and was going to schedule her for a repeat cesarean. When she objected he told her that labor would be risky to her and her baby and allegedly compared her uterus to a "hydrogen bomb." The women developed severe gastrointestinal complications after the repeat cesarean and required prolonged hospitalization. She sued, claiming that the cesarean was unwanted and unnecessary. The doctor was found negligent and the patient was awarded $1.53 million by a superior court jury.[20,21]

Case B2: The patient has had a prior cesarean and demands a cesarean before labor but there is no definite medical indication (*an elective repeat cesarean with possible medical indications*).

A typical example of a B2 type of case would be a woman who planned to have a trial of labor but changes her mind near term because it appears that her baby is too large. She then requests or demands an elective repeat cesarean. This creates a very difficult dilemma. Studies have shown that both clinical and sonographic estimates of fetal macrosomia are inaccurate.[22–27]

Even if we could accurately predict fetal macrosomia, it has been shown that the majority of trials of labor will result in uneventful vaginal births. We retrospectively reviewed 301 trials of labor in patients with prior cesareans in which the actual documented birthweight for the current pregnancy was more than 4000 g (8 pounds, 13 ounces). In the 4000–4499 g weight range, 58% of the patients had vaginal births after cesarean, and the perinatal morbidity and mortality was no higher than for the infants with birthweights less than 4000 g.[28] These data are reassuring, but are no panacea for the obstetrician's dilemma. If a woman with a prior cesarean requests a repeat operation but is strongly encouraged to have a trial of labor and then goes on to experience a traumatic birth of a 4700-g infant, litigation is likely to occur. On the other hand, if a physician agrees to operate in any case in which the baby appears to be a little large, many unnecessary cesareans will be performed for normal-sized infants.

Another example of a B2 type of case is a woman who had a prior cesarean performed for failure to progress after a long and difficult labor. Although studies have shown that many of these patients can go on to have subsequent normal vaginal births, this outcome can certainly not be guaranteed. If such a patient demands an elective repeat cesarean, should her request be denied? On the other side of the coin, it must be remembered that the single most common reason for *primary* cesarean in the United States is failure to progress, or dystocia. If all these women had elective repeat operations, there would be few VBAC candidates and overall cesarean rates would continue to rise.

While the B1-type cases can strain the doctor–patient relationship, the B2-type cases can destroy it. There are currently no easy solutions. A management plan often used in cases in which there is a possible but not definite medical indication for repeat cesarean is to recommend a trial of labor with the understanding and agreement that cesarean will be performed "expeditiously" if labor does not progress or if there is any sign of fetal distress. This reasonable plan should not be confused with a "perfunctory" trial of labor in which the real plan is to perform a repeat cesarean an hour or two after admission to the hospital regardless of how the labor progresses.

Type "C" Cases

Case C: The patient has no prior cesarean and demands a cesarean during labor (*a primary cesarean during labor*).

Because of the enormous physical and emotional stresses of labor, type C cases are perhaps the most common of all the cesarean decision conflicts. However, it should emphasized that only cases that involve little or no medical indication for cesarean are discussed. For example, assume a woman in labor has a fetus with a suspicious fetal heartrate monitor pattern. Further assume that her cervical dilatation has not changed from 6 cm despite several hours of adequate oxytocin augmentation. If this woman were to request or even demand a cesarean, her doctor would almost certainly agree. There would be no conflict, the interaction would rank near the middle of the cesarean decision spectrum, and everyone would be happy. Unfortunately, things rarely are so simple in the practice of obstetrics.

Case C1: The patient has no prior cesarean and demands a cesarean during labor although there is no medical indication (*a primary cesarean during labor with no medical indication*).

An obstetric intern quickly learns that many women demand cesarean delivery in the midst of labor. The obstetric intern also learns that most of these women will go on to experience normal vaginal births. If a cesarean was performed every time an exhausted woman screamed "Take my baby now!," the cesarean rate would probably be more than 50%. Although some attorneys would argue that even during the most difficult moments of labor a woman has the right to demand a cesarean operation, most physicians would disagree. For example, a survey of 112 obstetricians found that only 2% would agree to a laboring patient's request for cesarean if she were at term with no medical problems.[8] Interestingly, if it was assumed that the same patient was married to a lawyer, the number of doctors who said they would agree to perform a cesarean rose to only 3%. To be sure, this was a survey and reflected only what individual physicians said they might do in hypothetical situations! In any case, the current state of affairs seems to be that the majority of doctors would not agree to perform a primary cesarean section in the absence of medical indications.

Case C2: The patient has no prior cesarean and demands a cesarean during labor yet there is no definite medical indication (*a primary cesarean during labor with possible medical indications*).

The survey of obstetricians mentioned earlier (case C1) also addressed the issue of making the decision to perform a primary cesarean during labor with possible medical indications.[8] The method of the study can be described briefly as follows. A survey requesting opinions about the decision to perform a cesarean section in several hypothetical situations was sent to obstetricians in three cities, and 112 of the doctors responded (97 university staff and 15 in private practice). Four of the situations were clearly type A1 cases (elective primary cesarean with *no* medical indication). However, several scenarios involved the decision-making process when there were possible (but not definite) indications for surgery. For example, 7% of respondents said that they would agree to a patient's request for primary cesarean if she had a history of stress urinary incontinence since her last delivery. An additional 12% of respondents said that they might agree to performing the cesarean. This was true even though only 1% of respondents felt that stress incontinence represented an actual medical indication for cesarean. Similarly, 8% of respondents said that they would agree to a laboring patient's request for primary cesarean if she had a history of postpartum hemorrhage with a previous birth. An additional 15% of respondents said that they might agree to performing the cesarean in this situation. But only 1% of respondents thought that postpartum hemorrhage experienced after a previous vaginal birth represented an actual medical indication for cesarean. Clearly, nonmedical factors do influence the decision-making process.

Type "D" Cases

Case D: The patient has had a prior cesarean and demands a cesarean during labor (*a repeat cesarean during labor*).

Case D1: The patient has had a prior cesarean and demands a cesarean during labor yet there is no medical indication (*a repeat cesarean during labor with no medical indications*).

This is a unique situation. At least in the United States, it is a common practice in many hosptials to perform a repeat cesarean at any time during a trial of labor if the patient requests it. This may explain why some reports have been unable to reproduce the rate of vaginal birth after cesarean of 75% or more found in large studies. Although most doctors would not even ponder an affirmative response to such a request made by a woman with no prior cesarean (type C1 case), many will rush to the operating room if she has had one. Most programs with high rates of vaginal birth after cesarean have adopted the policy of treating trial-of-labor patients essentially the same as any other patients and using similar indications for primary and repeat cesarean. However, in the survey of 112 obstetricians discussed here, if a woman was progressing well during a trial of labor after prior cesarean and changed her mind at 9 cm of dilatation and requested a repeat operation, 30% of the physicians said they would definitely comply and an additional 24% said they might comply. This was true in spite of the fact that only 4% believed that a medical indication for the repeat operation actually existed.

In another type of survey study addressing the decision to choose trial of labor or repeat cesarean section, questionnaires were sent to women rather than physicians.[29] Of those respondents who opted for repeat cesarean, about 25% listed concerns about the risks of labor as important in their decision-making process. Far more women listed such factors as convenience of timing the birth, knowing what to expect, and avoidance of the pain of labor. Also, 73% of these patients reported that their physicians either discouraged vaginal birth after cesarean or had neutral attitudes. The authors concluded that the physician has a central role in the decision-making process.

Case D2: The patient has had a prior cesarean and demands a cesarean during labor yet there is no definite medical indication (*a repeat cesarean during labor with possible medical indications*).

Everything said about the C2 type of cases (primary cesarean during labor with possible medical indications) applies to the D2 situations. However, when the additional risk factor of a prior cesarean birth is added, the scale tips toward repeat operation. For example, if a laboring patient with no prior cesarean requested a primary cesarean because she had a history of postpartum hemorrhage with a previous birth, 8% of doctors surveyed said they would agree to perform the operation even though only 1% thought the prior event had any bearing on the current labor.[8] If a laboring patient with a history of prior cesarean operation requests that her trial of labor be terminated because a prior postpartum hemorrhage, how many doctors would refuse? In contrast to the D1 case where everything is going fine with the trial of labor but the patient simply changes her mind and requests a cesarean, this situation involves additional risk factors. As another example, what if variable decelerations of the fetal heart rate are noted during a trial of labor and the concerned patient (or her partner) demands an immediate repeat cesarean? In the majority of cases, these decelerations are innocuous but at times they could be an indication of uterine rupture. What is the solution to these dilemmas? It may be a reasonable policy to have a "lower threshold for intervention" during trial of labor than for patients with no prior cesarean. This is certainly not the same as saying that every woman who changes her mind during a trial of labor should be rushed to the operating room.

How the Cesarean Operation Differs from Other Surgical Procedures

Part of the reason why the cesarean decision-making process can become so complicated is that the cesarean operation differs somewhat from all other operations. In a strictly anatomic sense, cesarean involves cutting through and then repairing the same or similar layers of tissue as those encountered during many abdominal operations. But when nonanatomic factors are considered, several differences become apparent. These differences may explain why doctor–patient conflicts may arise in obstetrics that would not occur in other surgical fields.

1. There are very limited alternatives. When an individual presents with symptoms or signs of illness, diagnostic testing may lead to the conclusion that surgery could be beneficial. In many cases there is often an alternative to surgery such as

administration of appropriate medications and waiting to see if the symptoms improve or resolve. With the cesarean operation there are generally no such alternatives. When a woman is in labor the baby must be born and there are only two possible routes. The woman can not go home and come back in a couple of weeks after she has had a chance to ponder her options or try other treatment regimens.

2. The decision to operate is often made on an emergent basis. With the exception of procedures performed on trauma patients, the vast majority of operations that take place in most surgical disciplines are scheduled elective procedures. In contrast, the decision to perform a cesarean operation (except for elective repeat procedures) is often made only minutes before the surgery actually begins.

3. The patient is often under great physical and emotional stress. No surgeon would consider jogging along with a patient during a marathon to discuss the potential risks versus benefits of a proposed major operation. Yet this is analogous to the situation that commonly occurs when one obtains the consent of a woman in the midst of active labor.

4. The health of two patients must be considered simultaneously. Yet another difference between the cesarean operation and all other operations is the fact that two patients are always involved. An individual who might be quite comfortable considering risks versus benefits involving their own health may be far less comfortable handling decisions that also affect the health of their unborn child. In addition to the patient's opinions about the welfare of her fetus, the actual rights of the unborn child also must be considered. These fetal rights and the issue of maternal versus fetal beneficence have been extensively debated by specialists in medical ethics. The opinion of the unborn child's father may also have an impact on the decision as to whether or not a cesarean is performed. For example, if a woman has gallstones and is unclear as to whether or not she wants a cholecystectomy, her husband's opinion may be of little significance to the surgeon. However, if a woman in labor has a fetal monitor pattern indicating possible fetal distress and she is unclear as to whether or not she wants a cesarean, her husband's opinion can become quite important. If he demands a cesarean delivery to protect the health of his unborn child, can his request be ignored?

Conclusions

1. Approximately 1 million cesareans are performed in the United States each year. It has been claimed that 500,000 of these operations are unnecessary.[30] While most experts would argue that this estimate is vastly inflated, many would agree that cesareans are indeed being performed that are not medically indicated.

2. The decision to perform a cesarean may involve some degree of disagreement or conflict between a patient and her physician. Although media attention has focused on rare cases where a patient *refuses* a medically indicated cesarean, the opposite situation, where a patient *demands* a medically unnecessary cesarean, occurs much more frequently. Yet the "patient-demanded cesarean" has been virtually ignored in both medical literature and media coverage of rising cesarean rates.

3. The entire concept of a chapter devoted to "the patient who demands a cesarean" may infuriate members of the natural childbirth movement along with some childbirth educators and midwives. The idea that large numbers of women are requesting or demanding cesareans may seem absurd to them. However, these individuals may not realize that they are dealing with a select portion of the population. Women who seek out midwifery care may do so because they strongly desire natural childbirth. Women who desire cesareans do not contact midwives. Women who are upset after cesarean delivery may contact cesarean-related support groups to ventilate negative feelings.

Women who are pleased with their operations do not contact support groups.

4. Although this chapter focused on patients who demand or strongly request cesareans, all the eight cases discussed could also be applied to a discussion of patients who prefer not to have a cesarean (left side of the cesarean decision-making spectrum), for example, the woman who is very interested in natural birth but who is told by her doctor early in her pregnancy that the baby "looks too big." It must be acknowledged that, while cases of court-ordered cesareans are rare, situations where women are subtly urged to accept a cesarean occur frequently.

5. This chapter only considers cases in which there is some degree of conflict in the cesarean decision-making process. In most cases the patient and her doctor agree on the decision to perform, or not to perform, a cesarean. The fact that a patient and her doctor agree on the decision to perform a cesarean does not necessarily imply that the procedure is medically indicated. The agreement does imply that no conflict exists.

6. Conflicts that occur before labor allow time for many options to be considered (see case A2), whereas conflicts which occur during labor allow for few options. This raises the question of whether the "cesarean issue" should be discussed during routine prenatal care. Contemporary prenatal care involves the discussion of a plethora of subjects, many of which will have little bearing on the pregnancy outcome. Whether the patient should take one or two prenatal iron tablets a day is probably of far less significance than whether the patient will end her pregnancy on an operating table. With one of every four women currently being delivered by cesarean in the United States, a prenatal discussion about why a cesarean operation will or will not be recommended would be far from esoteric.

7. The issue of who actually decides whether a cesarean will be performed needs further discussion in the ethical, legal, and medical literature. The prevailing opinion in the legal field, that a woman has the right to request or demand the operation at any time, raises ominous implications. This was discussed under case A1.

8. The cesarean operation differs from other surgical procedures in several important ways. These differences may explain why the cesarean decision-making process can become quite complicated.

References

1. Kolder VE, Gallagher J, Parsons MT. Court ordered obstetrical interventions. N Engl J Med 1987;316:1192–1197.
2. Elkins TE, Brown D, Barclay M, Anderson HF. Maternal-fetal conflict: a study of physician concerns in court-ordered cesarean sections. J Clin Ethics 1990;1:316–319.
3. Curran WJ. Court-ordered cesarean sections receive judicial defeat. N Engl J Med 1991; 324:272–273.
4. Elkins TE, Andersen HF, Barclay M. Court-ordered cesarean section: an analysis of ethical concerns in compelling cases. Am J Obstet Gynecol 1989;161:150–154.
5. Kluge EW. When cesarean section operations imposed by a court are justified. J Med Ethics 1988;14:206–211.
6. Cunningham GF, MacDonald PC, Gant NF. Williams obstetrics, 18th Ed. Norwalk, Connecticut: Appleton & Lange, 1989:445.
7. Ryding EL. Investigation of 33 women who demanded a cesarean section for personal reasons. Acta Obstet Gynecol Scand 1993; 72:280–285.
8. Johnson SR, Elkins TE, Strong C, Phelan JP. Obstetric decision-making: responses to patients who request cesarean delivery. Obstet Gynecol 1986;67:847–850.
9. Feldman G, Freiman J. Prophylactic cesarean section at term. N Engl J Med 1985; 312:1264–1267.
10. Brodsky A. Letter to the editor: prophylactic cesarean section at term. N Engl J Med 1985;313:753.
11. Flamm B, Lim O, Jones C, Fallon D, Newman L, Mantis J. Vaginal birth after cesarean section: results of a multicenter study. Am J Obstet Gynecol 1988;158: 1079–1084.
12. Flamm B, Neuman L, Fallon D, Thomas S, Yoshida M. Vaginal birth after cesarean sec-

tion: results of a five-year multicenter study. Obstet Gynecol 1990;76:750–754.
13. Phelan J, Clark S, Diaz F, Paul R. Vaginal birth after cesarean section. Am J Obstet Gynecol 1987;157:1510–1515.
14. Smolowe J. Waiting for the verdicts. Time Magazine. 1993;142:26:48–49 (December 20).
15. ACOG. Guidelines for vaginal delivery after a previous cesarean birth. Committee opinion number 64. Washington, DC: American College of Obstetricians and Gynecologists, 1988.
16. Kline J, Arias F. Analysis of factors determining the selection of repeated cesarean section or trial of labor in patients with histories of prior cesarean delivery. J Reprod Med 1993;38:289–292.
17. Flamm B, Goings J, Liu Y, Wolde-Tsadik G. Elective repeat cesarean section versus trial of labor: a prospective multicenter study. Obstet Gynecol 1994;83(6):1–6.
18. Joseph G, Stedman C, Robichaux A. Vaginal birth after cesarean section: the impact of patient resistance to a trial of labor. Am J Obstet Gynecol 1991;164(6):1441–1447.
19. Abitbol M, Castillo I, Taylor U, Rochelson B, Shmoys S, Monheit A. Vaginal birth after cesarean section: the patient's point of view. Am Fam Physician 1993;47:129–134.
20. Lehman B. Woman wins $1.53 million suit on unwanted cesarean. The Boston Globe Newspaper, June 16, 1993 (Metro section), p 1.
21. Laska LL. Malpractice casebook. OBG Management 1993 December: 49–50.
22. Ong HC, Sen DK. Clinical estimation of fetal weight. Am J Obstet Gynecol 1972;112:877–880.
23. Simon NV, Levisky JS, Shearer DM, O'Lear MS, Flood JT. Influence of fetal growth patterns on sonographic estimation of featl weight. JCU 1987;15:376–383.
24. Miller JM, Kissling GA, Brown HL, Gabert HA. Estimated fetal weight: applicability to small and large-for-gestational-age fetus. JCU 1988;16:95–97.
25. Miller JM, Brown HL, Khawli OF, Pastorek JG, Gabert HA. Ultrasonographic identification of the macrosomic fetus. Am J Obstet Gynecol 1988;159:1110–1114.
26. Benacerraf BR, Gelman R, Frigoletto FD. Sonographically estimated fetal weights: accuracy and limitation. Am J Obstet Gynecol 1988;159:1118–1121.
27. Deter RL, Hadlock FP. Use of ultrasound in the detection of macrosomia: a review. JCU 1985;13:519–524.
28. Flamm B, Goings J. Vaginal birth after cesarean section: is suspected fetal macrosomia a contraindication? Obstet Gynecol 1989; 74:694–697.
29. Kirk E, Doyle K, Leigh J, Garrard M. Vaginal birth after cesarean or repeat cesarean section: medical risks or social realities? Am J Obstet Gynecol 1990;162(6): 1398–1405.
30. Silver L, Wolfe SM. Unnecessary cesarean sections: How to cure a national epidemic. Washington, DC: Health Research Group, 1989.

16
The Impact of Midwifery Care, Childbirth Preparation, and Labor Support on Cesarean Section Rates

JANICE R. GOINGS

> The last few years have witnessed the development of radical surgical obstetrics and brilliant the results have been. Long lists of Cesarean sections have been published with little if any mortality, and the indications for the operation have been extended.... The question before us is: Are we in our enthusiasm over radical obstetric surgery neglecting the fundamentals of obstetrics; the routine precautionary methods which may make the resort to radical obstetric surgery unnecessary?[1]
>
> Edwin B. Cragin

In 1916, when Dr. Cragin published his paper, which was titled "Conservatism in Obstetrics," the cesarean section rate was less than 1%.[2] Three-quarters of a century later, it has risen dramatically and has, through the intervening years, been the subject of much debate. In 1965 the rate was only 4.5%.[3] By 1980 the rate had risen so high that the National Institutes of Health convened a task force to study the topic. When this task force issued its report in 1981, the national rate was 17.6%.[4] It has since escalated to 23.5%,[5] an increase of 33.5% in 10 years.

Almost 80 years later, Dr. Cragin's question remains germane. Some have argued that the current cesarean rate has resulted in healthier mothers and babies.[6] However, many industrialized countries with cesarean rates much lower than ours also have much lower perinatal mortality rates than we do.[7] During the first half of this century, maternal and neonatal morbidity and mortality did decrease dramatically, but much of this decrease can be attributed to nonmedical factors, such as improved nutrition and sanitation. Further decreases resulted from medical, not surgical, advances such as the availability of type-specific blood transfusions, antibiotics, and the use of aseptic techniques.[8]

If, in fact, we believe that cesarean section is overutilized, how can we intervene to lower this rate? Should we employ the active management model? Will more liberal use of external cephalic versions and trial of labor after prior cesarean have sufficient impact? Is the diagnosis of fetal distress made too frequently?

The vast majority of literature devoted to this topic analyzes outcomes of the medical model of care. In this model, the laboring woman is seen as a patient and the process of labor as abnormal until proven otherwise.[9] But other models of caring for pregnant women also have the potential for impacting the cesarean rate. This chapter explores these models, which view pregnancy as a normal physiologic state, one that is only pathologic in retrospect. Included in this discussion are the midwifery model of care, childbirth education, and labor support.

Midwifery

The midwifery model of care is common in industrialized European countries. One of the best known and well-studied examples is at the National Maternity Hospital in Dublin. There, midwives are responsible for the management of all laboring women, including private patients who are delivered by the midwives with their personal physician in attendance. Midwives are closely involved with the education and training of resident physicians. The active management of labor, one of the components of this system, is partially credited with the impressive outcomes of care there, but the inclusion of midwives in their maternity service is, undoubtedly, an important factor.[10]

In this country midwives consist of two groups of practitioners: nurse-midwives who receive certification through the American College of Nurse-Midwives and who may legally practice in all 50 states; and a second group consisting of licensed midwives, lay midwives, and foreign-trained midwives, who have varying degrees of education and training and, depending on the state in which they practice, may or may not have legal standing for the practice of midwifery.

The philosophy of care given by the midwife differs from that of the medical practitioner, although both have a common goal, that each birth end with a healthy mother and baby. Midwives have historically been viewed as noninterventionists in the birth process, whereas physicians have been seen as much more likely to intervene, even in a normal, low-risk birth. Midwifery philosophy views pregnancy and childbirth as a normal physiologic female function, to be considered pathologic only in retrospect. The practice of midwifery has been described as a "combination of science, craft, art, talent, knowledge and experience".[11]

The safety, quality, cost-effectiveness, and patient acceptance of midwifery care in a variety of settings have been well documented.[12–14] But just how does this care differ from that provided by physicians?

A case-comparison retrospective review was conducted in 1985 in an attempt to answer this question. Twenty-nine laboring women attended by certified nurse-midwives (CNMs) were matched case by case with 29 women attended by physicians. Patients included in each group were low risk and were delivered in a tertiary-care hospital. Results showed no differences in complication rates, but very different care practices by the two provider groups. Intravenous fluids were administered to 38% of the CNM patients, compared to 72% of the physician (MD) patients. Electronic fetal monitoring was used 100% of the time by physicians, compared to 34% of the time by CNMs. Oxytocin was administered to 56% of the MD patients, but only 22% of the CNMs. Amniotomy was performed more often by physicians than midwives. Pain-relieving medication was administered less frequently to patients cared for by CNMs. The use of episiotomy was much higher in the physician group. Two of the women in the midwife group sustained third-degree lacerations, compared to 3 third- and 3 fourth-degree lacerations in the physician group. More women in the midwife group planned to breast-feed their infants at discharge.

This study was limited by its small sample size and retrospective nature; women were not randomly assigned to providers, a factor that might have accounted for some of the findings. With patients self-selecting for CNM or MD care, the differing patterns of care might have been a reflection of dissimilar needs in each group.[15]

Another study addressing the philosophic difference in pregnancy and childbirth care given by physicians and CNMs was conducted at the School of Nursing at the University of Michigan at Ann Arbor. A sample of 15 obstetricians and seven CNMs in a private service within a large tertiary-care medical center completed questionnaires on various aspects of their philosophy of care. These questionnaires were constructed by clinical experts familiar with the functioning of both obstetricians and nurse-midwives. Self-reports of amount of clinic time spent with patients were randomly verified by auditing provider office schedules.

Results of this study showed consistent differences between the CNMs and MDs.

Midwives spent significantly more time on initial and return prenatal visits and were more likely to do teaching in the areas of nutrition, infant feeding, exercise, parenting, and minor maternal illnesses than were physicians. Midwives reported that they personally did almost all the teaching, while physicians relied on ancillary staff members, such as nurses.

While there were differences in the two groups in their rating of the importance of management of intrapartum care, such as the use of pitocin, intravenous (IV) fluids, continuous electronic fetal monitoring (EFM), use of lithotomy position, etc., only the importance of ambulation in labor reached statistical significance. The authors noted that differences which did not reach statistical significance were, nonetheless, all in the expected directions, with CNMs placing more importance on management techniques such as not using IVs, not limiting oral fluids in labor, etc.

Study limitations included the small sample size and the volunteer nature of the study participants. In addition, the non-participation of five of the physicians in the group who failed to return their questionnaires might have influenced the outcomes, because there could have been differences in those choosing to participate and those choosing not to participate.[16]

These two studies found some significant differences in the philosophy of and care given by midwives and physicians. Other studies have affirmed these differences, including less instrumentation used in deliveries,[17,18] lower rates of epidural anesthesia and episiotomy,[19] and significantly more third- and fourth-degree lacerations in physician deliveries.[18]

Although small and limited in design, the foregoing studies basically describe much of the midwifery philosophy. Through such interventions as health education, midwives encourage women to accept responsibility for their own care. They have long embraced an opposition to routine interventions in childbirth and have strongly encouraged family-centered care, to include not only husbands but also siblings and other family members in the birth process. Physicians and nurses who work side by side with midwives in a variety of settings are familiar with such midwifery techniques as ambulation or nipple massage as an alternative to the administration of oxytocin to stimulate labor; employing massage and water therapy as alternatives to the use of medication and regional anesthesia; encouraging the intake of oral fluids in labor instead of routinely administering IV solutions; the avoidance of routine episiotomies, and the use of alternative laboring and birthing positions.

An important question arises when considering these midwifery interventions. Does this philosophic divergence and differing practice style between medicine and midwifery result in different outcomes for patients, especially when considering their impact on cesarean section? If midwifery care does lead to a lower cesarean section rate, what components of midwifery care are responsible?

Midwifery Care and Cesarean Rates

One of the largest studies to address this issue, conducted in the Netherlands from 1969 to 1983, is known as the Wormerveer study. It was a prospective study of 8480 women who were booked for delivery either at home, in a small maternity unit, or in hospital, with deliveries under the supervision of a midwife. After exclusion for risk factors, spontaneous abortions, ectopic pregnancies, and those lost to follow-up, the study group consisted of 4000 nulliparous and 3980 multiparous women.

The majority of births occurred at home, in the maternity unit, or in the hospital under the supervision of the midwives (74.3%); 17.3% births occurred in a hospital after referral of the woman during pregnancy, and the remainder (7.9%) occurred after referral during delivery.

The women were divided into two groups: the "selected group" were low-risk women selected for labor and delivery under care of the midwife, and the other group consisted of women referred to obstetricians because of complications during pregnancy or childbirth.

The overall cesarean section rate of 1.4% for the total group was very low, but was even lower (0.4%) for the selected, low-risk group cared for exclusively by the midwives. This was in comparison to a national rate of 4.2% in 1969 and 9.7% in 1980. The perinatal mortality rate was low, especially, as anticipated, in the low-risk group.

The authors concluded that selection of pregnant women into high- and low-risk groups is possible and felt that within the scope of the Dutch system of obstetric care it is possible to achieve very good results with midwifery care for selected women.[20]

In a prospective evaluation study of the effectiveness of nurse-midwives conducted at the University of Mississippi from October 1972 to April 1973, low-risk maternity patients were randomly allocated to the nurse-midwifery or house staff service; the study was not blinded. There were 298 patients in the midwife group and 140 patients in the house staff group. Although differences were observed between the two groups for other parameters, the cesarean section rate for patients cared for by the midwives was 3.7% as compared to 4.3% for the resident staff, a number that did not reach statistical significance.[17]

In a report on the development of the midwifery service that was initiated in 1973 at the Hennepin County Medical Center, Hewitt analyzed the outcomes of their experience with more than 2400 births occurring at an in-hospital birth center. High-risk patients were comanaged with obstetric residents, both antepartum and intrapartum, and the delivery was done by the midwife when appropriate. From July 1973 to January 1982, the nurse-midwives cared for 2418 women. Of these, 117 (4.8%) required cesarean section.[21]

Sharp and Lewis[22] reviewed the experience of nurse-midwives at a tertiary-level, university-affiliated hospital from 1973 to 1982. In caring for this indigent population, two models of care, comprehensive and episodic, were established. In the comprehensive model, patients were assigned to the midwives in the prenatal period, comanaged with physicians if complications arose, or transferred to medical management if complications arose that required ongoing medical management. Episodic care applied to patients who had not been cared for prenatally by the midwives but who received such care in the intrapartum period. Of the 6671 patients who received episodic care (low-risk, screened in labor), only 0.2% required cesarean section. For the remaining 4095 patients in the comprehensive group, the cesarean rate was 8.2%.[22]

In 1981 a study was conducted retrospectively on the first 1000 births at the midwifery service of the San Francisco General Hospital (California). The population was screened to exclude very high risk women but included some with risk factors such as adolescence and advanced maternal age. The overall cesarean section rate was 9.0% for the study period between 1975 and 1979. It was noted that the hospital rate was 12.6% in 1977 and 12.2% in 1978. The perinatal mortality rate of 9 in 1000 compared favorably to 13 to 1000 for the hospital.[23]

Stewart and Clark[24] described the outcomes of patients admitted to an in-hospital birthing center in Douglas County, Georgia. Their experience spanned 5 years from 1976 to 1981, and included 2050 births in a private CNM-MD practice. Patients were accepted into the practice regardless of risk factors; the only exclusions were patients admitted in active labor before 32 weeks of gestation. Eighty percent of the deliveries were performed by the nurse-midwives. The authors reported a primary rate of 7.9% and a repeat rate of 1.5%, which they believed to be low in comparison to national averages.[24]

The Tucson Nurse-Midwifery Service reported the outcomes of patients to whom the CNMs provided prenatal, intrapartum, and postpartum care from 1978 to 1982. Their population was low risk, with physician consultation being obtained in the antepartum period by a physician at the

Arizona Health Sciences Center. If physician consultation was needed once patients were admitted in labor, it was provided by resident staff from the University of Arizona. Consultation was required on almost 50% of their patients. The midwives delivered 711 (86%) of 823 patients on their service; 39 women required cesarean section, for an overall rate of 5%.[25]

The Nurse-Midwifery Service at Women's Hospital of the Los Angeles County/University of Southern California Medical Center has been in existence since 1974. In a 1985 report on the role of the nurse-midwife in a large teaching hospital, although not specifically addressing the topic of cesarean section, they noted that only 2% of the patients cared for primarily by the CNMs during the last 6 months of their study required a cesarean, compared with an primary rate of 8.7% for the hospital during that period.[26]

A retrospective chart review was conducted on 175 consecutive patients who received care from the Yale Nurse-Midwifery Practice, a nurse-midwifery service run by the Yale School of Nursing students and faculty. Their population was predominantly middle class, with insurance or self-paying, married, and with a planned pregnancy. Women with prior cesareans were accepted into their practice. Patients were followed by the midwives from their initial visit through the postpartum period. If complications arose requiring MD consultation or management, the midwives remained with the patients throughout their labors and births. The nurse-midwives delivered 76.4% of their patients; all vaginal birth after cesarean (VBAC) candidates had a successful trial of labor. Their cesarean section rate was 5.1%.[27]

Scupholme[28] analyzed the outcomes of 6313 births conducted by nurse-midwives at Jackson Memorial Medical Center (Florida) from 1977 to 1981. Group I, which consisted of 690 women, received both prenatal care and intrapartum management by CNMs. The remaining 5623 patients were assigned to midwifery care when admitted in labor. The total cesarean rate for all 6313 patients was 2.1%, with group I having a rate of 8.2%, compared to group II, which had a rate of 1.3%. The overall cesarean rate for the hospital was noted to be 17% at the time of publication of the study.[28]

At the University of Vermont between 1980 and 1985, nurse-midwives cared for 1966 women and retrospectively reported on their outcomes. Their data included all women followed prenatally by the midwives, regardless of intrapartum complications. Physician consultation was required for 46.5% of the women in labor. The authors reported a 10.4% primary cesarean section rate and noted that this was somewhat less than the hospital rate of 15%.[29]

Cavero reported on the outcomes of the first 1000 births at the Nurse-Midwifery Group of Fresno (California), where the population served was at mixed obstetric risk; more than 82% of the patients served were Medicaid recipients. Data were retrieved through a retrospective review of the charts of all women admitted to nurse-midwifery care. The primary cesarean rate was 6.7% with an overall rate of 7.7%, which was less than half that for the hospital in which the midwives practice and which was lower than the county and state statistics.[30]

An interesting comparison of midwife versus physician management was made using data obtained in a retrospective review of pregnant adolescents attending a tertiary-care center clinic. Nurse-midwives were on call for their patients from Monday through Friday; on the weekend, when patients from the midwifery case load were admitted in labor, they were managed by resident physicians. Fifty-three patients were cared for by the CNMs and 32 by the physicians. The population was essentially low risk, except that they were adolescents. The cesarean section rate in the CNM group was 1.9% and in the MD group it was 3.1%, a difference that was not statistically significant.[19]

In a retrospective descriptive study of midwife-attended home births in Toronto from January 1983 through July 1988,

Tyson described the outcomes of 1001 low-risk patients delivered by 26 midwives. The total number of cesareans was 35 for a rate of 3.5%.[31]

In one of the largest studies ever conducted on out-of-hospital births, Rooks et al.[32] analyzed the outcomes of care in U.S. birth centers from 1985 to 1987, a study that was undertaken to evaluate the safety of care in such centers. While physicians, primarily obstetrician, provided some of the labor (9.9%) and delivery (16%) care, the majority of care was given by certified nurse-midwives, student nurse-midwives, and midwives who were not CNMs.

A total of 11,814 women were admitted for labor and delivery to 84 free-standing birth centers. There were no maternal deaths, and the overall perinatal mortality rate was 1.3 of 1000 births, excluding lethal anomalies. There was no delineation of cesarean section rate by type of practitioner, but overall only 4.4% of patients required cesarean section. The authors then compared their outcomes to those of five other studies of low-risk hospital births. This included almost 40,000 births from hospitals such as Parkland Memorial in Dallas and the University of Illinois during the period from 1969 to 1985. Cesarean section rates for low-risk patients at these hospitals ranged from a low of 8.3% to a high of 17.6%.[32]

Most of the literature on outcomes of midwifery practice involves low-risk patients. However, at the North Central Bronx Hospital in New York City, nurse-midwives care for low- and high-risk patients with approximately 70% of the patients being at risk or high risk. Haire and Elsberry[33] reported on the outcomes of 3287 patients delivered in 1988. Midwives were the primary providers of antepartum, intrapartum, and postpartum care for all low-risk women and comanaged high-risk women with attending obstetricians. The CNMs delivered 86.1% of these patients, including 38 vaginal breech deliveries performed with the physician assisting or in attendance. Their cesarean rate was 11.8%, which was one half the national average.[33]

A study on the effect of midwifery interventions in lowering the cesarean section rate was conducted at Los Angeles County/University of Southern California (LAC/USC) Medical Center in 1991. Chambliss et al.[18] compared cesarean birthrates between physician- and nurse-midwife-attended births. Low-risk patients were randomly assigned to either MD or CNM management with the birth attendant being unable to determine the patient's group assignment. Patients assigned to the midwifery service were cared for during their labor and delivery experience on the in-hospital birth center, while those assigned to physician care were placed in the regular labor and delivery area. A total of 492 patients were included in the study.

Outcomes related to cesarean section showed a 0.4% cesarean rate on the MD service, compared to 2.0% on the CNM service, a difference that did not reach statistical significance. The authors believed that the selection criteria used for triage was responsible for this outcome, not midwifery management per se. The authors concluded that in selected low-risk populations, a low cesarean rate can be obtained with either physician- or midwife-managed services.[18]

In a Finnish study on the rate of interventions by midwife, researchers obtained data from a population-based trial on iron prophylaxis during 1986–1987. A research assistant abstracted data from 2135 births involving 25 midwives. Patients admitted in labor were routinely assigned to midwives without selection, but if possible women who were anticipated to have a difficult birth were assigned to the more experienced midwives. The hospital in which the midwives practiced had a low overall cesarean rate of 9.8% in 1987, compared to 15% for all of Finland. The combined cesarean rate for patients managed by the midwives was 8.4%, with division of cesareans into elective and emergency not possible with the data retrieval system used.[34]

The Menlo Park Birth Place (California) registered 898 women during a 7-year period ending in 1985. Of this number, 208 were not admitted because of medical and other factors, leaving a total of 690 who were followed through delivery. The population was predominantly Caucasian, married, and well educated; 82% gave birth at the center, with 18% requiring intrapartum transport. Of the 127 women who required transfer, 18 (3%) were delivered by cesarean section.[35]

In 1990 Schimmel et al.[36] analyzed the outcomes of patients registered at a midwifery service established to serve low-income women who previously had no access to prenatal care. Their patients were prospectively enrolled, and data were collected using a standardized form obtained from the American College of Nurse Midwives. Their population included high-risk women who were cared for collaboratively with attending physicians. Women who were planning trials of labor after prior cesarean were accepted into their practice. Comparison data for the private obstetricians was obtained from the hospital's delivery logbook. There were 496 patients in the midwifery group and 669 patients in the physician group. The primary and repeat cesarean rates for the midwives were 3.7% and 3.0%, respectively, for a total rate of 6.7%. For physicians, the primary rate was 11.4% with a total rate of 20.6%. The nulliparous cesarean rates for the midwives and the physicians were 7.6% and 18.5%, respectively.[36]

A study that found a statistically significant difference in the cesarean rates in low-risk patients cared for by midwives compared to physicians was conducted by the University of California at San Francisco. Using computerized records, researchers selected patients cared for by resident physicians or nurse-midwives at Moffitt Hospital, which is affiliated with the University. They then excluded all patients who were not delivered at term (37–42 weeks), did not have at least five prenatal visits, had a multiple gestation, were admitted in labor with any fetal presentation other than an cephalic or position other than occiput, and who had infants who were not liveborn or who had congenital anomalies. They further excluded pregnancies with 1 or more of 52 medical or obstetric conditions that they felt predisposed to cesarean delivery. The final study group consisted of 4607 low-risk women; 3551 were in the physician group and 1056 were in the nurse-midwife group.

The overall incidence of cesarean section during their study period was 19.8%. After the exclusions described, the cesarean section rate for the low-risk cohort included in their study was 11.8%. When analyzed by type of provider, the nurse-midwives had a 9.75% cesarean rate and the physicians 12.3%. This difference was statistically significant.

Because the two groups differed significantly by age, race, parity, and year of delivery, multivariate analysis was used to estimate the risk of cesarean while controlling for these potential confounding variables, as well as birthweight. Even after this analysis, the differences remained statistically significant.

Because this was a low-risk cohort, all cesarean sections were performed for either labor abnormalities or fetal distress. In analyzing the frequency of these two diagnoses by provider, the authors found that physician patients were 50% more likely to be diagnosed with fetal distress and 25% more likely to be diagnosed with labor abnormalities.[37]

Discussion

Of the 21 studies analyzed, the majority were retrospective reviews of the outcomes of midwifery management in a variety of settings, both in and out of hospital (Table 16.1). Of the 6 prospective studies, only 2 were randomized, and only 1 was blinded. The study at the University of Mississippi, which was randomized but not blinded, consisted of a group of 438 women. The differences in the cesarean section rates of CNMs and resident physicians were not statistically significant.[17] The randomized, blinded study at LAC/USC Medical Center of 492 women showed a similar result.[18] The

TABLE 16.1. Studies on the impact of midwifery care on cesarean section rate

Study	Limitations	Type	Cesarean section rate	Comparison	Population
Wormeveer (1969–1983), n = 7,980	Not randomized No control group	Prospective	1.4%, total 0.4%, select	4.2%, 1969[a] 9.7%, 1980[a]	Low risk
University of Missippi (1972–1973), n = 438	Not blinded Small sample	Prospective Randomized	3.7% —	4.3%[b] —	Low risk —
Minnesota (1973–1981), n = 2,418	No control group	Retrospective	4.8%	—	Low risk
Grady Memorial (1973–1982), n = 10,776	No control group	Retrospective	0.2%[c] 8.2%[d]	—	Mixed
San Francisco General Hospital (1975–1979), n = 1,005	No control group	Retrospective	9.0%	12.6%, 1977 12.2%, 1978	Mixed risk
Georgia (1977–1982), n = 2,050	No control group	Retrospective	7.9%, primary 1.5%, repeat	—	Mixed risk
Tucson (1978–1982), n = 838	No control group	Retrospective	5.0%	—	Low risk
LAC/USC (1979–1980), n = 863	Cesarean section rate incidental finding	Retrospective	2.0%	8.7%	Low risk
Yale (1985), n = 175	No control group Small sample	Retrospective Consecutive	5.1%	—	Low risk
Jackson Memorial (1977–1981), n = 6,313	No control group	Retrospective	2.1%	17%	Mixed risk
University of Vermont (1980–1985), n = 1,852	No control group	Retrospective	10.4%	15%	Low risk
Fresno (1980–1985), n = 1,000	No control group	Retrospective	7.7%	15%[a]	Mixed risk
Yale (1984) n = 85	Small sample No random assignment	Retrospective	1.9%, CNM	3.1%, MD[b]	Low risk
Toronto (1983–1988), n = 1,001	No control group	Retrospective	3.5%	—	Low risk
National Birth Center (1985–1987), n = 11,814	No control group	Prospective	4.4%	8.3%–17.6%	Low risk
Bronx Hospital (1988), n = 3,287	No control group	Retrospective	8.1%, primary 3.4%, repeat	1/2 the national rate	High risk
LAC/USC (1992), n = 492	Small sample; important variables not analyzed	Prospective Randomized Blinded	2.0%	0.4%[b]	Low risk

TABLE 16.1. *Continued*

Study	Limitations	Type	Cesarean section rate	Comparison	Population
Finland (1988–1987), $n = 2{,}135$	No random assignment	Prospective, population based	8.4%	9.8%, hospital 15%[a]	Low risk
Menlo Park (1979–1985), $n = 690$	No control group	Retrospective	3.0%	—	Low risk
Yolo County (1990), $n = 496$	Data on control group obtained from delivery logbook	Prospective	6.7%	20.6%	Mixed risk
Moffitt Hospital (1981–1988), $n = 4{,}607$	Self-selection of patients	Retrospective cohort	9.75%	12.3%[e]	Low risk
Booth Medical Centers, Thomas Jefferson University Hospital (1977–1978), $n = 1{,}600$	Self-selection of patients	Retrospective cohort	4.3%	12.9%[e]	Low risk

[a] National rate.
[b] Not statistically significant.
[c] Episodic care.
[d] Comprehensive care.
[e] Statistically significant.

study by Beal[19] in which the same caseload of women were cared for intrapartum by nurse-midwives during the week and residents on the weekend also showed no statistical difference in cesarean rates between the two types of practitioners.[19]

Does this mean that midwifery care has no effect on the cesarean section rates? Are the low rates achieved by midwives simply a reflection of selection bias? Do either physicians or midwives achieve low cesarean rates when caring for low-risk women? Risk status of patients is an important variable to consider, because it is known from a Canadian study that women with an absence of risk factors at the beginning of their pregnancy have a cesarean rate that is half the rate of the general hospital population.[38]

How do physicians caring for patients in low-risk settings that are not large tertiary-care centers with residency programs compare with midwives in their incidence of cesarean section? Goyert et al.[39] analyzed the method of delivery of 1533 women who were cared for by 11 board-certified obstetrician-gynecologists at a community hospital in an affluent suburb of Detroit (Michigan). Their study was conducted during a 12-month period from 1986 to 1987. Only very low risk private patients gave birth at this 320-bed hospital; patients with risk factors were transferred to tertiary-care centers. The mean primary cesarean rate for this low-risk population was 17.2%, with a range of 9.6% to 31.8%.[39]

In the randomized, blinded LAC/USC study, Chambliss et al.[18] concluded that the low cesarean rates achieved by the nurse-midwives resulted from selection bias. However, nurse-midwives are an integral part of the health care team at this hospital and have been so for approximately 20 years. When physicians are exposed to the midwifery style of care, they will often adopt many midwifery management techniques. Could the results at this hospital be a reflection of the philosophy of care there, rather than selection bias? The authors stated that physician management at this hospital rarely includes ambulation, the presence of a support person for the laboring woman, or the use of alternative laboring and birthing positions, but does include a high rate of use of epidurals. However, none of these variables, which have the potential to impact outcomes, were analyzed. And, as was pointed out by Butler et al.,[37] the small sample size of the USC study did not have sufficient power to detect the 30% lower incidence found in their own study.

In a comparison of the differences in intrapartum care of low-risk women, Reid et al.[40] found lower intervention rates by family physicians when compared to obstetricians. They thought this reflected a more expectant style of practice by family physicians. Could this principle of expectant practice style be extrapolated to midwifery care? Do midwives in general have more of a commitment to achieve a normal birth when possible than do physicians, especially when compared to obstetricians?

This topic was addressed by Baruffi et al.[41] in a 1990 study investigating the differences in cesarean section rates at two institutions in Philadelphia. Booth Maternity Center (BMC) was staffed by CNMs with a consulting physician present on the premises. Thomas Jefferson University Hospital (TJUH) was staffed by resident and attending physicians. A sample of 796 women who received care at BMC was randomly selected and then matched to 804 women delivering at TJUH. After adjusting for demographic factors and medical conditions that might explain any differences, the authors found a significantly higher primary cesarean rate at TJUH. The rate at TJUH was 2.5 fold higher than at BMC, being 12.9 for the former and 4.3 for the latter. Importantly, the higher cesarean rate at the tertiary hospital was not associated with improved newborn outcomes.

After complete analysis of their data Baruffi et al. believed that the most likely explanation for the differences in the cesarean rates related to "differences in maternity care practices that may arise from different philosophies about the management of labor and delivery." They noted that the model of care at Booth Maternity Center was Similar to that in many European countries with low intervention rates and perinatal mortality rates lower than the U.S. rate.[41]

As was discussed, it is often the contention that midwives have low cesarean rates only because they care for low-risk patients. But midwives also care for high-risk patients, usually in tertiary-care centers. The review by Haire and Elsberry[33] at the North Central Bronx Hospital in 1988 was conducted in a midwifery service in which 70% of the women were at risk or high risk. Risk factors present in patients in their study population included medical induction of labor, diabetes (A-R), pregnancy-induced hypertension, abruptio placenta, placenta previa, maternal addiction, lack of prenatal care, etc. Even in this high-risk population, the midwives managed to achieve a cesarean rate of only 11.8% with a neonatal mortality rate of 9.2 per 1000 live births for all birthweights and 3.7 per 1000 for infants over 1000 g.[33] At Jackson Memorial, where nurse-midwives care for high-risk patients intrapartum, Schupholme[28] reported an overall cesarean rate of 2.1% in the study conducted there. Five-minute Apgars of 7–10 were present in 99.65% of their infants.

Many of the studies reviewed were conducted in large teaching hospitals, where the patients, although ostensibly low risk, were from the lower socioeconomic strata, a group that generally has all the medical problems associated with being disadvantaged such as poor nutrition, lack of contraception (resulting in closely spaced conceptions), and poor compliance with prenatal care.

Mayes et al.[15] pointed out that when patients have a choice between midwifery and physician care, such women may choose providers according to their own internal needs, resulting in a self-selection bias. In many of these busy, tertiary-care centers, however, patients frequently are assigned to midwifery or physician care solely on the basis of the demands of the labor and delivery unit, not patient preference.

In all the studies reviewed, covering the years from 1969 to 1982, cesarean sections were infrequently performed for patients cared for by midwives. Rates ranged from 0.2% to 11.8%, with the highest rate occurring in a high-risk population. Although many of the studies were methodically flawed, one cannot ignore the universally low cesarean rates reported, which often have been achieved despite the fact these midwives work on labor and delivery units in busy, tertiary, high-risk medical centers. In a study on the effects of hospital environment on the practice of low-risk obstetrics, Carroll et al.[42] found that caring for low-risk patients in a high-risk setting is associated with higher intervention rates for family physicians. And yet midwives have achieved low intervention rates in such high-risk settings.

In examining some of the most successful programs discussed, it becomes apparent that these programs have one thing in common, that of a strong working relationship between physicians and midwives. This team approach to obstetric practice is a model that has been utilized in a variety of settings to achieve good maternal and neonatal outcomes. One such setting in Kaiser-Permanente, the largest and oldest health maintenance organization (HMO) in this country. In this HMO setting the midwife-physician team approach has been used successfully since its inception in the Southern California region in 1980.[14,43] Nurse-midwives and nurse-practitioners are involved in the antepartum care of low- and high-risk women. At several of the medical centers in the region, all laboring women with low-risk pregnancies are managed independently by nurse-midwives with in-house physician consultation available. Physicians are, thus, freed to care for patients with high-risk pregnancies and complicated deliveries.

Nurse-midwives also collaboratively care for selected at-risk patients, such as those with mild PIH, gestational diabetes, postdates inductions, mild abruptio placenta, previous cesarean section, etc. At the Riverside Medical Center, nurse-midwives annually deliver almost 70% of the patients. For the 3-year period from 1991 to 1993, the CNMs have managed and comanaged 5812 women of mixed risk status. Their overall cesarean section rate is 5.5%.[44]

Many theories have been postulated as to how or why midwives are successful in lowering the incidence of operative intervention. Butler[37] proposed two possible mechanisms. The first is that care provided by the midwife empowers women, enabling them to cope more successfully with childbirth. Second, she explored the possibility that the midwife provides a mechanism for delaying the decision to interfere (physician consultation), thus lengthening the decision-making time, during which some women may make progress in their labor or be delivered.[37] Hemminki[34] believed that midwives may influence cesarean rates in three ways: that the manner in which they provide care may influence the mother's attitude and her progress in labor; that the frequency and style of consultation may influence the physician's decision regarding the need for cesarean section; and that how actively the midwife intervenes in each labor (such as labor augmentation, analgesia) may influence its outcome.

The possibility exists that other personnel who elect to work in maternity care settings where midwives function may influence the outcomes of laboring women. Nurses who support the midwifery philosophy of care frequently will choose to work in hospitals and birth centers with midwives. The type of supportive care that they provide in labor may additionally impact cesarean section rates.

The available evidence seems to suggest that the midwifery model of care can lower cesarean section rates. Utilization of cesarean section is decreased in maternity services in which midwives provide intrapartum care in industrialized countries around the world. Large randomized trials are needed to assess the specific ways in which midwives achieve their outcomes. An interesting finding of the Finnish study was the wide variation in cesarean rates among their midwives. While the overall cesarean rate was low for patients cared for by midwives, individual rates ranged from 0.0% to 17.6%.[34]

More investigation is needed to determine if the low cesarean rates of midwives are achieved by means of their philosophy of care, types of care given, professional standards, practice environments, or other factors that perhaps have yet to been considered.

One final area that should be investigated in future research concerns physicians who ally themselves with midwives. It is important to remember that a midwife may care for a laboring woman, but it is always a physician who decides to perform a cesarean section. While the practice style of the midwife who is obtaining the consult may have an effect on the decision, the ultimate judgment to resort to surgical intervention lies with the consultant. Midwives obtain physician consultation frequently,[25,29] but rarely is surgery needed for their patients. More information is needed about this decision-making process. Additionally, more data on the characteristics and motivations of these physicians would be helpful in determining the reasons for their support of midwives. Possibilities might include such factors as personal belief systems that value the birth process or prior exposure to midwives and their style of practice. The latter could have profound implications for the manner in which students in medical school and residency programs are educated about the normal birth process.

In the final analysis, it is physicians and midwives together who can have a positive impact on lowering the cesarean section rate in this country.

Childbirth Education

Historical Perspective

There is evidence to show that attempts to manage the pain of childbirth date back to ancient times. Recently discovered evidence in Jerusalem from the fourth century A.D. indicates that marijuana was used to strengthen labor contractions and reduce the pain of labor. Skeletal remains in an undisturbed tomb were identified as those of a 14-year-old with a term fetus in the pelvic area. The young girl, who was found lying on her back in an extended position, apparently in the last stages of pregnancy or of giving birth, had very contracted pelvic measurements. Residue of burnt cannabis was present near the abdominal area. Researchers concluded that this drug was inhaled by the pregnant girl to possibly increase uterine contractions and/or relieve the pain of labor. They also cite the use of cannabis during childbirth in several nineteenth-century publications.[45]

In the nineteenth century, methods introduced to reduce the pain of childbirth were surrounded by controversy. When choloroform was used in 1847 by Simpson in Scotland to alleviate the pain of childbirth, it was condemned by the clergy as being contrary to biblical teaching that women should suffer in childbirth to atone for original sin.[46]

In 1848, in Boston, Dr. Walter Channing introduced the use of ether for childbirth, and encouraged other physicians to try it. Most were reluctant to do so and, thus, the use of anesthesia was infrequent in the latter half of the century. Various reasons for this reluctance included uncertainty about the necessity and safety of anesthesia for most births, and moral objections, with some doctors fearing that the drugs might excite women into erotic behavior. When used, it was most often employed for first births or when internal versions or instruments were needed.[46]

With the movement to hospital birth in the early part of the twentieth century, the use of analgesia and anesthesia became commonplace, with women feminists and suffragettes spearheading the efforts to introduce "twilight sleep' into this country from Germany. A hospital devoted to the utilization and promotion of this method of pain relief was established in Boston by a female physician in 1914. It was called the Twilight Sleep Maternity Hospital and eventually was the site of 3000 births attended by Dr. Eliza Ransom. Mrs. John Jacob Astor's picture even appeared in newspaper articles extolling the virtues of this method.[46]

However, concerns for such extensive use of sedation during hospital childbirth were being voiced in the 1920s and 1930s. Dr. Grantly Dick-Read, a British physician, introduced the concept of natural childbirth into this country with the publication of his classic text in 1933. He advocated teaching relaxation techniques to handle what he termed the fear–tension–pain cycle that occurred with laboring women.[47] Other methods of managing the pain of childbirth by relying on preparation during the prenatal period instead of the use of medications in labor were gradually introduced. These included the methods of Lamaze and Vellay in France,[48] which were introduced by Marjorie Karmel in 1959[49] and Elizabeth Bing in 1967,[50] and Bradley's husband-coached childbirth.[51] Many of these techniques of childbirth preparation remain popular today and are an accepted part of prenatal care. Various other methods of handling the pain of childbirth have been employed in prenatal classes in this country, including such techniques as hypnosis, self-hypnosis, acupressure, water therapy, and biofeedback.

In the United States, there are various organizations, many of which date back to the early years of the natural childbirth movement, devoted to the education and certification of childbirth educators. One of the first of these, founded in 1960 by Elizabeth Bing and Marjorie Karmel, is the American Society for Psychoprophylaxis in Obstetrics (ASPO/Lamaze), a national organization that promotes the principles of Lamaze childbirth and family-centered maternity care. Also founded in 1960, the International Childbirth Education Association (ICEA) is a nonprofit organization devoted to the support of family-centered maternity care and the encouragement of freedom of choice based on the knowledge of alternatives available to childbearing families.

One of the first organizations in this country to focus on the issue of cesarean section was the Cesarean Support Education and Concern or C/SEC, which was founded in 1973. Its purpose was to increase awareness of the rising cesarean section rates, prevent the unnecessary utilization of the cesarean operation, and to humanize the cesarean childbirth experience. In 1982 another cesarean awareness group, the Cesarean Prevention Movement (CPM), was founded, and later became the International Cesarean Awareness Network (ICAN). Their goals have been to provide a forum for the exchange of ideas about birth, to provide a childbirth support network for women, and to lower the rising cesarean section rate through education.

Limitations of Research Studies

There are many problems inherent in attempting to assess the effectiveness of childbirth education programs. The most apparent is that of selection bias. The type of patient who chooses to attend childbirth classes can have a significant effect on outcomes. The attitudinal factors of these women may make it difficult to ascertain if outcomes result from the classes themselves or occur because of psychologic and sociologic differences. Women who choose to attend childbirth classes seem to differ from those who choose not to attend such classes. Lumley and Brown[52] found that nonattenders were younger, less well educated, poorer, and more likely to be unmarried. They were also more likely to miss clinic appointments, had higher rates of alcohol and tobacco use, and were less likely to initiate and maintain breast-feeding. Davis and Morrone[53] found that women who attended classes were of a higher socioeconomic group, older, and better educated than those who did not attend. Leonard [54] found that women attending childbirth education classes were older, with higher educational levels of both the women and their spouses, were of a higher socioeconomic level, and had a higher incidence of planned pregnancies.[54]

Prenatal classes vary greatly in their content, with some focused on early pregnancy and covering such topics as nutrition, fetal growth and development, exercise, and maternal body changes during pregnancy. Later classes can focus on any subject from hospital policies to sibling preparation, breast-feeding, and parenting. Even classes that are supposedly devoted to labor preparation can differ greatly in the their approach. Standardization of class content, as well as the focus of such educational efforts, is an important factor. An additional variable that must be considered is the expertise and effectiveness of the childbirth educator. Also, problems arise when trying to quantify instruction that is given; simply attending class does not ensure internalization of the material presented.

Goals of prospective parents vary greatly, and certainly may differ from that of their childbirth educators and caregivers. Hetherington[55] conducted a study in an inner-city hospital and found that motivating factors for the majority of women were simply to be better prepared for delivery and to share their childbirth experience with their parner. Charles et al.[56] related that most women who take childbirth classes do so not for obstetric reasons but to increase enjoyment of their birth experience and to reduce and control pain.

Early Studies

Reports examining the effectiveness of childbirth preparation seem to have begun in the 1940s. Early investigators seemed to have been concerned with the safety of mothers and neonates and addressed such issues as the amount of time required of the physicians and nurses who worked with prepared women; the effects on forcep rates, which were quite high in that era; and the "success or failure" of the laboring woman.

Goodrich and Thoms[57] introduced the concept of childbirth preparation to Yale University Hospital in the late 1940s. In their initial publication from 1948, they looked at the outcomes of 156 women trained by their method, which was based on the work of Read. The patients they selected to receive childbirth education were from two groups: volunteers, who came to the clinic requesting natural childbirth; and every third woman registering for prenatal care at their clinic, providing they had completed at least 28 weeks of gestation and were "deliverable from below." These women then attended classes on anatomy/physiology, relaxation and breathing techniques, the course of labor, and prenatal exercises. Outcomes were only evaluated for the amount of analgesia/anesthesia required by the patient in labor, being characterized from excellent to poor. This study, while very flawed by today's standards, was important however because it showed that childbirth education could successfully be taught to patients in a clinic service.[57]

These authors expanded their study in 1949[58] by combining the earlier group with another study, reporting on a total of 546 women with childbirth preparation. The additional women were recruited consecutively from their clinic population; apparently after their initial report, all patients were given the opportunity to attend classes. In analyzing their data, only vaginal deliveries were considered, but they did report an operative vaginal delivery rate of 8.4%. They again concluded that natural childbirth could be taught successfully in a teaching ward obstetric service, and that the method had advantages for mother and child and was psychologically desirable for most women.[58]

Interestingly, they never even considered the impact that this method might have on cesarean sections. In their total series, it is difficult to determine their cesarean section rate, because they only reported on patients who delivered vaginally. The cesarean rate in this country in 1948–1949 was approximately 3%.[2] One can only surmise that to the authors this was not an important variable to analyze when looking at the effects of childbirth preparation on labor and delivery.

In 1951, Thoms, this time in conjunction with Wyatt, analyzed the outcomes of 1000 deliveries under a childbirth training program and in 1954 expanded this study with Karlovsky[59] to include 2000 consecutively selected deliveries. Although interesting, these studies were quite flawed, because not all the women in the study group participated in classes; 4% of the women included in the analysis of their data were unregistered and seen for the first time in labor. In addition, there was no control group for comparison. Of the patients whom they evaluated in their first series of 1000, they noted a 2.3% cesarean rate for primigravidas and 1.8% for multiparas. In the second series of 1000, the rate for primigravidas was 0.8% and for multiparas 2.4. The primary rate was 1.8%. There were 14 repeat cesareans in the second group, but it is not clear if these were failed trial-of-labor (TOL) patients, unlikely given the obstetric philosophy at the time. And if they were not failed TOL patients, their inclusion in this study of the effects of childbirth preparation on labor and delivery outcomes is puzzling.[59]

One of the first studies to use a control group when assessing the effects of childbirth training was reported by Van Auken and Tomlinson in 1953.[60] They made a comparative study of 400 primigravidas in their practice in Troy, New York. One-half received prenatal training consisting of a hospital tour and four classes conducted at their office; the other half of the group acted as controls. It does not appear that their subjects were randomly assigned, however. This might have shown some interesting comparisons, but they specifically excluded patients who were delivered by cesarean section from their data. They found that women who received childbirth classes had shorter labors, less instrumented deliveries, less perineal trauma, and a decreased use of drugs and anesthetics in labor.[60]

Davis and Morrone in 1962[53] studied 463 patients receiving childbirth preparation and labor support. The subjects were divided into four groups: 320 received both preparation and support, 35 received preparation but no support, 85 had support but no preparation, and 23 had no preparation or support. They concluded that the type of person selecting childbirth preparation is more important than the preparation itself. In considering this influence on delivery factors, they only noted an increased use of elective low forceps in women who were not supported in labor, but failed to comment on cesarean sections.

Beck and Hall[61] in their 1978 review article on natural childbirth examined approximately a dozen studies from the 1940s to 1977, including some of those discussed here. They believed that these studies tended to show such effects as decreased pain perception, increased cooperativeness on the part of the mother, reduced use of analgesia/anesthesia, decreased rate of episiotomies and operative deliveries, etc. But they concluded that a variety of errors in study design, such as absence of control groups, failure to provide a group undergoing attention-placebo treatment, failure to report statistical methods used to analyze data, and nonrandom assignment when there were control groups confounded all the measured treatment outcomes.[61]

Recent Research

It was not until the 1970s that researchers began to consider the impact of childbirth preparation on cesarean rates. Scott and Rose[62] from the University of Iowa retrospectively analyzed the data from 129 primiparous patients who completed Lamaze classes and delivered at the University Hospital in 1967–1974. These patients were matched with a control group of 129 patients of the same parity who had not received preparation and were of the same age and economic status as the study group. The cesarean rate was 6.2% in the study group and 4.7% in the controls, a finding that did not reach statistical significance.

Charles et al., in 1977,[56] analyzed the obstetric and psychologic outcomes of patients who had received childbirth preparation in a large urban teaching hospital. 249 women who delivered at this hospital were interviewed 1–3 days postpartum, and information about obstetric complications was obtained by reviewing their medical records. Of this group, 95 had received training in the Psychoprophylatic method of childbirth education. It appears that there were more cesareans in the prepared group, but this was probably related to their selection method, thus making it difficult to interpret their data on cesarean sections.

In a 1978 retrospective study of the maternal and fetal outcome of Lamaze-prepared patients, Hughey and McElin[63] analyzed the outcome of 500 consecutive Lamaze deliveries at the Evanston Hospital in Illinois and compared this group with 500 controls, who were consecutively matched for age, race, parity, and educational level. The authors reported a statistically significant decrease in the cesarean rate of the Lamaze group, 3% versus 12% in the controls. While acknowledging that analyzing the effects of consistency of class attendance might have altered their outcomes, the authors did not collect data on this aspect.[63]

An Australian study conducted from 1981 to 1982 consisted of a random sample of 398 low-risk women who gave birth in five Sydney teaching hospitals. Data were obtained by questionnaire-based interviews 3 weeks after birth and by abstracting information from hospital charts. The women were divided into groups based on the number of hours of class attendance: none, low (1–12 h), medium (13–19 h), and high (20 or more h). The overall cesarean rate was 5% and was not significantly related to the extent of antenatal preparation.[64]

Sturrock and Johnson[65] reviewed the records of 207 primigravid patients who attended childbirth education classes at a military hospital from December 1986 through July 1987. Patients were divided into two groups: the treatment group, which attended two to four classes, and the comparison group, which attended one or no prenatal classes. Family practice residents delivered the majority of the patients. The cesarean section rate in the group who attended classes was 38% compared with 29% in the group who did not attend classes; this did not reach statistical significance.

A population-based study comparing the differences in women attending childbirth classes and those who did not attend such classes was conducted in Australia in 1989. Data were obtained through analysis of a questionnaire mailed to a representative sample of women 8 months after they had given birth for the first time. The questionnaire contained 66 precoded questions about such aspects as reproductive history, duration of breast-feeding, sociodemographic data, and emotional well-being. The emergency cesarean section rate for those women attending classes was not significantly different from those who did not attend class (9.8% versus 10.6%). Interestingly, the difference in women who chose elective cesarean section was slightly higher in the attenders (11.0% versus 8.5%), but this also was not a significant difference.[52]

Discussion

When the literature related to childbirth preparation is reviewed, results are conflicting. Some have thought that childbirth preparation results in substantial benefits and has positive outcomes on birth experiences.[55,66] Others believe that attendance at such classes is not associated with differences in emotional well-being, satisfaction with care received, or differences in birth procedures.[52,53] All authors do, however, seem to agree that the process is not harmful to mother or infant.

As Beck and Hall[61] pointed out, to evaluate the outcomes of childbirth classes, large randomized trials are needed that include three groups: a control group, which receives no treatment; the experimental group; and a third group consisting of those who receive placebo-attention. Shearer[67] believes that even this might not be adequate, because so much of intrapartum management, including use of EFM, labor induction, use of analgesia and anesthesia, and cesarean section is based on practice habits within hospitals and communities. Intrapartum care, thus, conforms mostly to physician peer practices and not to patients' wishes, diluting whatever effect is present from the classes.

Perhaps we are attempting to measure the wrong outcomes when evaluating the "success" or "failure" of prenatal classes. As Enkin[68] stated, "It is important to remember . . . that these classes are an educational rather than an obstetrical intervention, and the outcomes used to evaluate them must reflect their objectives." Certainly, prenatal and childbirth preparation classes seem to fill a perceived need on the part of expectant parents. While concrete data are not available, estimates indicate that between 30% and 70% of all expectant women in English-speaking countries attend such classes.[69] Perhaps it would be more relevant to evaluate the effect of classes on the parent's actual goals, rather than on outcomes that health care providers perceive as important.

One of the initial aims of the childbirth education in this country was to help women cope with their labors by using nonpharmalogic means, so it is not unexpected that researchers have paid relatively little attention to the impact of childbirth classes on cesarean section rates. They have focused, instead, on analyzing such factors as use of pain medication, use of instrumented deliveries, incidence of complications, length of time in labor, incidence of episiotomy, use of forceps and vacuums, and patient satisfaction with their birth experience.

Of the available research that addresses the question of potential impact of childbirth education on cesarean section, none were randomized, prospective studies. The two studies reviewed with matched controls, both of which were retrospective, revealed differing results; one showed a statistical difference in the incidence of cesarean section,[63] and the other did not.[62] The retrospective review by Charles[56] was designed in such a manner as to make interpretation of data uncertain. In neither the retrospective reviews by Bennett[64] and Sturrock[65] nor the population-based study by Lumley[52] were there significant difference in cesarean rates between attenders and nonattenders of classes (Table 16.2).

If the desired goal is to reduce the incidence of cesarean section through educating women and their partners, the first requirement would be specific instruction in cesarean prevention techniques. Classes such as these are already provided through the International Cesarean Awareness Network (ICAN) in their Birthworks program. Research is needed on the effectiveness of classes that are specifically focused on reducing cesarean section rates by empowering women to cope with their labors.

It is ironic that an educational movement which began in response to the overuse of sedating drugs during childbirth seems to have failed, somewhat, in its purpose. All too frequently childbirth preparation classes in this country simply prepare women to accept hospital policies while reviewing the types of analgesia and anesthesia available. While not negating the efforts of dedicated childbirth educators or denying that childbirth preparation is beneficial, it is, perhaps, too simplistic to believe that attending sessions which review the process of labor, describe support and comfort measures for the laboring woman, and discuss hospital procedures can significantly impact a woman's ability to avoid a major surgical procedure.

Educational efforts that are directed more specifically at cesarean prevention can undoubtedly play a large role in reducing the number of cesarean sections performed in this country. More effort is required on

TABLE 16.2. Comparison of cesarean section rates between women who did or did not attend birth preparation classes

Study	Type	Cesarean section rate	Comparison
Scott (1967–1974),[62] $n = 359$	Retrospective, matched controls	6.2%	4.7%
Charles (1978),[56] $n = 249$	Retrospective, control group	? (increased)	—
Hughey (1978),[63] $n = 1000$	Retrospective, control group	3%	12%[a]
Bennett (1981–1982),[64] $n = 398$	Random sample, questionnaire-based interview	—	—
Sturrock (1986–1987),[65] $n = 207$	Retrospective	38%	29%
Lumley (1989),[52] $n = 292$	Population based	9.8%	10.6%

[a] Statistically significant.

the part of childbirth educators to promote this type of childbirth education.

Labor Support

Before this century, women were always supported in labor by other women—midwives, female friends, and relatives. When hospital births became the norm, this form of labor support all but vanished. If support was given, it was usually by the bedside nurse who was providing care to the laboring woman.

An outgrowth of the childbirth education and to some extent the feminist movement was an increased awareness that support was an important element of care for women in labor. Starting with the natural childbirth movement in the 1930s and 1940s, first husbands and then other family members and friends joined laboring women in hospital (and later birth) rooms across America. Today, it is unusual for a woman who is laboring in a hospital to do so alone.

Midwives and nurses have traditionally supported laboring women. In addition, doulas, or companions whose sole responsibility is to support the laboring woman, are increasingly being utilized in hospital births. These doulas are not necessarily professionals, but function as an ombudsman for women in labor and after birth.[70]

That labor support might be an important variable in outcome studies was touched upon by Goodrich and Thoms[57] as early as 1948, when they advocated constant attention by an understanding nurse who would help the laboring woman adapt her relaxation exercises. Van Auken and Tomlinson[60] in their 1953 study commented that "labor patients should be out of bed and with their husbands as much as possible during labor."

In 1980 a study was conducted in Guatemala by Sosa et al.[71] on the effects of a supportive labor companion. Hospital policy did not allow the presence of husbands, other family members, or friends in the labor or delivery rooms. Primigravid low-risk women were randomly assigned to either an experimental (labor support) or control (no labor support) group when admitted in active labor. Women in the experimental group received constant support from an untrained lay woman from admission to delivery. There was a 27% rate of cesareans in the control group, compared to 15% in the experimental, a rate that was statistically significant.[71]

Klaus and Kennell[72] expanded that study to include low-risk primigravid women at the same hospital, using the same methods described by Sosa et al., and reported the outcome in 1986. There were 249 women who labored alone and 168 who labored in the presence of a supportive female companion. Once again, there was a statistically significant difference in the cesarean section rates of the two groups, with the supported group having fewer cesareans than the unsupported group (7% versus 17%).[72]

The two foregoing studies were criticized on the grounds that the care environment in which they were conducted in Guatemala was vastly different from that in this country. In an attempt to see if this type of care would have similar results in a hospital with modern obstetric practices, these authors conducted a study at Jefferson Davis Hospital in Houston, which is affiliated with Baylor College of Medicine. Hospital policy prohibited the presence of support persons during labor and delivery. Nulliparous low-risk women admitted in active labor who consented to participate in the study were assigned to either a doula support group (n = 212) or an observed group (n = 204) in which an inconspicuous observer remained in the laboring woman's room. A control group (n = 204) was added to examine the potential supportive effects of the nonparticipant observer. Their findings in relation to cesarean section rates were similar to those observed in Guatemala. Of the patients in the supported group, 8% required a cesarean, compared to 13% in the observed group and 18% in the controls. It is interesting to note that the presence of someone who simply remains with the mother at all times (observed group) can lower the cesarean rate.[73]

Because nurses provide much of the care given to laboring women in this country, it would seem important to consider the effects of their care on cesarean rates. In 1991 Radin et al.[74] studied labor and delivery nurses in a nonprofit, tertiary referral hospital with approximately 6700 deliveries per year. A staff of 45 registered nurses provided care for patients throughout their labor and birth. These nurses were ranked according to the cesarean rates of their nulliparous, low-risk women who went into spontaneous labor. Cesarean rates differed markedly between nurses in the lowest quintile (4.9%) and the highest quintile (19%) and were not explained by various factors, such as maternal age, gravidity, childbirth preparation, insurance status, reliance on public assistance, physician who attended the birth, labor augmentation, epidural anesthesia, infant weight/gestational age, or dilatation when the nurse assumed care. The authors concluded that the care given by nurses during labor has an important influence on cesarean birthrates.[74]

Hodnett and Osborn[75] conducted a randomized trial on the effects on labor support on childbirth outcomes. Low-risk women who were attending childbirth preparation classes were enrolled in the study, and randomly assigned to either the experimental group, who received prenatal and intrapartum support from a monitrice, or the control group, who received prenatal and postpartum support from the same monitrice. The study's purpose was to evaluate the effect of labor support on women having vaginal births, so those requiring cesareans were excluded from the analysis of the data. The final sample consisted of 103 subjects. The authors did comment that the cesarean section rate in the experimental group was 17%, compared to 18% in the controls. Results failed to show that continuous supportive care had any evident direct effect on cesarean delivery rates. Obstetric practices at the hospital in which the study was conducted may have been an important factor in outcomes; there was a high incidence of a forcep deliveries (31 of 103), and epidurals, both continuous (89%) and terminal.[75]

Discussion

The foregoing studies show the significant impact that labor support can have in lowering cesarean section rates (Table 16.3). However, support for many women laboring in hospitals is this country is often minimal. Most commonly, it is the baby's father who assumes the supporting role, a role for which he may be ill equipped. Many fathers can function quite well in assuming an active role in caring for their partners in labor; in fact, this is the basis of the Bredley method of childbirth preparation. Other fathers find the experience distressing and overwhelming, and may, perhaps, contribute to the decision to perform a cesarean by their inability to cope with their partner's pain.

Other family members or friends of the laboring woman may be involved in supportive care. The usual practice during home or birth center births is to have several people present. As birth rooms have become more common in hospitals, there has been an increase in the number of women who choose to have more than one person present during their births. But here are some voices of dissension over this practice. Odent[76] believes that this is disruptive to the laboring woman and may, in fact, result in a long, difficult labor.

Nurses have increasingly become involved with the technologic aspects of birth, activities that are more highly valued and which consume more of their time than labor support. Using work-sampling, McNiven et al.[77] found that the proportion of time that nurses spent on supportive care on the day shift at a Toronto teaching hospital was only 9.9%. And yet patients view nurses who remain with them for longer periods of time as more helpful.[78]

Midwives have traditionally been seen as a category of caregiver who remains constantly with the laboring woman. However, midwives in this country, who are working in increasing numbers in busy, tertiary medical centers, may fail to provide adequate labor support. Not infrequently, a midwife in such a setting may care for more than one patient at a time.

Doulas are assuming an increasing role in supporting women during labor and after birth. One organization offering certification for these caregivers is Doulas of North America. It is estimated that there are 800–1000 birth doulas and 500–1000 postpartum doulas in the United States.[79]

Thus, although it appears that labor support decreases obstetric interventions, lowers cost, and has a positive impact on maternal and neonatal health,[73] providing such support does not seem to be a priority in this country. This is not the case in other industrialized countries with low cesarean section rates. At the National Maternity

TABLE 16.3. Effect of support during labor on cesarean rate

		Cesarean section rate	
Study	Type	Experimental	Controls
Sosa (1980)[71]	Prospective-randomized	15%; $n = 20$	27%[a]; $n = 20$
Klaus (1986)[72]	Prospective-randomized	7%; $n = 168$	17%[a]; $n = 249$
Kennell (1991)[73]	Prospective-randomized	8%; $n = 212$	13%, observed group; $n = 204$ 18%, control group[a]; $n = 204$
Hodnett (1988)[75]	Retrospective	17%; $n = 72$	18%; $n = 73$

[a] Statistically significant.

Hospital in Dublin, where the cesarean section and perinatal mortality rates are very low, laboring women are never left alone. Student midwives or medical students sit continuously with patients to provide emotional support. Such activity is considered to be an important part of the education and training of both midwives and physicians.[10]

If we are to lower the cesarean section rate in this country, we need to reassess the practice, that is so common in many hospitals, of leaving the laboring woman alone with her significant other, who may be poorly prepared to be her sole labor support. Nurses, midwives, doulas, and physicians together must investigate ways in which continuous support can be given to each and every laboring woman in this country.

References

1. Cragin E. Conservatism in obstetrics. N Y Med J 1916;CIV:1–3.
2. Levine W. Thirty-year survey of cesarean section. Obstet Gynecol 1962;19:42–49.
3. Marieskind H. Cesarean section in the United States: has it changed since 1979? Birth 1989;16:196–201.
4. National Institute of Health: Consensus Development Conference on Cesarean Childbirth, September 1980, sponsored by the National Institute of Child Health and Human Development. Bethesda, maylands: NIH publication no. 82–2067,1981.
5. Taffel S, Placek P, Rosary C. U.S. Cesarean section rates 1990: an update. Birth 1992; 19:21–22.
6. Williams RL, Hawes WE. Cesarean section, fetal monitoring, and perinatal mortality in Callifornia. Am J Public Health 1979;69: 864–870.
7. O'Driscoll K, Foley M. Correlation of a decrease in perinatal mortality and increase in cesarean section rates, Obstet Gynecol 1984;63:205–208.
8. Stewart D, Stewart L. Compulsory hospitalization: freedom of choice in childbirth?, Vol. II. Missouri: Napsac Reproductions, 1979.
9. Klein M. Active management of labor: whose agenda? Birth 1993;20:97–99.
10. Boylan P. Active management of labor: results in Dublin, Houston, London, New Brunswick, Singapore, and Valparaiso. Birth 1989;16:122.
11. Swinnerton T. Traditional midwifery skills. Nurs Times 1990;86:74–75.
12. Levy B, Wilkinson F, Marine W. Reducing neonatal mortality rate with nurse-midwives. Am J Obstet Gynecol 1971;109:50–58.
13. Reid M, Morris J. Perinatal care and cost effectiveness: changes in health expenditures following the establishment of a nurse-midwife program. Med Care 1979;17: 491–500.
14. Bell KE, Mills, JI. Certified nurse-midwife effectiveness in the health maintenance organization obstetric team. Obstet Gynecol 1989;74:112–116.
15. Mayes F, Oakley D, Wranesh B, et al. A retrospective comparison of Certified Nurse-Midwife and physician management of low risk births. J Nurse-Midwifery 1987; 32:216–221.
16. Yankou D, Petersen B, Oajkley D, et al. Philosophy of care: a pilot study comparing Certified Nurse-Midwives and physicians. J Nurse-Midwifery 1993;38:159–164.
17. Slome C, Wetherbee H, Daly M, et al. Effectiveness of certified nurse-midwives: a prospective evaluation study. Am J Obstet Gynecol 1976;124:177–182.
18. Chambliss L, Daly C, Medearis A, et al. The role of selection bias in comparing cesarean birth rates between physician and midwifery management. Obstet Gynecol 1992;80: 161–165.
19. Beal M. Nurse-midwifery intrapartum management. J Nurse-Midwifery 1984;29: 13–17.
20. Van Alten D, Eskes M, Treffers PE. Midwifery in the Netherlands. The Wormerveer study; selection, mode of delivery, perinatal mortality and infant morbidity. Br J Obstet Gynecol 1989;96:656–662.
21. Hewitt M, Hangsleben KL. Nurse-midwives in a hospital birth center. J Nurse-Midwifery 1981;26:21–29.
22. Sharp ES, Lewis LE. A decade of nurse-midwifery practice in a tertiary university-affiliated hospital. J Nurse-Midwifery 1984; 29:353–365.
23. Mann R. San Francisco General Hospital nurse-midwifery practice: the first thousand births. Am J Obstet Gynecol 1981;40: 676–682.

24. Stewart RB, Clark L. Nurse-midwifery practice in an in-hospital birthing center: 2050 births. J Nurse-Midwifery 1982;27: 21–26.
25. Schreler A. The Tucson Nurse-Midwifery Service: the first four years. J Nurse-Midwifery 1983;28:24–30.
26. Platt LD, Angelini DJ, Paul RH, et al. Nurse-midwifery in a large teaching hospital. Obstet Gynecol 1983;66:816–820.
27. Nichols C. The Yale Nurse-Midwifery practice: addressing the outcomes. J Nurse-Midwifery 1983;30:159–165.
28. Scupholme A. Nurse-midwives and physicians: a team approach to obstetrical care in a perinatal center. J Nurse-Midwifery 1982;27:21–27.
29. Keleher K, Mann L. Nurse-midwifery care in an academic health center. J Obstet Gynecol Neonatal Nurs 1986;September/October: 369–372.
30. Cavero C, Fullerton J, Bartlome J. Assessment of the process and outcomes of the first 1000 births of a nurse-midwifery service. J Nurse-Midwifery 1991;36:104–110.
31. Tyson H. Outcomes of 1001 midwife-attended home births in Toronto, 1983–1988. Birth 1991;18:14–19.
32. Rooks J, Weatherby N, Ernst E, et al. Outcomes of care in birth centers: the national birth center study. N Engl J Med 1989;321: 1804–1811.
33. Haire D, Elsberry CC. Maternity care and outcomes in a high-risk service: the North Central Bronx Hospital experience. Birth 1991;18:33–37.
34. Hemminki E, Kojo-Austin H, Malin M, et al. Variations in obstetric interventions by midwife. Scand J Caring Sci 1992;6:81–86.
35. Eakins P, O'Reilly W, May L, et al. Obstetric outcomes at the Birth Place in Menlo Park: the first seven years. Birth 1989;16:123–129.
36. Schimmel L, Hogan P, Boehler B, et al. The Yolo County Midwifery Service: a descriptive study of 496 singleton birth outcomes, 1990. J Nurse-Midwifery 1992;37:398–403.
37. Butler J, Abrams B, Parker J, et al. Supportive nurse-midwifery care is associated with a reduced incidence of cesarean section. Am J Obstet Gynecol 1993;168:1407–1413.
38. Moutquin J, Gagnon R, Rainville C, et al. Maternal and neonatal outcome in pregnancies with no risk factor. Can Med Assoc J 1987;137:728–732.
39. Goyert G, Bottoms S, Treadwell M, et al. The physician factor in cesarean birth rates. N Engl J Med 1989;320:706–709.
40. Reid A, Carroll J, Ruderman J, et al. Differences in intrapartum obstetric care provided to women at low risk by family physicians and obstetricians. Can Med Assoc J 1989;140:625–633.
41. Baruffi G, Strobino D, Paine L. Investigation of institutional differences in primary cesarean birth rates. J Nurse-Midwifery 1990; 35:274–281.
42. Carroll J, Reid A, Ruderman J, et al. The influence of the high-risk care environment on the practice of low-risk Obstetrics. Fam Med 1991;23:184–1881.
43. Goings JR. Success in a health maintenance organization. In: Rooks J, Haas JE, eds. Nurse-Midwifery in America. Washington, DC: The American College of Nurse-Midwives Foundation, 1986:36–37.
44. Unpublished data. nurse-midwifery service, Kaiser-Permanente, Riverside, California, 1993.
45. Zias J, Stark H, Seligman J, et al. Early medical use of cannabis. Nature 1993; 363:215.
46. Wertz R, Wertz D. Lying-in: a history of childbirth in America. New York: Free Press, 1977.
47. Dick-Read G. Natural childbirth. London: Heinemann, 1933.
48. Lamaze F. Painless childbirth. New York: Pocket Books, 1972.
49. Karmel M. Thank you, Dr. Lamaze. Philadelphia: Lippincott, 1959.
50. Bing E. Six practical lessons for an easier childbirth. New York: Grossop & Dunlap, 1967.
51. Bradley R. Husband-coached childbirth. New York: Harper & Row, 1965.
52. Lumley J, Brown S. Attenders and non-attenders at childbirth education classes in Australia: how do they and their births differ? Birth 1993;20:123–129.
53. Davis CD, Morrone FA. An objective evaluation of a prepared childbirth program. Am J Obstet Gynecol 1962;84:1196–1201.
54. Leonard RF. Evaluation of selection tendencies of patients preferring prepared childbirth. Obstet Gynecol 1973;42:371–377.
55. Hetherington S. A controlled study of the effect of prepared childbirth classes on obstetric outcomes. Birth 1990;17:90.

56. Charles AG, Norr KL, Block CR, et al. Obstetric and psychological effects of psychoprophylactic preparation for childbirth. Am J Obstet Gynecol 1978;131:44–52.
57. Goodrich FW, Thoms H. A clinical study of natural childbirth: a preliminary report from a teaching ward service. Am J Obstet Gynecol 1948;56:875–883.
58. Thoms H, Goodrich FW. Training for childbirth. JAMA 1949;140:156–158.
59. Thoms H, Karlovsky ED. Two thousand deliveries under a training for childbirth program: a statistical survey and commentary. Am J Obstet Gynecol 1954;68: 279–285.
60. Van Auken WB, Tomlinson DR. An appraisal of patient training for childbirth. Am J Obstet Gynecol 1953;66:101–105.
61. Beck NS, Hall D. Natural childbirth: a review and analysis. Obstet Gynecol 1978;52: 371–379.
62. Scott JR, Rose NB. Effect of psychoprophylaxis (Lamaze preparation) on labor and delivery in primiparas. N Engl J Med 1976; 294:1205–1207.
63. Hughey MJ, McElin TW. Maternal and fetal outcome of Lamaze-prepared patients. Obstet Gynecol 1978;51:643–647.
64. Bennett A, Hewson D, Booker E, et al. Antenatal preparation and labor support in relation to birth outcomes. Birth 1985; 12:9–16.
65. Sturrock WA, Johnson JA. The relationship between childbirth education classes and obstetric outcome. Birth 1990;17:82–85.
66. Cogan R. Effects of childbirth preparation. In: Clinical obstetrics and gynecology. New York: Harper & Row, 1980.
67. Shearer MH. Commentary: effects of prenatal classes cannot be measured by obstetric management. Birth 1993;20:130–131.
68. Enkin MW. Commentary: are the correct outcomes of prenatal education being measured? Birth 1990;17:90–91.
69. Simkin P, Enkin M. Antenatal classes. In: Chalmers I, Enkin M, Keirse MJN, eds Effective care in pregnancy and childbirth. Oxford: Oxford University Press, 1989:318–333.
70. Raphael D. New patterns in doula-client relations. Midwife, Health Visitor and Community Nurse 1988;24:376–379.
71. Sosa R, Kennell J, Klaus M, et al. The effect of a supportive companion on perinatal problems, length of labor, and mother-infant interaction. N Engl J Med 1980;303:597–600.
72. Klaus M, Kennell J, Robertson S, et al. Effects of social support during parturition on maternal and infant morbidity. Br Med J 1986;293:585–587.
73. Kennell J, Klaus M, McGrath S, et al. Continuous emotional support during labor in a US hospital: a randomized controlled trial. JAMA 1991;265:2197–2201.
74. Radin TG, Harman JS, Hanson DA. Nurses' care during labor: its effect on the cesarean birth rate of healthy, nulliparous women. Birth 1993;20:14–21.
75. Hodnett ED, Osborn RW. A randomized trial of the effects of monotrice support during labor: mothers' views two to four weeks postpartum. Birth 1989;16:177–183.
76. Odent M. Birth reborn. London: Souvenir Press, 1984.
77. McNiven P, Hodnett E, O'Brien-Pallas LL. Supporting women in labor: a work sampling study of the activities of labor and delivery nurses. Birth 1992;19:3–7.
78. Klein RPD, Gist NF, Nicholson J, et al. A study of father and nurse support during labor. Birth 1981;8:161–164.
79. Queenan JT, ed. New mothers add to their support staff. Contemp Ob/Gyn 1994;39:145.

17
Cesarean Projects at the State and National Level

BRUCE L. FLAMM

The 450% increase in the U.S. cesarean rate over 20 years did not go unnoticed by governmental agencies. In response to persistently increasing rates, several states embarked upon cesarean-related projects. Related committees have been formed to address rising cesarean rates in other industrialized nations. Summaries of some of these programs are outlined here. This information will be of value to physicians, hospitals, and others who are interested in developing quality assurance programs to evaluate the utilization of cesarean operations. Other nations and other states within the United States have no doubt formed similar committees, and hence the following examples are not meant to be an exhaustive listing of such projects.

United States

California

The California Collaborative Cesarean Appropriateness Review and Education (CARE) Project

California State Assembly Concurrent Resoution 199 (Resolution Chapter 124, 1984) requested studies and reports that focused on the rising costs of health care. In 1985, responsibility for these projects was assigned by Governor George Deukmejian to the California Health Policy and Data Advisory Commission CHPDAC.

The third biennial report to the governor and the state legislature focused on cesarean section.[1] Cesarean was chosen for investigation because of rising cesarean rates and a very wide variation of rates observed in different parts of the state.[2] Data were analyzed from the Department of Health Services Birth Cohort Data File for 1987, which included information on more than 500,000 births. The report noted that the cesarean delivery rate for 1987 in California was 24.4% but that there were wide variations in different areas. For example, in one area of Los Angeles (including 10 hospitals) the cesarean rate was 32.9%. By comparison, an area (including 13 hospitals) with similar demographics and economics in another part of the state had a cesarean rate of 22.9%. In spite of the very different cesarean rates, perinatal outcomes for the two areas were found to be similar. There was no obvious explanation why cesareans were used 44% more frequently in one area than in the other.

The report recommended that various organizations be brought together to address "what appears to be unnecessarily high C-section rates in certain parts of California." The CHPDAC subsequently initiated the formation of a Cesarean Work Group, which convened for the first time in January 1992. This group included representatives from The American College of Obstetrics an Gynecology (ACOG), the California Association of Hospitals and Healthcare Systems

(CAHHS), the California Medical Association (CMA), the California Office of Statewide Health Planning and Development (OSHPD), Kaiser Permanente (America's largest health maintenance organization, or HMO), Blue Cross of California, the California Health Information Association (CHIA), the Bay Area Business Group on Health, SRI International, and the University of California San Francisco.

The California Cesarean Work Group reviewed state and national data along with the work that had already been done in this area in other states (see following). It was noted that one of the U.S. Public Health Service's goals was to reduce the national cesarean rate from 24.4% in 1987 to no more than 15% by the year 2000. It was also noted that the U.S. Public Health Service's goals for the year 2000 were based on the work of a consortium of nearly 300 national organizations and all state health departments. To reach that national target, California would have to perform about 44,000 fewer cesarean operations each year.

During 1992–1993, the California Cesarean Work Group developed the The California Collaborative CARE (Cesarean Appropriateness Review and Education) Project. This was designed to be a voluntary program in which cesarean-related data would be collected at voluntarily participating hospitals. Medical records staff at participating hospitals would abstract relevant data from the charts of all patients who underwent cesarean operations onto a specifically designed data-collection form. The data would then be analyzed and sent back to the participating hospital's obstetric staff in a series of confidential reports. Individual physicians would then be able to compare their practice patterns to those of their peers (with names deleted) both locally and across the state. At the time of this writing, the California Collaborative CARE project is in its pilot phase.

California State Assembly Bill No. 686

A fundamental principle of the California Collaborative CARE project just described is that detailed information about cesarean rates for specific physicians and specific hospitals would be confidential. These data would be compiled and fed back to the voluntarily participating hospitals and physicians to be used in confidential educational and quality assurance programs. As a member of the California Cesarean Work Group and the California Collaborative CARE project, I can verify that public dissemination of abstracted data was not a consideration.

In contrast, California State Assembly Bill No. 686 would make cesarean rates public. First introduced by Assembly Member Roybal-Allard in 1989, AB 686 would add Chapter 6.6 to Division 2 of the Health and Safety Code, relating to cesarean sections. The legislative counsel's digest pointed out that existing law provides specific informed consent requirements which must be satisfied before a hysterectomy may be performed but that no similar requirements exist for the cesarean operation. AB 686 would mandate that a specific information about cesarean delivery be given to all women during the first 6 months of pregnancy. This would be accomplished by means of a standardized written summary in layman's language that would discuss risks, benefits, and alternatives to cesarean along with several other topics. Other topics would include a discussion of vaginal birth after cesarean section.

The Bill would also require documentation that standardized written information was given to the patient. Perhaps the most controversial aspect of AB 686 was that it would also allow for public disclosure of cesarean rates. "This bill would require every hospital that provides obstetrical services to, upon request, provide information to any person regarding the rate of cesarean sections performed at that hospital." Perhaps because of this controversial aspect, the Bill faced opposition from the California Medical Association (CMA) and the American College of Obstetricians and Gynecologists (ACOG). AB 686 has been amended several times, but thus far it has not been ratified.

California State Assembly Bill No. 524

AB 524 (California Hospital Outcomes Assessment) was approved by the state legislature and signed by Governor Pete Wilson in late 1991. AB 524 requires that the California Office of Statewide Health Planning and Development (OSHPD) begin *Risk-Adjusted Monitoring of Outcomes* (RAMO) of hospital inpatient care. A summary of the evolution of AB 524 and a description of the RAMO approach to outcome analysis has been recently published.[3] AB 524 established a 12-member technical advisory committee to make recommendations about production of RAMO reports. The committee chose cesarean section, the most common major operation performed in California, as one of the RAMO projects.

One of the benefits of the RAMO process is that it is apparently better suited to the analysis of statewide hospital discharge abstract databases than "severity adjustment" methodologies which have traditionally been applied to situations in which the medical record of each patient is available. However, it must be noted that the data contained in the statewide hospital discharge database related to indications for cesarean delivery is limited. In contrast, the data abstraction form developed by the California Collaborative CARE project was designed specifically to address the indications for cesarean. Several members of the Clinical Advisory Panel on Obstetric Outcomes (of the AB 524 California Hospital Outcomes Assessment Project), including myself, cautioned that any report on statewide cesarean rates must therefore be interpreted with caution. This is particularly true because the public may be misled by confusing results. For example, analysis may reveal a slightly higher cesarean rate at one hospital than at another hospital in the same area. This difference may persist even after "risk adjustment" by the RAMO process. However, a small difference in the two neighboring hospital's cesarean rates might be explainable by factors that are not contained in the state's discharge abstract database. Newspaper reports touting one hospital's cesarean rate over the others may thus be somewhat misleading. Even with these limitations, RAMO reports on cesarean section would be of great value if fed back to hospitals and physicians for use in their educational and quality assurance programs.

Florida

In 1991 the Florida state legislature passed a bill which requires that hospitals receiving public funds for maternity care must comply with practice parameters that include the assessment of the feasibility of attempting vaginal birth after a previous cesarean delivery.

Maine

Although often not discussed in publications on rising cesarean rates, fear of malpractice litigation is clearly one of the most important factors. The general rule is that the only cesarean a doctor is sued for is the one he or she does not do. In other words, if a cesarean is done early in labor, it would be difficult to blame any neonatal or childhood problems on the obstetrician. This type of thinking and the resultant practice of "defensive medicine" has no doubt resulted in countless cesarean operations.

In April of 1990, Maine Governor John R. McKernon, Jr. signed into law Public Law 1990, Chapter 931, which included subchapter IX, the 5-year Medical Liability Demonstration Project (MLDP). With this law, Maine will become the first state to formally use medical practice guidelines for malpractice defense. If an obstetrician elects to participate in this program, and if he or she practices in accordance with the practice guidelines, the guidelines may be used as an affirmative defense in a medical malpractice suit. By proving compliance with the guidelines, a physician will be able to win dismissal of a suit. In other words, the MLDP protects physicians from liability in exchange for following specific practice guidelines.[4]

Five obstetric protocols developed for the project potentially affect cesarean rates: cesarean for failure to progress, breech, fetal distress, repeat cesarean, and management of prolonged pregnancy. There is often confusion about what actually constitutes a legally defensible standard of care. If guidelines are accepted as the standard of care, and if a physician adheres to these guidelines, it follows that a bad outcome does not imply negligence. The Public Health Research Institute of Portland, Maine plans to evaluate this project. If successfully implemented, the Maine MLDP Law could serve as a model for the rest of the naiton.

Maryland

Cesarean rates varied from 17.8% to 42.6% at hospitals in Maryland in 1987. An inner-city hospital, presumably handling many high-risk obstetric cases, had a cesarean rate of 14.5% while a rural hospital had a rate of 37.6%. In 1987 the Maryland Hospital Association's Council for Quality Health Care began to investigate cesarean rates in the state and to search for reasons for interhospital variations. The Council established an obstetrics advisory committee that included representatives from the Maryland Medical Society and the Obstetrical and Gynecological Society of Maryland. The Council developed guidelines and data collection forms that hospital quality assurance (QA) staff and obstetrics chiefs could use. The Maryland Hospital Association is currently helping hospitals perform internal review of records to investigate reasons for large interhospital variations in cesarean rates.[5,6]

Massachusetts

In 1985 legislation was passed in Massachusetts which mandated that every pregnant woman has the right to know a hospital's cesarean section rate. This includes the total rate, the primary rate, the repeat rate, and the vaginal birth after cesarean rate. This is very similar to the controversial California State Assembly Bill No. 686 that was described earlier.

Michigan

The Greater Detroit Area Health Council in conjunction with the ACOG began a vaginal birth after cesarean section (VBAC) educational project in 1990. The program involves 13 hospitals and includes presentations to the medical staff along with patient education in the form of specific pamphlet.

Minnesota

The Minnesota Clinical Comparison and Assessment program developed cesarean-related guidelines and distributed them to all providers of obstetric care in 1991. More than 40 Minnesota hospitals began abstracting medical records in 1992 to evaluate the impact of the guidelines.

New York

In 1986, cesarean rates for hospitals in New York State varied from 11% to 40%. This huge variation could not be explained by demographic or geographic factors. In 1987 a joint Task Force was formed by ACOG District II and the New York State Department of Health. The Task Force developed a two-tiered review program in which were included both internal and external (site visit) reviews of cesarean charts.[7]

Similar to the California CARE project, participation in the program was voluntary. In 1989–1990, 24 of 176 hospitals in the state with obstetric services had site visits. Perhaps partially as a result of this program the New York State cesarean rate fell from more than 25% in 1987 to 23.4% in 1991. National rates also leveled off at about 24% during the same time interval, but in New York State the downward trend was greater in those hospitals that participated than in those which did not.

Vermont

In 1990 the Hospital Data Council accepted the Vermont Program for Quality in Health Care's proposal to develop a cesarean section quality improvement project.[8] The project focused specifically on VBAC. Data were collected from all 12 hospitals with obstetric services. Results indicated that 72% of patients with prior cesarean were delivered by repeat cesarean. Overall cesarean rates varied widely from less than 12% to more than 30% at some hospitals in Vermont. A program was developed that included both patient and professional education.

U.S. National Cesarean Programs and Reports

U.S. National Consensus Conference on Cesarean Birth

Between 1970 and 1980, the U.S. cesarean rate increased from 5% to 15%. Because the rate had tripled in a single decade, the National Institutes of Health (NIH) formed a 19-member Task Force on Cesarean Childbirth.[9] In 1980 a major conference was held at the NIH headquarters to review the Task Force findings and bring together experts from many fields to evaluate the rising cesarean rate. The Task Force findings were published in 1981 in a 537-page report.[10] A summary of the Task Force recommendations was also published in several major medical journals. A subsequent study demonstrated that the cesarean project received more media coverage than any of the seven other NIH consensus development projects even though other topics included controversial issues such as breast cancer and coronary bypass surgery.[11] In spite of the widespread dissemination of recommendations aimed at lowering the utilization of this operation, the national cesarean rate continued to rise.[12]

The Washington Business Group on Health

The Washington Business Group on Health (WBGH) is a national nonprofit health policy and research organization that represents more than 150 corporations representing all segments of industry. In 1991 the group researched the topic of VBAC and published a 23-page document aimed at the purchasers of health care.[13]

The Public Citizen Health Research Group

The Public Citizen Health Research Group is a division of the Washington-based Public Citizen, a group founded in 1971 by consumer advocate Ralph Nader. The Health Research Group obtained and analyzed statewide hospital discharge databases that included information on cesarean operations. In 1989, the group released a controversial report that listed cesarean rates at more than 2400 hospitals across the United States.[14] The report concluded that 475,000 of the 934,000 cesareans performed in 1987 were unnecessary resulting in more than 1 million extra hospital days and extra costs of more than $1 billion. The report generated much media attention and was sharply criticized by the American College of Obstetricians and Gynecologists.[15] Although many experts felt that the report greatly overestimated the number of unnecessary operations, it was difficult to argue with the conclusion that at least some unnecessary cesareans were being performed.

Canada

The Canadian National Consensus Conference on Cesarean Birth

In 1985 a project similar to the cesarean consensus development program of the U.S. NIH was developed in Canada. The committee reviewed the relevant literature and held a meeting to develop recommendations. They then circulated an interim statement to more than 30,000 individuals and organizations. A final statement (Canadian Consensus Statement on Cesarean Birth), which included cesarean reduction guidelines, was then developed and published in the *Journal of the Canadian Medical Association.*[16] The Society of Obstetricians and Gynecologists of Canada and the Association of Professors of Obstetrics and Gynecology fully endorsed the final statement.

The Ontario Cesarean Birth Quality Assurance Committee

As had been the case in the United States, there was evidence that the Canadian Consensus Statement on Cesarean Birth had little effect on cesarean rates. Ontario, Canada's largest province, contains 35% of the country's population and 44% of its obstetricians; 154 hospitals offer maternity care in Ontario. A study conducted in Ontario in 1987–1988 showed that although about 90% of obstetricians were aware of the cesarean guidelines, only one-third reported changing their practice because of them. Although self-reported data from physician surveys indicated that cesarean rates declined after publication of the guidelines, actual data showed that rates of cesarean were 15%–49% higher than rates reported by the obstetricians.[17]

In 1989 the Minister of Health announced the formation of the Ontario Cesarean Birth Quality Assurance Committee. The Minister of Health suggested a 15% cesarean rate as a target goal to be achieved within 2 years. The committee consisted of members from the Ontario Medical Association, the Ontario Section of the Society of Obstetricians and Gynecologists of Canada, the Ontario chapter of the Canadian College of Family Practice, the Ontario College of Physicians and Surgeons, the Ontario Nurses' Association, the Ontario Hospital Association, childbirth consumer groups, and the Ministry of Health. In 1991 the committee published a 76-page report that analyzed national and international cesarean rates and made many specific recommendations.[18]

Conclusion

Rising cesarean rates have clearly captured the attention of agencies at state and national levels. Because cesarean section is the most frequently performed major operative procedure in the United States and because it has held this position each year for the past decade, governmental interest is not surprising. In fact, in the era of health care reform it would be surprising if commonly performed operations were not subjected to governmental scrutiny. Physicians and hospitals that voluntarily develop quality assurance programs for cesarean delivery will undoubtedly fair well in any future health care environment.

Finally, some physicians are fearful that governmental or third-party agencies, perhaps motivated by cost-containment issues, will attempt to dissuade obstetricians from performing cesarean sections that truly are medically indicated. The American College of Obstetricians and Gynecologists is working to avoid this potentially dangerous scenario by ensuring that ACOG representatives participate in cesarean projects at the state and national levels.

References

1. Kerr C, et al. Variations in cesarean section rates in California, Biennial report on cost containment. Sacramento, California Health Policy and Data Advisory Commission, 1991.
2. Dumbauld S. Trends in cesarean births in California, 1970–1986. Sacramento: State of California Department of Health Services, Health Data and Statistics Branch, No. 88-05171, July 1988.
3. Johns L. State Report: measuring quality in California. Health Affairs 1992;Spring:266–270.
4. Smith GH. Maine's liability demonstration project: Relating liability to practice parameters. Washington, DC: American Medical Association, AMA State Health Legislation Report, Fall 1990.
5. Kazandjian VA, Summer SJ. Evaluating the appropriateness of care: a study of cesarean section rates. Qual Rev J Qual Assur 1989; 15:206–214.

6. Kazandjian VA, Summer SJ. Cesarean section guideline: the Maryland experience. JAMA 1990;263:1491.
7. Dillon WP, Choate JW, Nusbaum ML, McCarthy MA, McCALL M, Rosen MG. Obstetric care and cesarean birth rates: a program to monitor quality of care. Obstet Gynecol 1992;80:731–737.
8. Vermont Program for Quality in Health Care, Inc. (VPQHC), 136 Main Street, Box 1356, Montpelier, Vermont 05601.
9. Jacoby I. The consensus development program of the National Institutes of Health. Current practices and historical perspectives. Int J Tech Assess Health Care 1985; 2:420–432.
10. U.S. Department of Health and Human Services. Cesarean childbirth. Washington, DC: Government Printing Office, 1981. (National Institutes of Health publication No. 82-2076.)
11. Winkler JD, Kanouse DE, Brodsley L, Brook RH. Popular press coverage of eight National Institutes of Health Consensus Development Topics. JAMA 1986;255:1323–1327.
12. Gleicher H. Cesarean section rates in the United States: the short-term failure of the National Consensus Development Conference in 1980. JAMA 1984;252:3273–3276.
13. Vaginal birth after cesarean section: information for health care providers. Quality Resource Center Report. Washington, DC: Washington Business Group on Health, 1991.
14. Silver L, Wolfe SM. Unnecessary cesarean sections: how to cure a national epidemic. Washington, DC: Health Research Group, 1989.
15. Half the cesarean operations in U.S. called unnecessary. Los Angeles Times, January 27, 1989.
16. Lomas J, et al. Consensus Conference Report. Indications for cesarean section: final statement of the panel of the National Consensus Conference on Aspects of Cesarean Birth. Can Med Assoc J; 134:1348–1352.
17. Lomas J, Anderson G, Domnick-Pierre K, Vayda E, Enkin M, Hannah W. Do practice guidelines guide practice? N Engl J Med 1989;321:1306–1311.
18. Cesarean Birth Quality Assurance Committee. Appropriate use of cesarean section: recommendations for a quality assurance program. Ministry of Health Ontario, Canada: 1991;ISBN 0-7729-8614-2.

18
Guidelines for Appropriate Utilization of Cesarean Operations

Bruce L. Flamm

The U.S. Department of Health objectives for the 1990s call for decreasing the national cesarean rate to no more than 15% of all deliveries by the year 2000.[1] This would mean performing about 400,000 fewer cesareans each year in the United States. To some this may seen like an impossible goal. However, by using guidelines, many hospitals have already lowered their overall cesarean rates to less than 15%. Importantly, this has been achieved with no increase in perinatal or maternal morbidity or mortality.

In today's medical-legal and social climate, it may be unreasonable to assume that a set of guidelines aimed solely at physicians will lead to a major reduction of the U.S. cesarean rate. If the U.S. Department of Health's objectives are to be achieved and sustained, obstetricians will need the support of nurses, childbirth educators, attorneys, governmental agencies, and the lay public. Similar cooperation will be necessary in any country that plans to address the cesarean issue on a national level.

Guidelines for Physicians

1. Decrease the Rate of Primary Cesarean Section for "Failure to Progress" to 5% (or less) of All Deliveries

This category should also include all cesareans for "dystocia," "failure to descend," "cephalopelvic disproportion," and "failed induction."

A. Avoid "latent-phase" cesareans. Cesareans for "failure to progress" performed before the cervix has reached 4 cm of dilatation are generally not indicated. Make sure that the patient is truly in labor before admitting her to the hospital. This is one of the key components of Dublin's "active management" program, and may contribute more to the lowering of cesarean rates than the aggressive use of oxytocin. Patients with reassuring fetal monitoring results who are not in labor should be sent home to await the onset of labor. For those patients who are experiencing a prolonged latent phase, therapeutic rest with morphine sulfate or a similar medication is often very beneficial.

B. Avoid failed inductions. If the cervix is not favorable and does not respond to oxytocin, consider delaying the induction for a few days if fetal monitoring is reassuring. Alternatively, consider using prostaglandins for cervical ripening. Above all, confirm that the indication for induction is medically valid. Many patients strongly request elective induction near term. Such patients should be reminded that awaiting a more favorable cervix and the onset of natural labor may result in an easier and less complicated birth.

C. A lack of change in cervical dilatation after 1 h (multiparous patient) or 2 h (primiparous patient) in the active phase of labor is an indication that some type of intervention may be necessary. This finding is not an indication for cesarean delivery. Appropriate interventions may include ambulation, placement of an internal uterine pressure catheter to assess the labor pattern, oxytocin augmentation of labor, or continued observation.

D. Avoid placing arbitrary time limits on the second stage of labor. This is especially true if epidural anesthesia is being used. Consider alternative positions for pushing, such as squatting.

E. The performance of a primary cesarean will place the patient at increased risk during future pregnancies. Repeat cesarean section and its alternative, vaginal birth after cesarean, are each associated with specific risks. This fact should be part of the informed-consent process at the time of the decision to perform an initial cesarean, particularly if there is no emergent need to proceed with the operation. The prime example is a cesarean performed solely because of "slow progress in labor."

2. Decrease the Rate of Primary Cesarean Section for All Indications *Other Than* "Failure to Progress" to 5% (or less) of All Deliveries

A. Fetal distress: Use electronic fetal heartrate monitoring appropriately and avoid overinterpretation of subtle monitor findings. In recent years there has been a proliferation of for-profit educational courses on the interpretation of electronic fetal monitormg. Some of these meetings have had unfortunate effects on both physicians and nurses. Presentations linking minimally suspicious monitor strips with subsequent poor outcomes leave attendees with the mistaken impression that almost any change in the fetal heartrate is an indication for immediate cesarean delivery. Although such cases do occur they are the exception, not the rule. Most experts agree that fetal monitoring is an *in*exact science and that true cases of "fetal distress" necessitating immediate operative delivery are infrequent. One potential benefit of electronic fetal monitoring is that, if the pattern is reassuring, there is no need for intervention. This is particularly important when a patient is making slow progress in labor.

B. Genital herpes: Assuming a 5% prevalence of genital herpes in the reproductive-aged population, approximately 200,000 women with a history of herpes will deliver each year in the United States. Studies from Washington State revealed that increasing the cesarean rate to 75% for women with a history of recurrent genital herpes failed to reduce the incidence of neonatal herpes. Recent reports indicate that cesarean delivery is indicated only for those women with lesions or symptoms of a recurrence during labor. Specific details are presented in Chapter 10.

C. Breech presentation: The prevalence of breech presentation at term is approximately 3%–4%. Currently about 85% of breech presentations are delivered by cesarean section in the United States. Many of these 100,000 operations per year could be prevented by external cephalic version while others could be safely avoided by allowing vaginal breech delivery in selected cases. Specific details are presented in Chapter 6.

D. Twin Gestation. Twins comprise approximately 1% of all pregnancies. Cesarean section is generally not indicated when both twins are in vertex presentation, regardless of gestational age. Although external cephalic version of the second nonvertex twin has recently become popular,[2] breech extraction remains a reasonable alternative.[3] Specific guidelines for the intrapartum management of twin gestation are presented in Chapter 9.

3. Decrease the Rate of Repeat Cesarean Section to 5% (or less) of All Deliveries

A. Follow the American College of Obstetricians and Gynecologists (ACOG) guidelines for vaginal delivery after previous cesarean birth.[4] Approximately 75% of women who are candidates for trial of labor will deliver vaginally if labor is allowed.

B. Most women with a prior cesarean *are* appropriate candidates for trial of labor. Women who have contraindications for labor with the current pregnancy such as complete placenta previa or transverse lie are obviously not candidates for trial of labor.

C. Several studies indicate that trial of labor with two prior cesareans is relatively safe; however, the risk of uterine rupture appears to be higher than for patients with one prior cesarean (approximately 1% compared to 0.5%). Minimal data are available on trial of labor with more than two prior cesareans; therefore, it may be prudent to perform a repeat cesarean unless the patient has a very strong desire for vaginal birth and is willing to assume a potentially increased risk.

D. Unknown uterine scar type is definitely *not* a contraindication to trial of labor. The majority of patients in several large studies on trial of labor had undocumented uterine scar types.[5] However, if the records cannot be obtained and the primary cesarean was performed because of transverse lie, placenta previa, or premature breech, elective repeat cesarean may be indicated. Although some centers have allowed trial of labor with a known prior classical or vertical uterine incision, experts generally agree that these patients are at significantly increased risk for uterine rupture.

E. Prior cesarean for "failure to progress" or "cephalopelvic dysproportion" is not a contraindication to trial of labor. The majority of patients in several large studies actually had "failure to progress" listed as the indication for their primary cesarean. Current data do not support the use of x-ray pelvimetry to screen such patients before trial of labor.

F. Many women request repeat cesarean operations. However, many women request primary cesarean operations. It is a double standard to treat these requests differently. The decision to perform a major operation must be made on the basis of medical indications.

Guidelines for Midwives

1. Reaffirm That Pregnancy Is a Normal Condition, Not a Disease

A. Nurse-midwifery educational programs usually require that registered nurse (RN) candidates for admission have a minimum of 2 years of experience working in a labor and delivery unit, which is generally in a medical center environment. Thus, nurses accepted to nurse-midwifery schools may have only experienced birth in a high-tech, high-risk environment. Furthermore, some nurses apply to midwifery school because they want to "do" more. Admissions panels may want to focus on those candidates who have a great interest in normal childbirth and a lesser interest in advanced ultrasound and electronic fetal monitoring techniques.

B. Nurse-midwifery educational programs are commonly located in university medical centers. This may expose midwifery students to a continuous stream of complicated pregnancies. Nurse-midwifery educators must continue to emphasize that midwives are experts in normal childbirth. They should strive to make educational experiences available, perhaps at out-of-hospital birth centers, which promote the commitment that midwifery has to normalizing the birth process.

Guidelines for Nurses

1. Recognize the Impact of Intrapartum Nursing Care on Cesarean Rates

A. Often a nurse is the only health care professional at the bedside during most of labor. Labor can be physically and emotionally exhausting not only for the patient and her partner but for the labor and delivery nurse as well. In particular, assisting the couple with the pushing stage of labor makes great demands on the nurse. Anyone who has been with a woman during the second stage of labor knows that this can be a difficult time. If the patient becomes angry or "gives up," it may be tempting for the nurse to call the doctor and say, "she's just not going to make it." Such comments are often taken very seriously by physicians, and at times the response is to order that the patient be prepared for surgery. Clearly, this may end a difficult situation for the nurse but may not be in the patient's best interest. The bedside nurse must be aware of this potential scenario and request to be relieved by a colleague if she feels that she is becoming emotionally or physically exhausted while caring for the patient.

B. Nursing administrators must be aware of the importance of a nurse at the bedside during active labor. This one-on-one labor support is a key component of Dublin's "active management" program and undoubtedly contributes to their low cesarean rates. If a nurse is assigned to care simultaneously for several women in active labor, it will not be physically possible for her to offer the support needed by each individual patient. Although one-on-one labor support may appear to be an inefficient use of limited nursing resources, it must be remembered that every patient who "gives up" during labor and demands an unnecessary cesarean will require additional nursing care during her postoperative hospitalization.

C. Charting duties that require extensive paperwork also detract from the labor and delivery nurse's primary role. Every effort should be made to streamline unnecessary paperwork so that the nurse can spend more time with the patient and less time with the chart.

D. In settings where one-on-one nursing is not possible, consideration should be given to trained "doula" support. The patient's husband or partner, who may be attending a birth for the very first time, is not equipped to handle alone the physical and emotional needs of a laboring woman. (See point number four under guidelines for the lay public, following.)

Guidelines for Childbirth Instructors

1. Recognize the Impact of Childbirth Education on Cesarean Rates

A. Studies on the impact of childbirth education on cesarean rates have had mixed results (see Chapter 16, this volume). However, this may simply reflect the difficulty in conducting such studies. Experienced doctors and midwives agree that women who have no preparation for labor and delivery are far more likely to decompensate and demand unnecessary cesareans than women who have a good understanding of what to expect.

B. Present an optimistic but realistic picture of the normal birth process. Prominent childbirth educators have pointed out that some of their colleagues strictly avoid the use of the word "pain" in their classes. This may be a disservice to those women who find labor to be far worse than anything they expected.

C. Avoid the use of comprehensive birth plans that specifically list each of the interventions that a woman will (or will not) allow during labor. These point-by-point birth plans often alienate both the physician and nursing staffs. It is my opinion that it is far more fruitful for a couple to stress their desire for natural childbirth during the prenatal visits and to reiterate these feelings during labor.

Guidelines for Medical School and Residency Faculty

1. Emphasize Normal Pregnancy and Birth Before Presenting Detailed Discussions of Pathologic Conditions

Professors involved in the development of medical school and residency curriculum should place more emphasis on the normal aspects of pregnancy and childbirth. Current teaching often leads medical students to believe that pregnancy is an extremely dangerous condition that often culminates in a series of life-threatening catastrophes. In reality, many if not most pregnancies would have good outcomes with little or no medical intervention. Students should be taught that the goal of obstetric care is to screen for risk factors and to avoid interventions when no problems exist. Overemphasis on pregnant patients with insulin-dependent diabetes and other serious medical diseases may lead students to the mistaken conclusion that every pregnancy is fraught with danger. In addition to lectures by perinatologists on high-risk obstetrics, medical students and obstetric residents should be exposed to the nurse-midwifery care of uncomplicated pregnant women.

2. It Is Far Easier to Teach a Young Physician *How* to Do a Cesarean Than *When* to Do One

In the past, cesarean-related education has focused almost entirely on discussions of subtle differences in surgical technique. Entire grand rounds programs are often dedicated to topics such as antibiotic prophylaxis for cesarean delivery. Clearly, more attention must be devoted to helping residents understand when the operation is really necessary rather than simply teaching them how to best perform the procedure or what antibiotics to use.

Guidelines for Attorneys

1. Ponder the Impact of Your Actions and Those of Your Colleagues on Society

Malpractice attorneys now commonly advertise in newspapers and telephone books, seeking out cases that involve alleged "birth injuries." These ads generally offer incentives such as a "no recovery, no fee" guarantee and occasionally include testimonials by clients who have successfully won millions of dollars. To many individuals such ads are an open invitation to obtain wealth far beyond their wildest dreams. These ads will certainly attract some individuals who have actually been harmed by medical malpractice. Unfortunately, they will also attract fortune hunters. Initial review of most cases solicited in this manner will probably lead the attorney to doubt if there was any wrongdoing. If consultation with medical experts confirms this opinion, it would be unethical to proceed with litigation. Still, the potential for a huge verdict or out-of-court settlement must be very tempting. Attorneys faced with such temptations must understand that thousands upon thousands of unnecessary cesarean operations have been performed by doctors fearful of frivolous litigation.

Many obstetricians fear that any birth with a poor outcome will result in a lawsuit even if superior medical care is provided. This fear is not totally unjustified. Although a physician who has provided excellent care will typically prevail in such a lawsuit, the agony of being dragged through the litigation process cannot be ignored. Formulating a page of statements repetitively charging that a physician is guilty of gross negligence and incompetence may be all in a day's work for a malpractice attorney but such a document will be devastating for the physician who receives it. Likewise, taking a deposition from an accused physician may be a mundane matter for an experienced malpractice attorney, but it may be the most difficult day in the life of the doctor.

Attorneys may argue that induring such pain is an unavoidable consequence of committing "negligent" actions, but it must be remembered that standard-of-care issues are often confusing. Unintentionally missing a few subtle late decelerations on a fetal monitor recording is definitely not equivalent to robbing a bank, yet some plaintiff attorneys treat doctors like armed robbers. Worse yet, attorneys who knowingly pursue *frivolous* malpractice cases should be aware that they are committing legal malpractice and that their actions will have profound consequences not only for innocent physicians but also for society.

Guidelines for Medical Directors and Chiefs of Obstetric Services

1. Avoid Punitive Obstetric Meetings

It is common practice at many hospitals to review electronic fetal monitor strips at weekly or monthly obstetric conferences. Generally, cases with low Apgar scores are chosen for review. The doctor who attended the birth squirms in his or her seat as other physicians review every inch of the monitor strip. With the clarity of hindsight (and preordained knowledge that the Apgar scores were low because the case is being presented), participants call out their opinions and point out exactly at which point *they* would have done a cesarean if they had been managing the case. Eventually some obstetricians learn that they can avoid being placed on the "hot seat" by doing a cesarean any time there is any suspicious area on a fetal monitor tracing. This is the wrong lesson.

A more balanced meeting would include the review of suspicious monitor strips that culminated in the vaginal birth of vigorous infants with normal Apgar scores. Such cases are not difficult to find. In fact it is unusual to find a monitor strip that does not include at least some decelerations. Apgar scores should not be revealed until after everyone has been allowed to voice their opinions.

2. Initiate Cesarean Review Meetings

In contrast to fetal monitor review meetings, which tend to focus on cases in which cesareans should have been done (or should have been done sooner), cesarean review meetings focus on cases where cesareans might have been avoided. As for fetal monitor review meetings, these meetings should not be punitive lest obstetricians be placed in the double-bind situation of being criticized at one meeting if a cesarean is performed and at the other meeting if it is not. At some hospitals, every cesarean is briefly presented. This is unnecessary because many cases involve clear-cut or at least very reasonable indications for surgery. However, "failure to progress" cesareans performed at 2 cm and elective repeat cesareans (or repeat operations performed after an hour or two of labor) make for interesting discussions.

3. Perform a Confidential Audit of Individual Cesarean Rates

At some hospitals, the cesarean rates of one or two physicians are so high that the entire hospital cesarean rate is elevated above the national average. If an audit reveals that nine physicians have annual cesarean rates in the range of 12%–19% while one physician has a rate of more than 45%, it would be logical to focus educational efforts on that physician.

Guidelines for Third-Party Payers

1. Pay Physicians More for Normal Vaginal Delivery

Historically, physicians have been paid more for performing a cesarean than for attending a vaginal birth. In other words, a physician could stay in the hospital all night helping a woman through a difficult labor and make *less* money than for performing a 1-h operation and going home. This does not make sense.

2. Do Not Reimburse Patients for Operations That Are Not Medically Necessary

Some patients demand cesarean delivery solely as a matter of convenience. Whether it is reasonable to perform such operations is an ethical issue that is discussed elsewhere (see Chapter 15, this volume). Whether society or third-party payers should cover the expenses of such operations is a separate issue. A woman might not find the convenience of knowing her baby's exact birth date to be quite so appealing if this option was offered along with the condition that she agree to pay the $3000 in extra hospital bills generated by the operation and postoperative care.

3. Do Not Offer Additional Weeks of Paid Disability for Operations That Are Not Medically Necessary

In some states it is common practice to pay for 6 weeks of postpartum disability after a vaginal birth and 8 weeks after a cesarean delivery. The same woman who might insist on cesarean delivery because it is "easier on her body than going through labor" might also insist on the 2 additional weeks of paid disability. In reality, most women are sufficiently recovered by 6 weeks postpartum to resume full activities regardless of the mode of birth. The "traditional" 2 weeks of additional disability compensation offered after cesarean delivery is a tempting incentive for some women to lean toward operative delivery.

Guidelines for Hospital Administrators

Be Wary of Potential Conflicts of Interest

Hospital obstetric services are generally financially rewarding. In 1987, U.S. hospitals billed more than $10 *billion* for obstetric care (this did not include physician fees).[6] This makes obstetric services the greatest revenue generator for most hospitals. It is common knowledge that the hospital length of stay for cesarean delivery is approximately twice as long as for vaginal birth. For hospital administrators faced with difficult financial situations and empty beds, it may be difficult to support initiatives aimed at lowering cesarean rates.

Guidelines for Legislatures

Pass Legislation That Eliminates Frivolous Lawsuits

The Maine plan (see Chapter 17, this volume) could serve as a model for other states. The impact of legal issues on cesarean rates is difficult to measure, but many experts believe that it is of monumental importance. Fear of frivolous malpractice suits may be the single most important factor driving up cesarean rates, and this factor alone has probably resulted in hundreds of thousands of unnecessary cesarean operations.

Guidelines for the Lay Public

1. Inquire about cesarean rates for hospitals and physicians in your area. Small differences may simply reflect the fact that

some doctors care for more patients with complications than others. However, very high cesarean rates (greater than 30%) may be a warning if you hope to have a natural birth.

2. During prenatal visits and again on admission to the hospital, stress the fact that you would like to avoid a cesarean operation if possible.

3. Women should recognize that childbirth involves hard but valuable work and that a request for the attending physician to "just do a cesarean" is inappropriate when there is no medical indication for surgery.

4. Consider the help of a doula (trained labor support person). Although husbands can often be very helpful during labor, many have never seen a baby born except perhaps in a film shown during a childbirth education class. A trained support person who has attended dozens of births will be better equipped to reassure a woman who is having trouble with her labor. Also, the husband may be too emotionally involved to make objective decisions. The husband's response to seeing his wife in pain may be to beg the doctor to do a cesarean and just "get it over with."

References

1. Healthy People 2000. Washington, DC: U.S. Department of Health Services, Public Health Services (publication no. 91-50212:378–379).
2. Chervenak FA, Johnson RE, Berkowitz RL, et al. Is routine cesarean section necessary for vertex-nonvertex and vertex-transverse twin gestations? Am J Obstet Gynecol 1984;148: 1–5.
3. Gocke SE, Nageotte MP, Garite T, Towers CV, Dorcester W. Management of the nonvertext second twin: primary cesarean, external version, or primary breech extraction. Am J Obstet Gynecol 1989;151:111–114.
4. Guidelines for vaginal delivery after a previous cesarean birth. Committee Opinion Number 64. Washington, DC: American College of Obstetricians and Gynecologists, 1988.
5. Leung A, Farmer R, Leung E, Medearis A, Paul R. Risk factors associated with uterine rupture during trail of labor after cesarean delivery: a case control study. Am J Obstet Gynecol 1993;168:1358–1363.
6. The stork's share of America's hospital tab. OBG Management, May 1994.

Index

J

K

MIX
Papier aus verantwortungsvollen Quellen
Paper from responsible sources
FSC® C105338

If you have any concerns about our products,
you can contact us on
ProductSafety@springernature.com

In case Publisher is established outside the EU,
the EU authorized representative is:
Springer Nature Customer Service Center GmbH
Europaplatz 3, 69115 Heidelberg, Germany

Printed by Libri Plureos GmbH
in Hamburg, Germany